Biological Effects of Cholesterol Oxides

Editors

Shi-Kaung Peng, M.D., Ph.D.
Professor
Department of Pathology and Laboratory Medicine
University of California at Los Angeles School of Medicine
Harbor-UCLA Medical Center
Torrance, California

and

Robert J. Morin, M.D.
Professor
Department of Pathology and Laboratory Medicine
University of California at Los Angeles School of Medicine
Harbor-UCLA Medical Center
Torrance, California

CRC Press
Boca Raton Ann Arbor London

Library of Congress Cataloging-in-Publication Data

Biological effects of cholesterol oxides / editors, Shi-Kaung Peng and Robert J. Morin.
p. cm.
Includes bibliographical references and index.
ISBN 0-8493-6776-X
1. Cholesterol oxides--Physiological effect. 2. Cholesterol oxides--Pathophysiology. 3. Cholesterol--Oxidation. I. Peng, Shi-Kaung. II. Morin, Robert J.
[DNLM: 1. Cholesterol--analogs & derivatives. 2. Cholesterol--pharmacology. QU 95 B614]
QP752.C53B56 1991
612'.015731--dc20
DNLM/DLC 91-23092
for Library of Congress CIP

Direct all inquiries to CRC Press, Inc., 2000 Corporate Blvd., N.W., Boca Raton, Florida 33431.

International Standard Book Number 0-8493-6776-X

Library of Congress Card Number 91-23092

Printed in the United States of America 1 2 3 4 5 6 7 8 9 0

PREFACE

In recent years, researchers have accumulated a growing body of evidence linking cholesterol oxides with human diseases such as atherosclerosis and cancer. Since cholesterol is present ubiquitously in all cells, we may well ask "could it be cholesterol oxides rather than cholesterol per se that is the true culprit in human diseases?" The relative roles of cholesterol and cholesterol oxides are as yet unresolved, but the purposes of this book are to give the reader an overall indication of current knowledge on the biological activities of cholesterol oxides and to stimulate thinking in some new areas of diet-heart or diet-cancer research. Cholesterol oxides could be exogenously obtained from the diet as well as endogenously generated *in vivo* through free radical or enzymatic reactions. The oxygen-based reactions that convert cholesterol to its oxides could speed up dramatically once meats or other animal products are heated. Statistically, elevated serum cholesterol is associated with increased risk of coronary heart disease; however numerous exceptions occur at both ends of the spectrum: many persons with normal serum cholesterol do develop coronary heart disease, and vice versa. Is it possible that cholesterol oxides may be the missing piece of the puzzle? Furthermore, in developed countries, populations that consume high-fat diets with a high mortality rate for coronary heart disease also exhibit a similar high mortality rate for colon and breast cancers. Cholesterol oxides again may play a role in these disease processes. A remarkable diversity of biochemical and physiological effects of cholesterol oxides have been reported, particularly with regard to cholesterol metabolism, where some of them appear to have a regulatory role. Other significant effects are those on membrane function and on prostaglandin metabolism. In this volume, we have attempted to relate these basic cellular effects of cholesterol oxides to their pathophysiological effects to better understand the mechanisms of involvement of these compounds in disease processes. We have tried to make the book of use to scientific workers, not only those dealing directly with cholesterol and cholesterol oxides but to all concerned with the wider aspects of chemistry, biology, and medicine for the directions of future research which may lead to knowledge highly relevant to the understanding of human diseases and to development of preventive strategies such as improvement in food processing techniques or use of antioxidants. Finally, the hard work and forbearance of our secretaries, Carlida Miguel-Bocatija and Frances Barbara Frese, is greatly appreciated.

Shi-Kaung Peng, M.D., Ph.D.
Robert J. Morin, M.D.

EDITORS

Shi-Kaung Peng, M.D., Ph.D., is Chief of Surgical Pathology and Head of Cardiovascular Pathology at the Harbor-UCLA Medical Center and Professor of Pathology and Laboratory Medicine at the University of California at Los Angeles School of Medicine. He received his M.D. from the Medical College of National Taiwan University in Taipei, Taiwan in 1966 and his Ph.D. in experimental pathology from the Northwestern University Medical School in Chicago, Illinois in 1971. He was an Assistant Professor of Pathology at the Albany Medical College of Union University in 1972 and an Associate Professor of Pathology in 1977. In 1983, he joined the faculty at the UCLA School of Medicine in the Department of Pathology and Laboratory Medicine as Head of Cardiovascular Pathology in the Harbor-UCLA Medical Center. He is board certified in Anatomic and Clinical Pathology.

His major research interest involves studying the biological effects of cholesterol oxides and their role in atherogenesis and carcinogenesis. His other research interests include cholesterol and cholesteryl ester metabolism, the effects of vitamin D on atherosclerosis and laser application in vascular surgery. He serves on the Editorial Board of the *Journal of Cardiovascular Biomechanics, Technology and Surgery* and as the Editorial Reviewer for *Lipids, Atherosclerosis,* and *Archive of Pathology and Laboratory Medicine.* He has published more than 120 research papers, review articles, and book chapters.

Dr. Peng is a member of several national and international societies, including the International Academy of Pathology, the American Association of Pathologists, and the Society for Cardiovascular Pathology, and he is a fellow of the Council on Arteriosclerosis of the American Heart Association.

Robert J. Morin, M.D., graduated from the University of California at Los Angeles in 1956 with a degree in Zoology and received his M.D. from the University of California at San Francisco in 1960. He then took 2 years of post-doctoral training in lipid biochemistry at UCLA with Drs. Roslyn Alfin-Slater and James F. Mead, and additional training in Clinical Pathology at Harbor General Hospital. In 1967 he was appointed Assistant Professor of Pathology, and in 1978 attained the position of Professor of Pathology at Harbor-UCLA Medical Center. Since 1970 he has also been Head of the Chemical Hematology Section and Director of the School of Medical Technology at that institution. He is presently the Acting Chairman of Pathology Department and Director of the UCR-UCLA Biomedical Sciences pathology curriculum.

Dr. Morin's earlier research interests involved fatty acid, phospholipid, and cholesterol metabolism. He was the first to demonstrate the presence of 8,11,14-eicosatrienoic acid in animal tissues. His later research concerned the regulation of cholesterol esterification and hydrolysis, and most recently has focused on the effects of cholesterol oxidation derivatives on cultured cells in relation to atherogenesis and carcinogenesis. He has published 140 papers in these fields.

CONTRIBUTORS

Paul B. Addis, Ph.D.
Professor
Department of Food Science and Nutrition
University of Minnesota
St. Paul, Minnesota

Bing Hu, M.D.
Research Associate
Department of Pathology
Harbor-UCLA Medical Center
Torrance, California

Robert J. Morin, M.D.
Professor
Department of Pathology and Laboratory Medicine
UCLA School of Medicine
Harbor-UCLA Medical Center
Torrance, California

Peter S. W. Park, Ph.D.
Senior Scientist
Mead Johnson Research Center
Bristol-Myers Squibb Co.
Evansville, Indiana

Shi-Kaung Peng, M.D., Ph.D.
Professor
Department of Pathology and Laboratory Medicine
UCLA School of Medicine
Harbor-UCLA Medical Center
Torrance, California

Alex Sevanian, Ph.D.
Associate Professor
Institute for Toxicology
University of Southern California
School of Pharmacy
Los Angeles, California

Leland L. Smith, Ph.D.
Professor
Department of Human Biological Chemistry and Genetics
University of Texas Medical Branch
Galveston, Texas

C. Bruce Taylor, M.D.
Professor Emeritus
Department of Pathology
Albany Medical College
Union University
Albany, New York

TABLE OF CONTENTS

Chapter 1

INTRODUCTION AND HISTORICAL PERSPECTIVE

C. Bruce Taylor and Shi-Kaung Peng

TABLE OF CONTENTS

I. HISTORICAL BACKGROUND OF CHOLESTEROL OXIDATION

The compound cholesterol was first described in 1789 by De Fourcroy,[1] who isolated a crystalline substance from alcoholic extracts of human gallstones. Chevreul[2] named the substance *cholesterine* which was derived from the Greek words *chole* (bile) and *steros* (solid). Approximately a century later, the labile nature of cholesterol was discovered. In 1904, Shultze and Winterstein[3] reported that crystalline cholesterol could undergo spontaneous oxidation in air. Although Lifschütz had applied color tests involving acetic and sulfuric acid for the detection of sterols other than cholesterol in wool fat by 1907,[4] his further studies of cholesterol oxidation were clearly stimulated by the report of Schultze and Winterstein. Lifschütz observed the presence of "oxycholesterol" when a solution of cholesterol in acetic acid was treated with a little benzoyl peroxide and heated briefly; ill-defined products in this solution gave an intensive blue or green color on addition of sulfuric acid.[5] Many other methods were subsequently found for oxidizing cholesterol. In fact, mere heating of cholesterol in air resulted in products that gave a positive Lifschütz color reaction with acetic-sulfuric acid for oxycholesterol.[6] It was the work of Blix and Lowenstein[7] that firmly established the indispensable role of molecular oxygen in the spontaneous autoxidation of cholesterol.

Haslewood in 1939[8] was the first investigator to report the isolation of a known autoxidation derivative of cholesterol from animal tissue. He identified 7β-hydroxycholesterol in an extract of bovine liver. In the following year, Winterstein and Ritzmann[9] isolated the epimeric 7α-hydroxycholesterol from the nonsaponifiable fraction of pregnant mare's serum. This substance formed a blue color with antimony trichloride and showed a positive Lifschütz color test. Many autoxidation derivatives have since been isolated and identified in animal tissue.

In 1941 to 1942, Bergstrom and Winterstein[10-12] developed a model system for the study of cholesterol autoxidation. Colloidal cholesterol in aqueous sodium stearate dispersions is readily attacked by oxygen with formation of predominantly 7-ketocholesterol and a small portion of 7α-hydroxycholesterol and 7β-hydroxycholesterol. The results of Bergstrom and Winterstein in their studies of cholesterol autoxidation attracted much continuing interest and in essence clarified the earlier studies on cholesterol autoxidation by presenting a fairly complete concept of the process and by identifying the most frequently encountered oxidation products. Mosbach et al.[13] in 1953 were the first to identify cholestane-3β,5α,6β-triol as an autoxidation product formed in cholesterol-soap dispersions. Schenck et al.[14] in 1957 identified the formation of 6-cholesten-5α-hydroperoxide when cholesterol was irradiated in the presence of hematoporphyrin. This hydroperoxide is isomerized gradually in chloroform solution to 5-cholesten-7α-hydroperoxide and on dehydration yields 7α-hydroxycholesterol. In the presence of cupric salt 7α-hydroxycholesterol is converted to 7-ketocholesterol. Fieser et al.[15] demonstrated that cholesterol stored in the crystalline state in the laboratory could undergo autoxidation, and identified 25-hydroxycholesterol as a product of oxidation of one of the tertiary carbon atoms in the aliphatic side chain. They also showed that the amount of 25-hydroxycholesterol formed was proportional to the age of the specimen. Fieser and Fieser[16] in 1959 published a major text — *Steroids Oxidation* — with a comprehensive review of cholesterol oxidation.

Further advancement in the understanding of cholesterol autoxidation occurred in the 1960s with the introduction of advanced analytic techniques, such as thin-layer chromatography, gas-liquid chromatography, high-performance liquid chromatography, and mass spectrometry. Smith et al.[17] in 1967 demonstrated the complexity of natural autoxidation of cholesterol using two-dimensional thin-layer chromatography, and identified at least 32 oxidation products formed over 12 years of natural air aging of a cholesterol sample which

TABLE 1
Common Autoxidation Derivatives of Cholesterol

Chemical Name	Trivial Name
Cholest-5-ene-3β,7α-diol	7α-Hydroxycholesterol
Cholest-5-ene-3β,7β-diol	7β-Hydroxycholesterol
3β-Hydroxycholest-5-ene-7-one	7-Ketocholesterol
5α-Cholestane-3β,5,6β-triol	Cholestanetriol
Cholest-5-ene-3β,25-diol	25-Hydroxycholesterol
20S-Cholest-5-ene-3β,20-diol	20α-Hydroxycholesterol
5,6α-Epoxy-5α-cholestan-3β-ol	Cholesterol 5α,6α-epoxide
5,6β-Epoxy-5β-cholestan-3β-ol	Cholesterol 5β,6β-epoxide
3β,5-Dihydroxy-5α-cholestan-6-one	6-Ketodihydroxycholesterol
Cholest-4-ene-3-one	4-Cholestene-3-one

had been purified via the dibromide procedure of Fieser[18] and repeatedly recrystallized. Many of these compounds were only found in trace amounts. Smith[18] in 1981 wrote a major monograph on the subject of cholesterol autoxidation. Although the event initiating the formation of free radicals are not completely understood, many agents, including azo compounds, peroxides, hydroperoxides, transition metal ions, and excited molecular oxygen species could promote this step.[18] Photolysis by visible and ultraviolet light and ionizing radiation can also generate free radicals. The extent of autoxidation and the resultant products appear to be dependent on the physical state of cholesterol. In an aqueous phase at 85°C, pH 8.5, and aerated for 5 h, 80% of a sample of cholesterol is converted into oxidation products.[19] The susceptible site for oxidation of the cholesterol molecule in the dispersed aqueous state was shown to be the 5,6-double bond.[20] Predominant products identified were of ring-oxidized compounds, such as 7α- and 7β-hydroxycholesterol, 7-ketocholesterol, 3,5-cholestadien-7-one, and cholesterol 5β,6β-epoxide. In the presence of copper ion at 10^{-5} *M*, cholesterol showed maximal autoxidation, which could be inhibited by heavy metal complex formers such as sodium cyanide and diethyldithiocarbaminate. Hydrogen peroxide was found to be an important oxidizing agent in colloidal cholesterol. When cholesterol is in the crystalline state, the tertiary carbons in the side chain are also involved in autoxidation.[21] Biradical oxygen attacks at the preferred tertiary 20- and 25-carbon atoms forming 5-cholesten-3β-ol-20α-hydroperoxide and 5-cholesten-3β-ol-25-hydroperoxide, respectively, as intermediates before conversion to the dihydroxylated cholesterol.[22] The 3-hydroxysteroids crystallize similarly to the long-chain alcohols, which are in double layers with the hydroxyl groups in juxtaposition. The two major surfaces of the plate-like cholesterol crystals are composed, therefore, of molecules with their hydrocarbon side chain exposed. The molecules are evidently sufficiently closely packed to prevent the approach of oxidizing agent to the reactive position at, or adjacent to the 5,6-double bond, and the attack is directed to the tertiary 20- and 25-carbon positions. Table 1 lists the more common cholesterol autoxidation derviatives.

II. HISTORICAL BACKGROUND ON BIOLOGICAL EFFECTS OF CHOLESTEROL OXIDES

Since the discovery of the labile nature of cholesterol, studies have been undertaken to determine the biological effects of oxidized cholesterol derivatives. In the early twentieth century, Flury[23] tested the effects of oxycholesterol on isolated frog heart preparations. The oxycholesterol was effective in lengthening the duration of the heart beat and increasing the pulse frequency. Still other early investigations on the effects of oxycholesterol demonstrated that it had antirachitic properties.[24,25] In 1944, Haslewood,[26] suggested that systemic evaluation of the biological acitivites of oxidized cholesterol derivatives be undertaken. Although a thorough evaluation has not yet been accomplished, there is growing interest in the effects of cholesterol autoxidation derivates on biological systems.

Earlier studies on biological effects of cholesterol oxides were mostly concerned with the possibility that some of the cholesterol oxides may have carcinogenic properties. In 1931, Ruffo,[27] claimed that photoxidation of cholesterol in skin induced the development of skin tumors. Between 1946 and 1949, various investigations showed a high incidence of tumor development following subcutaneous injection of impure cholesterol (summarized by Fieser[28]). He speculated that cholesta-5-ene-3-one was the major carcinogenic agent. Bischoff[29] extended these earlier stuides by testing the carcinogenicity of several cholesterol oxidation products, and showed that when administered subcutaneously with sesame oil, they produced local fibrosarcomas in Marsh-Buffalo mice. Since then, many cholesterol oxidation derivatives have been identified as having carcinogenic effects.

Cholesterol has been implicated in the development of atherosclerosis since 1913 when Anitschkow[30] induced experimental atherosclerosis in rabbits by feeding them a diet containing cholesterol. Thus began the accumulation of massive amounts of data that dietary cholesterol and hypercholesterolemia were at the root of atherosclerosis and its clinical complications such as myocardial infarction and stroke. Most of the investigators in the field, however, were not aware of the labile nature of cholesterol. The possibility of atherogenic properties of cholesterol oxides was not explored until 1968 when Cook and MacDougall[31] produced experimental atherosclerosis in rabbits after feeding cholestane-3β,5α,6β-triol, one of the most toxic cholesterol oxides.

In 1971, Taylor rekindled a 20-year acquaintance with Werthessen, a steroid chemist and former co-worker of Smith, and expressed concern about the wide gap between the present level of knowledge of chemistry of cholesterol and its autoxidation products[17] vs. the biological effects of oxidized cholesterol derivatives. He reviewed two experiments in which hyperlipidemia in rabbits[32] and chickens[33] had been induced by dietary cholesterol along with endocrinologic manipulation. The animals with the diet-induced hypercholesterolemia had much more severe atherosclerosis than those with endogenously induced hypercholesterolemia by subcutaneous implantation of diethylstilbestrol. Werthessen also discussed a study by Schwenk et al.[34] in which he reported that purified cholesterol was more atherogenic than the impurities (oxidized cholesterol derivatives) in United States Pharmacopia (U.S.P.)-grade cholesterol. More recent studies by Smith et al.[17] and Engel and Brooks,[35] however, led to their mutual agreement that cholesterol purified via dibromidation was unstable when the method of Fieser[36] was used. During the conversation between Werthessen and Taylor, the question of a source of aged (possibly oxidized) cholesterol arose. Taylor reported the ownership of a 50-kg barrel of cholesterol purchased in 1966. The cardboard barrel of cholesterol had been opened to air and kept in a warm storage room for 5 years. Werthessen developed a methanol extraction method for separation of oxidized cholesterol derivatives from the pure cholesterol.[37] In 1976, Imai et al.[37] reported that purification of a stock of old U.S.P.-grade cholesterol (purchased in 1966) largely prevented its atherogenicity when fed to rabbits, whereas feeding 250 mg/kg of concentrated impurities in the aged cholesterol produced severe arterial damage with smooth muscle cell death and focal intimal edema within 24 h. Peng et al.[38] found the methanol-extractable impurities to be largely composed of cholesterol oxides. The major components identified were 25-hydroxycholesterol, cholestane-3β,5α,6β-triol, 7-ketocholesterol, 7α- and 7β-hydroxycholesterol, and 5-cholesten-7-hydroperoxide. Further evaluation of the biological activities of these cholesterol oxides in rabbit aortic smooth muscle cell cultures showed that 25-hydroxycholesterol and cholestane-3β,5α,6β-triol were the principal biologic toxins in the cholesterol impurities.[38]

The absorption and delivery of cholesterol oxides throughout the body are important considerations. Since cholesterol oxides have been shown to have significant biological activities, it seemed that there was a need for more knowledge concerning their absorption, distribution, and mode of action. There are, however, only a few reports on the dietary absorption of cholesterol oxides. Steinberg et al.[40] in 1958, reported a greater than 70% absorption of cholest-4-ene-3-one in rats. Kikuchi et al.[39] in 1968 showed that approximately 7% of a 25-mg oral dose of cholestane-3β,5α,6β-triol was absorbed in rats and monkeys. When cholesterol was administered simultaneously, its absorption rate could be enhanced almost twofold. Most of the absorbed cholestane-3β,5α,6β-triol exists as free sterol in the intestinal wall and is transferred to the lymphatic ducts without esterification. The mode of transport of cholesterol oxides in plasma has been investigated by Peng et al.[41] The oral administration of radiolabeled 25-hydroxycholesterol to squirrel monkeys resulted in its localization in low-density lipoproteins (LDL) (56%) and very-low-density lipoproteins (VLDL) (34%). High-density lipoproteins contained only a small amount of 25-hydroxycholesterol (10%). In contrast, radiolabeled cholesterol had a distribution in VLDL, LDL, and HDL almost identical to that of unlabeled cholesterol. The authors suggest that the preferential transport of 25-hydroxycholesterol by LDL and VLDL may be the basis for the more potent atherogenicity of these lipoprotein species. The potent cytotoxicity of LDL to endothelial cells *in vitro*[42] and the suggestion of *in vivo* oxidation of serum LDL based on its facile autoxidation *in vitro*[43] support this hypothesis.

In recent years, a significant and growing number of research studies concerning biological effects of cholesterol oxides have been reported, which provide the basis for the discussions in this book. It is apparent that these compounds have a wide variety of effects both *in vitro* and *in vivo* which may be linked to human diseases. The most important task, therefore, is to synthesize this knowledge toward a determination of whether or not they can pose a genuine threat to human health.

REFERENCES

1. **De Fourcroy,** De la substance feuilletee et cristalline contenue dan les calculs biliaires, et de la nature des concretions crystiques cristalisee's, *Ann. Chim.*, 3, 242, 1789.
2. **Chevreul, M. E.,** Sixieme Memoire. Examen des graisses d'homme, de mouton, de boeuf, de jaguar et d'oie., *Ann. Chim. Phys.*, 2, 339, 1816.
3. **Schültz, E. and Winterstein, E.,** Über das verhalten des cholesterins gegendas licht., *Physiol. Chem.*, 43, 316, 1904.
4. **Lifschütz, I.,** Die Oxydationsprodukte des cholesterins in den tierischen organen, *Physiol. Chem.*, 53, 140, 1907.
5. **Lifschütz, I.,** Quantitative Bestimmungen der Oxydationsprodukte des cholesterins, *Biochem. Z.*, 48, 373, 1913.
6. **Lamb, F. W.,** The oxidation of cholesterol, *J. Physiol.*, 48, 151, 1914.
7. **Blix, G. and Lowenstein, G.,** On the oxidation of cholesterol and molecular oxygen, *Biochem. J.*, 22, 1313, 1928.
8. **Haslewood, G. A. D.,** Metabolism of steroids. Isolation of 7-hydroxycholesterol and of "hepatols" from ox liver, *Biochem. J.*, 33, 709, 1939.
9. **Winterstein, O. and Ritzmann, J. R.,** The isolation of 7β-hydroxycholesterol from the serum of pregnant mares, *Biol. Chem.*, 136, 697, 1940.
10. **Bergstrom, S. and Winterstein, O.,** Autoxidation of sterols in colloidal aqueous solution in the nature of the products formed from cholesterol, *J. Biol. Chem.*, 141, 597, 1941.
11. **Bergstrom, S. and Winterstein, O.,** Autoxidation of sterols in colloidal aqueous solution. II. Δ^6-cholestene-diol-3(β),5α, a rearrangement product of 7(β)-hydroxycholesterol, *J. Biol. Chem.*, 143, 503, 1942.
12. **Bergstrom, S. and Winterstein, O.,** Autoxidation of sterols in colloidal aqueous solution. III. Quantitative studies on cholesterol, *J. Biol. Chem.*, 145, 209, 1942.

13. **Mosbach, E. H., Nienenberg, M., and Kendall, F. F.,** Separation of the air-oxidation products of cholesterol by column partition chromatography, *J. Am. Chem. Soc.*, 75, 2358, 1953.
14. **Schenck, G. O. Gollnick, K., and Neumüller, O. A.,** Photosensitized autoxidation of steroids preparation of steroid hydroperoxides by means of phototoxic photosensitizers, *Liebigs Ann.*, 603, 46, 1957.
15. **Fieser, L. F., Huang, W.-Y., and Bhattacharyya, B. K.,** Cholesterol and companion X. The diol fraction, *J. Oncol. Chem.*, 22, 1300, 1957.
16. **Fieser, L. F. and Fieser, M.,** *Steroids Oxidation,* Reinhold, New York, 1959, 189.
17. **Smith, L. L., Matthews, W. S., Price, J. C., Bachmann, R. C., and Reynolds, B.,** Thin-layer chromatographic examination of cholesterol autoxidation, *J. Chromatogr.*, 27, 187, 1967.
18. **Smith, L. L.** *Cholesterol Autoxidation,* Plenum Press, New York, 1981.
19. **Bergstrom, S. and Samuelson, B.,** *The Autoxidation of Cholesterol. Autoxidation and Antioxidants,* Vol. 1, Lundberg, W. O., Ed., Interscience, New York, 1961, 233.
20. **Winterstein, O. and Bloch, K.,** The products formed by the action of oxygen on colloidal solutions of cholesterol, *J. Biol. Chem.*, 137, 785, 1941.
21. **Beckwith, A. L. J.,** The oxidation of crystalline cholesterol, *Proc. Chem. Soc.*, 154, 1958.
22. **Van Lier, J. E. and Smith, L. L.,** Autoxidation of cholesterol via hydroperoxide intermediates, *J. Org, Chem.*, 36, 3627, 1970.
23. **Flury, F.,** Über die pharmakologischen Eigenschaften einiger saurer Oxydationsprodukte des Cholesterins, *Arch. Exp. Pathol., Pharmakol.*, 66, 221, 1911.
24. **Kramer, B., Shear, M. J., and Shelling, D. H.,** Fractionation of irradiated cholesterol. II. Antirachitic potency of the fractions, *J. Biol. Chem.*, 71, 221, 1926.
25. **Fosbinder, R. J., Daniels, F., and Steenbock, H.,** A quantitative study of the photochemical activation of sterols in the cure of rickets, *J. Am. Chem. Soc.*, 50, 923, 1928.
26. **Haslewood, G. A. D.,** Cholesterol metabolism in the animal body, *Nature (London),* 154, 29, 1944.
27. **Ruffo, A. H.,** The photoactivity of cholesterol in relation to cancer, *Am. J. Cancer,* 15, 915, 1931.
28. **Fieser, L. F.,** Some aspects of the chemistry and biochemistry of cholesterol, *Science,* 119, 710, 1954.
29. **Bischoff, F.,** Cocarcinogenic activity of cholesterol oxidation products and seame oil, *J. Natl. Cancer Inst.*, 19, 977, 1957.
30. **Anitschkow, N.,** Über die veranderungen der kaninchenaorta by experimenteller cholesterinsteatose, *Beitr. Pathol. Anat. Alig. Pathol.*, 56, 379, 1913.
31. **Cook, R. P. and MacDougall, J. D. B.,** Experimental atherosclerosis in rabbits after feeding cholestanetriol, *Br. J. Exp. Pathol.*, 49, 265, 1968.
32. **Seifter, J. and Baeder, D. H.,** Occurrence in plasma of an extractable lipid mobilizer, *Proc. Soc. Exp. Biol. Med.*, 91, 42, 1956.
33. **Chaikoff, I. L., Lindsay, S., and Lorenz, F. W.,** Production of atheromatosis in the aorta of the bird by the administration of diethystilbestrol, *J. Exp. Med.*, 88, 373, 1948.
34. **Schwenk, E., Stevens, D. F., and Altschul, R.,** Experiments in cholesterol atherosclerosis in rabbits, *Proc. Soc. Exp. Biol. Med.*, 102, 42, 1959.
35. **Engel, L. L. and Brooks, P.,** Cholesterol is stable, *Steroids,* 17, 531, 1971.
36. **Fieser, L. F.,** Cholesterol and companion. VII. Steroid dibromides, *J. Am. Chem. Soc.*, 75, 5421, 1953.
37. **Imai, H., Werthessen, N. T., Taylor, C. B., and Lee, K. T.,** Angiotoxicity and arteriosclerosis due to contaminants of U.S.P. grade cholesterol, *Arch. Pathol. Lab. Med.*, 100, 565, 1976.
38. **Peng, S. K., Taylor, C. B., Tham, P., Werthessen, N. T., and Mikkelson, B.,** Effect of auto-oxidation products from cholesterol on aortic smooth muscle cells — an in vitro study, *Arch. Pathol. Lab. Med.*, 102, 57, 1978.
39. **Kikuchi, S., Imai, Y., Suzuki-Ziro, S., Matsuo, T., and Noguchi, S.,** Biological studies of cholestane-3β,5α,6β-triol and its derivatives. III. The metabolic fate and metabolites of cholestane-3β,5α,6β-triol in animals, *J. Pharm. Exp. Ther.*, 159, 399, 1968.
40. **Steinburg, D., Fredrickson, D. S., and Avigan, J.,** Effects of 4-cholestenone in animals and in man, *Proc. Soc. Exp. Biol. Med.*, 97, 784, 1958.
41. **Peng, S. K., Taylor, C. B., Huang, W. Y., Hill, J. C., and Mikkelson, B.,** Distribution of 25-hydroxycholesterol in the lipoprotein and its role in atherogenesis. A study in squirrel monkeys, *Atherosclerosis,* 41, 395, 1982.
42. **Tauber, J. P., Cheng, J., and Gospodarowicz, D.,** Effect of high and low density lipoprotein on proliferation of cultured bovine vascular endothelial cells, *J. Clin. Invest.*, 66, 696, 1980.
43. **Schuk, J., Fairelough, G. F., and Haschemeyer, R. H.,** Oxygen mediated heterogeneity of apo low density lipoprotein, *Proc. Natl. Acad. Sci. U.S.A.*, 75, 3173, 1978.

Chapter 2

THE OXIDATION OF CHOLESTEROL

Leland L. Smith

TABLE OF CONTENTS

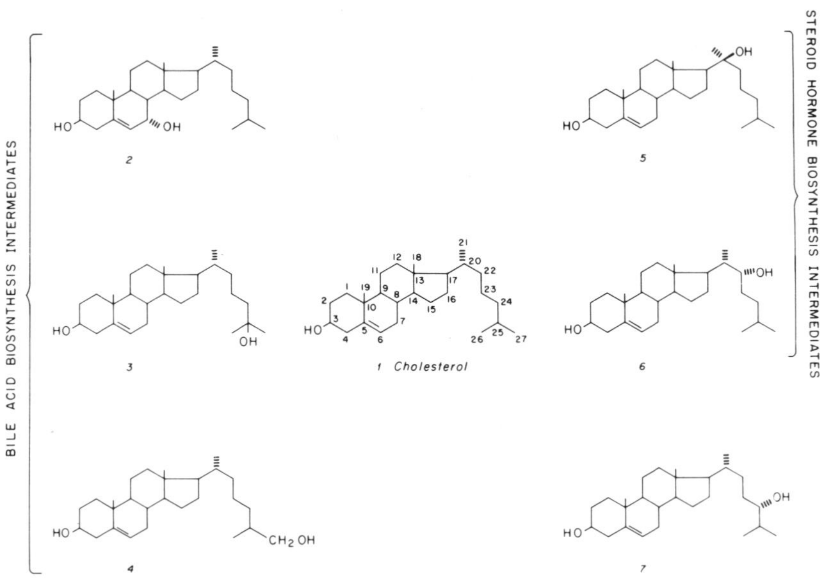

FIGURE 1. Monohydroxylated metabolites of cholesterol.

I. INTRODUCTION

Cholesterol (cholest-5-en-3β-ol, **1**) (Figure 1), essential to animal life via its stabilization of cellular membranes and to vertebrate life as a biosynthesis precursor of bile acids necessary for digestion and steroid hormones regulating cellular processes is readily oxidized by molecular oxygen in both enzymic and nonenzymic reactions. Indeed, the metabolism of cholesterol in animals is on balance oxidative, producing simple monooxygenated metabolites as precursors of more highly oxidized and degraded steroids. Additionally, nonenzymic processes such as autoxidation, lipid peroxidation, photochemical oxidation, and oxidation by environmental or *in vivo* reactive oxygen species generate analogous monooxygenated sterol derivatives with no presently recognized physiological role.[1-4]

These simple oxdation products, together with similar oxidized derivatives of other sterols and synthetic analogs, are the oxysterols, a set of oxidized sterol derivatives for which both *in vitro* and *in vivo* biological activities have been demonstrated,[1,5,6] biologically active oxysterols being the topic of the present monograph. More highly oxidized sterol metabolites such as the sapogenins, ecdysterols, antheridiols, polyhydroxylated marine sterols, and bile alcohols having recognized *in vivo* physiological roles, as well as oxidatively degraded steroids such as the bile acids and mammalian steroid hormones with established physiological functions, fall outside this definition.

Biologically active oxysterols derived from cholesterol have attracted the most attention for obvious reasons, cholesterol being the major sterol of human metabolism. However, analogous biologically active oxidized derviatives of other sterols formed by the same processes that act on cholesterol are also encompassed by the term oxysterol. For the sake of simplicity the origins of oxysterols derived from cholesterol and found in human tissues will be emphasized here. Chemical synthesis of oxysterol analogs will not be addressed. Also, references cited will be selected leading ones; full coverage of the pertinent literature is provided elsewhere.[1,2,5]

For satisfactory appreciation of the place oxysterols have in nature it is important to understand the processes by which oxysterols are formed. Some are established metabolites of cholesterol involved in the purposeful biosynthesis of steroids for which physiological roles are established. Other tissue oxysterols have less well-defined endogenous origins, whereas some have both an established metabolism origin and less well-defined origins; the origins of yet others remain uninvestigated. Dietary origins for many of these oxysterols have also been demonstrated.

With respect to the biological activities of dietary oxysterols it is now prudent to include among the oxysterols those nonoxidized sterol derivatives likely to be encountered in processed foodstuffs, including for example, stera-3,5-dienes and 3β-chloro-Δ^5-derivatives derived from acid hydrolysis of protein[7] and disteryl ethers formed during refining (bleaching) of edible oils.[8] Although these sterol derivatives are not formally oxysterols, their presence in processed foods together with the obvious potential for their oxidation to genuine oxysterols requires consideration, particularly as their biological properties are relatively unknown.

In addition, fatty acyl and sulfate esters of oxysterols are here encompassed by the generaly term oxysterol. Furthermore, inclusion of both oxidized and other sterol derivatives potentially formed *in vivo* under the action of environmental agents and xenobiotic substances must be considered for their relationships to the oxysterols (cf. Section III.C).

II. ENZYMIC OXIDATIONS

Endogenous sterol metabolism in which defined enzyme systems are associated generates a rich variety of oxysterols found as trace components of tissues. Among the enzymes involved are monooxygenases, dehydrogenases (and oxidases), and dioxygenases producing monohydroxysterols (stenediols), sterol epoxides, steroid ketones, and sterol peroxides. These oxysterols will be examined here in terms of the enzymes implicated in their formation, but as these enzymes are well-known types, details of mechanisms will not be treated.

The formation of oxysterols from cholesterol fatty acyl (or sulfate, etc.) esters by enzymic means, followed by hydrolysis of the 3β-ester, has not been demonstrated, although oxysterols esterified by fatty acids are present in freshly drawn human blood and oxysterol sulfates have been found in human blood, urine, and feces.[1,2]

A. MONOOXYGENASES

The monooxygenases implicated in oxysterol formation include the mixed-function oxidase (cytochrome P-450), hydroxylases, and epoxidases. Common tissue oxysterols such as the monohydroxylated cholesterols cholest-5-ene-3β,7α-diol (7α-hydroxycholesterol, **2**); cholest-5-ene-3β,25-diol (25-hydroxycholesterol, **3**); and (25R)-cholest-5-ene-3β,26-diol (26-hydroxycholesterol, **4**) (Figure 1) are endogenous cholesterol metabolites of established pathways to the bile acids. Others including (20S)-cholest-5-ene-3β,20-diol (20α-hydroxycholesterol, **5**) and (22R)-cholest-5-ene-3β,22-diol (22-hydroxycholesterol, **6**) are implicated in steroid hormone biosynthesis. Yet another is the cholesterol metabolite (24S)-cholest-5-ene-3β,24-diol (cerebrosterol, **7**), formed in central nervous tissue with no recognized physiological function there, but possibly also implicated in hepatic bile acid biosynthesis. The 7α-, 20-, 22-, 24-, 25-, and 26-monohydroxylated cholesterol derivatives **2** to **7** found in human tissues all have demonstrated enzymic origins and are formed purposefully. However, several (3β,7α-diol, **2**; 3β,20-diol, **5**; and 3β,25-diol, **3**) have established nonenzymic origins as well, to be discussed shortly.

The biosynthesis of cholesterol from lanosterol (5α-lanosta-8,24-dien-3β-ol, **10**) (Figure 2) also requires monooxygenase hydroxylations for removal of the 4- and 14α-methyl groups, thereby generating a set of C_{30}- and C_{27}-oxysterols bearing hydroxyl and carbonyl substituents. Some of these oxysterols, including 3β-hydroxy-5α-cholest-8(14)-en-15-one of current interest for its *in vitro* and *in vivo* biological activity, have been found in mammalian tissues and may contribute to any effects that tissue oxysterols may moderate.[9]

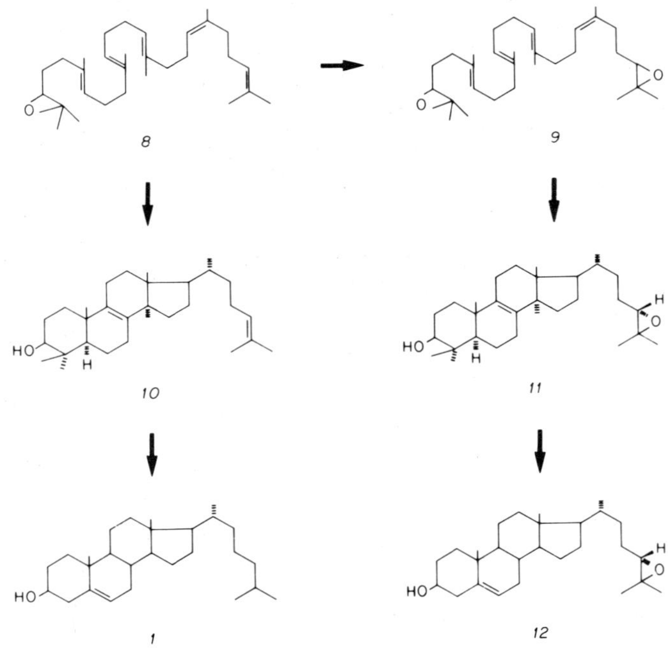

FIGURE 2. Sterol epoxides derived from (2S,22S)-squalene 2,3;22,23-bisepoxide.

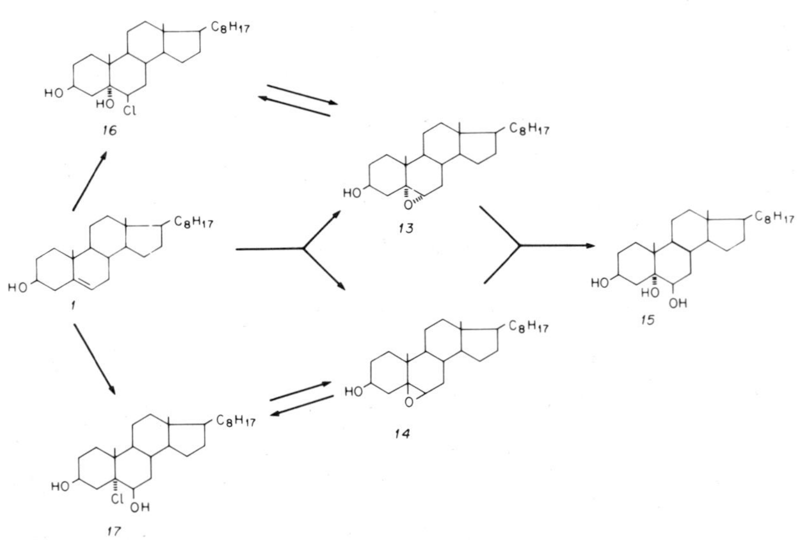

FIGURE 3. Cholesterol epoxides and chlorohydrins.

Also included among monooxygenases acting on sterols to generate oxysterols are epoxidases. The cholesterol 5α,6α-epoxide 5,6α-epoxy-5α-cholestan-3β-ol (**13**) (Figure 3) has a demonstrated enzymic origin as an *in vitro* metabolite of cholesterol in bovine adrenal cortex but also is found in human tissues along with the isomeric 5β,6β-epoxide 5,6β-epoxy-5β-cholestan-3β-ol (**14**). Both 5,6-epoxides **13** and **14** are also formed by nonenzymic action of several active oxygen species on cholesterol[10,11] Moreover, an alternative *de novo* biosynthesis pathway for oxysterol epoxides not involving cholesterol exists in which (3S,22S)-

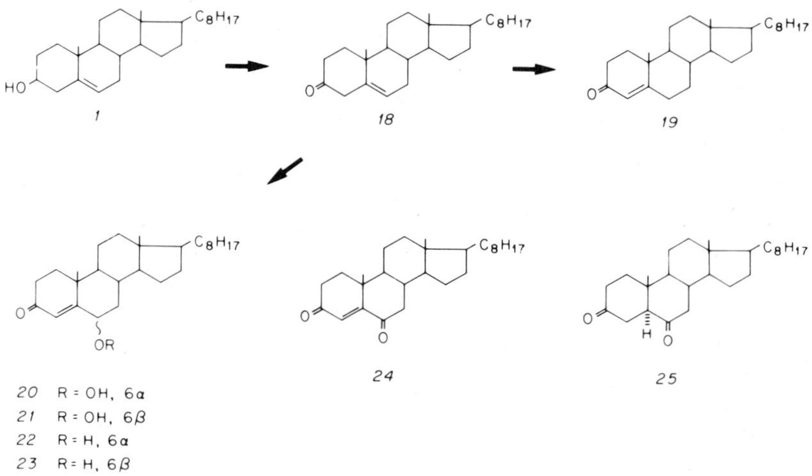

FIGURE 4. Oxysterol 3-ketones.

squalene 2,3;22,23-bisepoxide (**9**) (and not (2S)-squalene 2,3-epoxide, **8**) is cyclized to lanosterol 24,25-epoxide (24,25-epoxy-5α-(24S)-lanost-8-en-3β-ol, **11**) (Figure 2). Further enzymic oxidation of the C_{30}-epoxide **11** then yields desmosterol 24,25-epoxide (24,25-epoxy-(24S)-cholest-5-en-3β-ol, **12**).[12,13]

These specific *in vitro* enzymic 5,6- and 24,25-epoxidations may also occur *in vivo,* as the 5α,6α-epoxide **13** and 24,25-epoxides **11** and **12** have been detected in mammalian tissues.[14,15] However, epoxides **11** to **14** are also formed by nonenzymic processes,[16] and no physiological function has been demonstrated for any.

B. DEHYDROGENASES AND OXIDASES

Cholest-4-en-3-one (**19**) (Figure 4), found in mammalian tissues and implicated in the biosynthesis of the 5α- and 5β-cholestan-3β-ols, is a product of dehydrogenase action on cholesterol, via cholest-5-en-3-one (**18**). The same products are formed under the action of microbial cholesterol oxidases in which hydrogen peroxide (H_2O_2) is also generated. Both 3-ketones **18** and **19** are also formed nonenzymically in cholesterol autoxidation.

Dehydrogenases may also act on tissue oxysterols formed by enzymic or other means; for example, the 3β,7α-diol, **2**, is transformed by a liver dehydrogenase to 7α-hydroxycholest-4-en-3-one implicated in cholic acid biosynthesis. In this regard 7α-hydroxycholest-4-en-3-one is both an oxysterol (however one for which biological activity has yet to be discovered) and an endogenous cholesterol metabolite directed to bile acid and stanols biosynthesis.

C. DIOXYGENASES

In addition to hydroxylated, epoxidized, and dehydrogenated oxysterol metabolites, there is a set of tissue-peroxidized oxysterols that are formally products of the action of dioxygenases on the parent sterol. Thus, both atoms of dioxygen are present in these oxysterol metabolites. Soybean lipoxygenase (EC 1.13.11.12) and horseradish peroxidase (EC 1.11.1.7) oxidize cholesterol *in vitro*, yielding the epimeric cholesterol 7-hydroperoxides 3β-hydroxycholest-5-ene-7α-hydroperoxide (**28**) and 3β-hydroxycholest-5-ene-7β-hydroperoxide (**29**) (Figure 5) as initial products. Additionally, the hepatic lipid peroxidation system generates the same 7-hydroperoxides *in vitro*.[17] Also formed in these systems from the initial 7-hydroperoxides are the 3β,7α-diol, **2** (also a cholesterol monooxygenase metabolite); the epimeric cholest-5-ene-3β,7β-diol (7β-hydroxycholesterol, **36**) (Figure 6); and 3β-hydrox-

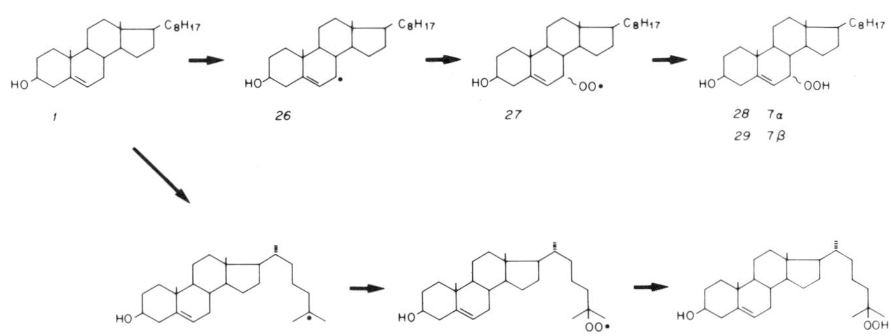

FIGURE 5. Derivation of cholesterol 7- and 25-hydroperoxides.

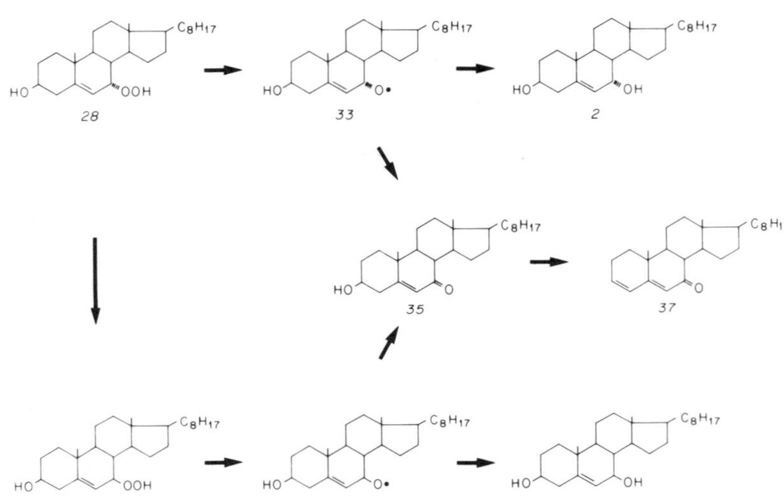

FIGURE 6. Derivation of the common 7-oxygenated oxysterols from the cholesterol 7-hydroperoxides.

ycholest-5-en-7-one (7-ketocholesterol, **35**); in some cases the isomeric 5,6-epoxides **13** and **14** and their common hydration product 5α-cholestane-3β,5,6β-triol (**15**) (Figure 3) may form.

In the absence of autoxidation, nonenzymic lipid peroxidation, and attack of hydroxyl radical (HO·), the product 7-hydroperoxides **28** and **29** infer dioxygenase action, and the product pattern of 3β,7-diols (**2** and **36**), 7-ketone (**35**), and 5,6-epoxides (**13** and **14**) implies generation of the 7-hydroperoxides, thus of dioxygenase action. This pattern of oxysterols (**2**, **13**, **14**, **35**, and **36**) is formed *in vitro* by human blood polymorphonuclear leukocytes and by rat lung macrophages. These results further suggest a possible intravascular lipid peroxidation of cholesterol *in vivo*.[18,19]

However, enzyme-substrate complex formation and other fundamental characterization of the reactions establishing genuine dioxygenase action on cholesterol have not been demonstrated, and the observed transformations may not utilize enzyme catalysis directly. Rather, these dioxygenases may merely oxidize cholesterol as a secondary matter in conjunction with their primary attack on polyunsaturated fatty acids as preferred substrates. The product pattern of 7-hydroperoxides **28** and **29** and their derivatives **2, 35, 36** and oxysterols **13** to **15,** as noted hereinafter, is exactly that evincing radical autoxidation of cholesterol.

The 7-hydroperoxides **28** and **29** have not been identified in fresh mammalian tissues, but their decomposition products 3β,7α-diol,**2**; 3β,7β-diol, **36**; and 7-ketone, **35** are present in freshly drawn human blood,[20-29] and a cholesterol fatty acyl ester hydroperoxide not otherwise identified has been detected in human blood.[30] These items imply an *in vivo* lipid peroxidation of cholesterol, whether involving dioxygenases or not.

The action of dioxygenases on sterols is more strongly supported in the case of numerous sterol Δ-5α,8α-peroxides found in marine invertebrates and microbial systems. However, the means by which these peroxysterols are formed remains uncertain, either by dioxygenases acting on sterol substrates or by active oxygen species, however derived, acting on $\Delta^{5,7}$-3β-alcohols. It is tempting to pose that Δ^{6}-5α,8α-peroxides be formed by action of haloperoxidases on appropriate sterol substrates in a fashion similar to the iodoperoxidase system forming ascaridole from α-terpinene.[31]

Yet other types of peroxysterols have been isolated from marine invertebrates[32,33] and plant tissue[34] under conditions suggesting likely metabolite status and dioxygenase action on sterol substrates.

The actions of monooxygenases, dehydrogenases, oxidases, and dioxygenases acting on cholesterol thus can generate oxysterols **2** to **7**, **11** to **15**, **18** and **19**, **28** and **29**, and **35** and **36** *inter alia,* as cholesterol metabolites potentially present in human tissues, oxysterols which are active in a variety of bioassays and for which there is current concern.

III. NONENZYMIC OXIDATIONS

Yet other important modes of formation in tissues of many of these same oxysterols exist. Among these are autoxidation involving ground-state dioxygen ($^3O_2$) and oxidation by other reactive oxygen species such as ozone (O_3), electronically excited (singlet) dioxygen ($^1O_2$), H_2O_2, dioxygen cation (O_2^+), and HO·. Of the recognized active oxygen species, only the dioxygen one-electron reduction product superoxide (O_2^-, HOO·) fails to oxidize cholesterol.[35] However, it must be recognized that O_2^- as a precursor of H_2O_2, in turn a precursor of HO·, may be a source of an oxidant attacking cholesterol, particularly in the presence of transition metal ions. Autoxidation is by far the most extensive process, yielding complicated mixtures of oxysterols only partially examined to date.

A. AUTOXIDATION

Autoxidation is the apparently uncatalyzed oxidation of cholesterol by the molecular oxygen of the air. Emphasis is placed on the apparent lack of catalysis, but one must posit that some agency promoting the initial events of autoxidation may be present nonetheless. No such agency has been established for the autoxidation of bulk cholesterol stored in the presence of air, but traces of O_3 or other active oxygen species in ambient air, of undetected peroxides or transition metal present in the sterol sample, or perhaps of exposure of the sample to heat or other radiation may be posed. For autoxidations in tissues the presence of iron or other transition metal ions as catalyst is presumed.

The autoxidation of cholesterol has been observed repeatedly for decades in a rich variety of experimental conditions. The mere storage of pure crystalline cholesterol in its original package from the manufacturer in the presence of air results in cholesterol autoxidation, generally easily confirmed by organoleptic means. Most newly opened bottles of cholesterol have a characteristic odor of oxidized side-chain fragments derived from carbon-carbon bond scissions concommitant with decomposition of sterol hydroperoxides, the earliest oxysterol products that can be isolated.

The development of odor and off-white to ivory to yellow coloration in bulk samples, both intensifying with air-aging, is characteristic of cholesterol autoxidation. Deeper tan and brown colors and the more acrid odor of destroyed sterol material ultimately may develop. The initially formed cholesterol hydroperoxides, easily detected by sensitive color tests on thin-layer chromatograms, are sequentially accompanied by thermal decomposition products and thereafter by a galaxy of degraded products. Hydrogen peroxide is also formed, which together with sterol hydroperoxides already present, may cause additional complexity to the spate of oxysterol products.

The oxysterol products of nonenzymic autoxidation and of enzyme-catalyzed oxidations just described are uncannily alike. The correspondence between products of enzymic peroxidation of cholesterol and nonenzymic autoxidation, including many secondary and higher-order transformation products of the initially formed sterol hydroperoxides, suggests the enzyme-catalyzed steps evolved simply from those that occur naturally by autoxidation. Indeed, the same three fundamental oxidation processes of oxygen insertion into C–H bonds, epoxidation of the Δ^5-double bond, and 3β-alcohol dehydrogenation to 3-ketone occur. Moreover, most of the oxysterol metabolites are also products of autoxidation, and in biological systems where both processes occur the oxysterols formed may have dual origins. Although the cholesterol metabolites are formed by monooxygenases using $^3O_2$, only one atom of dioxygen appears in the products. No added oxygen appears in the dehydrogenated 3-ketone metabolites, but the enzymic process requires dioxygen. Obviously both atoms of $^3O_2$ are incorporated into the dioxygenase products.

Reliable analysis data measuring the extent of artifact oxidations in tissues or *in vitro* experiments are not readily available, but the extent of the problem may be gauged in several cases. For example, freshly drawn human serum unprotected by added antioxidant having 24 ng/ml 7-ketone **35** had 63 ng/ml after 1 h. Other samples with 24 to 60 ng/ml were compared with 50- to 300-ng/ml-levels in samples frozen 1 to 6 months without antioxidant. Oxysterol fatty acyl esters of human plasma also increase in samples stored frozen, levels as high as 500 to 1000 ng/ml for the 3β,7-diols **2** and **36,** and 7-ketone **35,** being observed. Aged foodstuffs also show increased oxysterol levels, 30 μg/g 3β,7-diols **2** and **36** in butter oil increasing to 90 μg/g 3β,7α-diol, **2,** 60 μg/g 3β,7β-diol, **36,** after one year at −20°C.[20,21,24,36]

Despite rigorous attempts to limit adventitious sterol oxidation and estimation of artifact oxidation using added [4-^{14}C]-cholesterol, plasma levels of the three oxysterols 3β,7β-diol, **36,** 7-ketone, **35,** and 5β,6β-epoxide, **14,** most suggestive of *in vitro* lipid peroxidation of cholesterol (and thereby of cholesterol autoxidation in tissues *ex vivo*), were increased nonetheless during analyses. Levels were increased from 0 to 0.8 ng/ml to 2.8 to 3.6 ng/ml for 3β,7β-diol, **36;** from 0.4 to 2.8 mg/ml to 4.0 to 6.4 ng/ml for 7-ketone, **35;** and from 0 to 2.4 ng/ml to 7.2 to 31.7 ng/ml for 5β,6β-epoxide, **14**! The study further evinced basal levels of 3β,7α-diol, **2;** 3β,26-diol, **4;** and 3β,24-diol, **7.**[27]

Autoxidation also occurs during *in vitro* experiments involving cholesterol. The extent is measured by the relative amounts of 3β,7α-diol, **2** vs. 3β,7β-diol, **36** (680 ng/g vs. 145 ng/g in one case).[37]

It is of importance to recognize the insidious nature of cholesterol autoxidation so that mistaken views of sterol metabolism not be held. Except in those cases where enzymic metabolism is demonstrated by the usual sound methods, or in the case of side-chain hydroxylations where only one of two possible epimers are formed, the artifact nature of an oxysterol sample must be presumed.

As there shall always be new items discovered dealing with the intimate details of cholesterol disposition *in vivo,* so this generality must be revised as needed. As an example, a previous notion that cholest-5-ene-3β,4β-diol isolated from air-aged cholesterol be an autoxidation product of cholesterol[38] may have to be revised, for the oxysterol has been detected in human blood serum[22] and in butter[39] and the requisite isomeric cholesterol 4-hydroperoxides required of autoxidation have never been detected among cholesterol au-

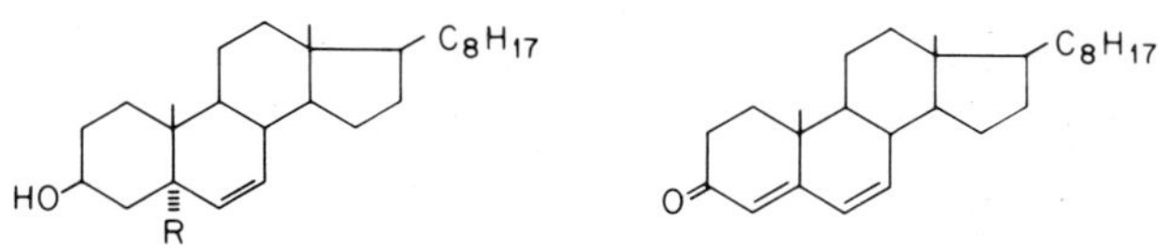

38 R = OOH
39 R = OO•
40 R = O•
41 R = OH

42

FIGURE 7. Oxysterols derived from singlet molecular oxygen attack on cholesterol.

toxidation products. Moreover, the recent discovery of 4β-hydroxylated bile acids in human infant bile and feces suggests hepatic 4β-hydroxylation is an important metabolic process.[40]

Exactly the same reconsideration must be given to the presence of cholesta-4,6-dien-3-one (**42**) (Figure 7), also found in air-aged cholesterol.[38] In this case, the $\Delta^{4,6}$-3-ketone, **42,** might be an autoxidation product of the 3β,7α-diol **2** (dehydrogenation of 3β,7α-diol, **2** to 7α-hydroxycholest-4-en-3-one, in turn dehydrated to $\Delta^{4,6}$-3-ketone, **42**), or a major decomposition product of 3β-hydroxy-5α-cholest-6-ene-5-hydroperoxide (**38**) (Figure 7), the product of the action of $^1O_2$ on cholesterol (see Section III.B.2), but more likely is a metabolite of the 3β,7α-diol, **2** implicated in hepatic bile acid biosynthesis, as the $\Delta^{4,6}$-3-ketone, **42,** has also been detected in human serum.[41]

The dilemma of whether an oxysterol recovered from biological material is metabolite or artifact thus remains a serious one in need of the most careful examination. The question of origin is all the more important as more becomes known about the biological activities of oxidized sterols. Whether tissue oxysterols are metabolites formed purposefully in connection with biosynthesis of other steroids, whether the oxysterols play a role per se in metabolism, or whether the oxysterols uniformly toxic in bioassays are deleterious *in vivo,* posing human health problems, require the utmost care be exercised in tissue analyses.

Cholesterol autoxidation is a radical process in which the same chemistry occurs as found for autoxidations of other unsaturated lipids and susceptible organic compounds. As previously noted, the lipid peroxidation of cholesterol in which dioxygenases are implicated appears to follow the same reaction pathway as autoxidation, and thus seems to be a radical process as well. The same initial oxidation products 7α- and 7β-hydroperoxides, **28** and **29,** (Figure 5) are formed in both processes, and the same secondary and higher-order transformation products are formed from the 7- hydroperoxides in both cases.

1. Initial Reactions

The initiation reaction of autoxidation is posited as involving abstraction of a reactive allylic 7-hydrogen atom, yielding the carbon-centered radical 3β-hydroxycholest-5-en-7-yl (**26**), which reacts with $^3O_2$ to yield the 3β-hydroxycholest-5-en-7-peroxyl radicals (**27**), stabilized in turn by hydrogen abstraction to provide the isolable product 7-hydroperoxides **28** and **29.** This process is thus the standard or classic representation of radical oxidation of the allylic C–H bond of olefins.

The epimeric 7-hydroperoxides **28** and **29** are the first-detected stable products of the process, but their precursors, C7 radical **26** and 7-peroxyl radicals **27,** are detected by electron spin resonance spectroscopy. Cholesterol irradiated in a vacuum with ionizing radiation gives rise to an electron spin resonance signal of the C7 radical **26;** introduction of $^3O_2$ into the system leads to the formation of 7-peroxyl radicals **27** at the expense of the carbon-centered radical **26.**[42,43] It is uncertain whether distinct epimeric C7 radicals **26** exist, but both epimeric 7α- and 7β-peroxyl radicals **27** are presumed to form. The spectroscopic detection of radicals **26** and **27** combined with isolation of the 7-hydroperoxides **28** and **29** defines the overall initial autoxidation process.

The cholesterol radicals **26** and **27** have not been sought in tissues or in cholesterol dispersions and solutions because of technical limitations, but the stable first product 7-hydroperoxides **28** and **29** have been observed in these circumstances, thus justifying extension of the described radical mechanism observed in the solid state to these other systems. Indeed, in every case where adequate product analysis has been used the presence of the 7-hydroperoxides as initial products is observed. Careful kinetics studies then show the decline in 7-hydroperoxides as secondary products incease, confirming in detail the processes advanced to account for most of the oxysterols formed in cholesterol-containing samples.

Other electron spin resonance signals of a tertiary carbon-centered C25 radical 3β-hydroxycholest-5-en-25-yl (**30**) are also observed, evincing the tertiary C25 carbon as the next most reactive radical oxidation site. Reaction of the C25 radical **30** with $^3O_2$ then yields 3β-hydroxycholest-5-ene-25-peroxyl radical (**31**) stabilized by hydrogen abstraction to give 3β-hydroxycholest-5-ene-25-hydroperoxide (cholesterol 25-hydroperoxide, **32**) as the third most prevalent peroxide autoxidation product of cholesterol (Figure 5).

The C25 radical **30** may form directly from cholesterol by the same primary hydrogen abstraction reaction described for formation of the C7 radical **26** but may also form by radical transfer reaction involving a radical already formed reacting with cholesterol, thus by hydrogen abstraction effected by C7 radical **26,** 7-peroxyl radical **27,** or 3β-hydroxy-cholest-5-en-7α- or 7β-oxyl radicals **33** and **34** (Figure 6). The C25 radical **30** then continues the radical chain reactions in the same manner as the C7 radical **26,** yielding the isolable 25-hydroperoxide **32.**

Yet other C20, C24, and C26 carbon-centered radicals must form in cholesterol subjected to autoxidations, as the additional cholesterol hydroperoxides 3β-hydroxy-(20R)- and (20S)-cholest-5-ene-20-hydroperoxides, 3β-hydroxy-(24R)- and (24S)-cholest-5-ene-24-hydroperoxides, and 3β-hydroxy-(25R)- and (25S)-cholest-5-ene-26-hydroperoxides have been isolated from air-aged cholesterol, along with their unique thermal decomposition products. Yet other degraded oxidation products found in air-oxidized cholesterol imply analogous oxidative attack at the C17, C22, and C23 sites, although the requisite hydroperoxides have not been discovered. For example, 3β-hydroxyandrost-5-en-17-one (dehydroepiandrosterone) infers C17 attack; 3β-hydroxy-22,23-bisnorcholenic acid infers attack at the C22 and/or C23 position. However, as these carboxylic acids are present at low levels in air-aged cholesterol, their origins as cholesterol metabolites in the liver biosynthesis of bile acids cannot presently be excluded. Other identified oxidized products establish that cholesterol is autoxidized predominantly at the allylic C7 position but also at every site in the side chain save the C21 carbon atom.[1,2] The autoxidation of cholesterol in the side chain is quantitatively much less than is autoxidation at the allylic C7 site and appears to be confined to the case of solid cholesterol exposed to air.

The suggested oxidative attacks at the C20, C22, C23, and C24 carbon atoms are confirmed by the presence of at least 14 volatile C_2- to C_6-components formed during autoxidation of cholesterol (cf. Section III.2). The characteristic odor of air-aged cholesterol arises from formation of these and other unidentified volatiles.

Whether autoxidations ocur at the allylic C4 site of the A ring or at nuclear tertiary carbon atoms (potentially with attendant formation of nuclear secosterols) has never been determined. Resolution of the very complex highly polar autoxidation products is a demanding matter attracting little attention, for which funding support does not exist. However, the highly polar cholesterol autoxidation products may be posited as having biological activity were bioassays conducted, witness weak mutagenicity of some of these fractions.[38]

Cholesterol autoxidation may occur wherever cholesterol is exposed to $^3O_2$, whether the cholesterol is solid, in solution, or in the dispersed state. The characteristic spate of

7-oxygenated products 7-hydroperoxides, **28** and **29;** 3β,7-diols, **2** and **36;** 7-ketone, **35;** isomeric 5,6-epoxides, **13** and **14;** and 3β,5α,6β-triol, **15** evince autoxidation, as no other naturally occurring oxygen species nor oxidation process has been described that generates all these oxysterols.[10,11] However, certain other oxidations of cholesterol mimic autoxidation with respect to the products formed; for example, platinum-catalyzed oxidation of cholesterol by air generates products of oxidation in the C7 allylic position as well as oxidation of the 3β-hydroxyl group to Δ^5-3-ketone, **18,** with its subsequent oxidation to the 6-oxygenated products 6α- and 6β-hydroperoxycholest-4-en-3-ones (**20, 21**); 6α- and 6β-hydroxycholest-4-en-3-ones (**22, 23**); and cholest-4-ene-3,6-dione (**24**) (Figure 4).[44]

The radical autoxidation of cholesterol at the allylic C7 site occurs in cellular membranes and in model systems that mimic biological membranes. The characteristic oxysterols 3β,7-diols, **2** and **36;** 7-ketone, **35;** 5,6-epoxides, **13** and **14;** and the 3β,5α,6β-triol, **15** are variously detected in model systems involving monomolecular film spread on water or solid support and in aqueous emulsions, micelles, and liposomes containing phospholipids. Reaction rates and specific oxysterol products formed depend on experimental conditions, and other system components affect results, but on balance the same processes and oxysterol products are implicated.[45-52] As the oxysterols formed are the same as those formed in model chemical systems, we may conclude that cholesterol autoxidation taking place in any of these media occurs by the common mechanism outlined for crystalline cholesterol.

Not every oxysterol of this set establishing autoxidation is reported in many studies; only selected members of the group are often detected. Particularly missing are the 7-hydroperoxides **28** and **29,** whose detection requires more care. Frequently the isomeric 5,6-epoxides **13** and **14** are not resolved, and other inadequacies of analysis also are recorded. In cases where the characteristic pattern is incomplete, it is probably that inadequate analysis methods were employed or that only one or a few of the oxysterols were of particular interest.

In addition to these autoxidation processes in which both atoms of dioxygen are introduced into the cholesterol molecule there are at least two other autoxidation processes recognized that involve oxidation of the two functional groups of cholesterol. Both processes require dioxygen and, putatively, formation of intermediate sterol hydroperoxides.

The cholesterol secondary 3β-hydroxyl group is dehydrogenated to the Δ^5-3-ketone **18** (Figure 4) and H_2O_2 is eliminated, yielding the 3-ketone, **18.** Oxygen-dependent dehydrogenations of sterol hydroperoxide to olefin has been observed in other cases (cf. Section III.A.2), and it is presumed that oxygen-dependent dehydrogenation of alcohol to ketone likewise involves transient hydroperoxide formation.

The initially formed Δ^5-3-ketone, **18,** is readily isomerized to the stable Δ^4-3-ketone, **19,** but (more significantly for generation of oxysterols) is subject to further autoxidation, yielding the epimeric 6α- and 6β-hydroperoxides, **20** and **21,** putatively by the same radical processes evinced for C7 and side-chain carbon autoxidations, thus via initial allylic C6 hydrogen abstraction followed by reaction with $^3O_2$ to give the corresponding 6-peroxyl radical, in turn stabilized by hydrogen abstraction yielding the epimeric 6-hydroperoxides **20** and **21.** These A-ring autoxidations occur with solid cholesterol[53] and with cholesterol in solution[54] but are not observed as frequently as is autoxidation at the allylic C7 carbon atom.

The second autoxidation process dependent on initial sterol hydroperoxide (or H_2O_2) formation is that of epoxidation of the cholesterol Δ^5-double bond, yielding the isomeric 5,6-epoxides **13** and **14** (Figure 3). The 5β,6β-epoxide **14** predominates in these cases, with **13**:**14** ratios of about 1:3 to 1:11. However, as noted hereinafter, other oxidants such as O_3 and HO· also epoxidize cholesterol with different proportions of 5α,6α-epoxide **13** and 5β,6β-epoxide **14.**[10,11] Moreover, the 5,6-epoxides are differentially unstable, both being hydrolyzed to the 3β,5α,6β-triol **15,** the 5β,6β-epoxide **14** being less stable at acid pH.[55] Accordingly, the ratio of 5α,6α-epoxide **13** to 5α,6β-epoxide **14** may be unreliable criterion for recognition of the oxidation process implicated in certain circumstances.

Cholesterol fatty acyl esters are also subject to autoxidation by radical processes. Cholesterol fatty acyl esters exposed in vacuum to ionizing radiation form initially an anion radical centered on the ester carbonyl group, but the C7 allylic and other radicals (probably including the C25 radical) are also formed.[42] Oxysterol esters recovered from the autoxidation of cholesterol 3β-acetate include 3β-acetoxycholest-5-ene-7β-hydroperoxide (**29** 3β-acetate), 3β-acetoxy-(20S)-cholest-5-ene-20-hydroperoxide, and 3β-acetoxycholest-5-ene-25-hydroperoxide (**32** 3β-acetate).[56] These items confirm that cholesterol fatty acyl esters are autoxidized by the same processes as the sterol.

The rates of autoxidation of cholesterol in comparison with its fatty acyl esters are different depending on conditions. In aqueous dispersions cholesterol is the more rapidly oxidized. Moreover, samples of cholesterol stored unprotected from air develop the characteristic odor of autoxidation within a year, but cholesterol fatty acyl esters stored similarly remain odorless for years. However, sterol fatty acyl esters are more rapidly autoxidized when heated in air, and the rate of cholesterol autoxidation is increased in mixtures of cholesterol and its fatty acyl esters.[57,58] These findings suggest caution in interpretation of analyses of tissues for oxysterols where cholesterol and its fatty acyl may both be present.

Other Δ^5-sterols are also subject to autoxidation by the same processes, yielding the analogous oxysterol products, some with similar biological activities. Sterols with double bonds at other sites may yield hydroperoxides at either or both allylic positions, witness the case for 5α-lanost-8-en-3β-ol where both 7- and 11-hydroperoxides are formed, and allylic rearrangements of the double bond may occur in conjunction with hydroperoxide formation. Moreover, sterols with alkyl substitution in the side chain may also undergo additional oxidations at the additional tertiary carbon centers.[1,2]

2. Subsequent Reactions

All of the initially formed peroxidic oxysterols are subject to further alterations, yielding a vast array of degraded oxysterols described shortly. However, it must be noted that the presently described autoxidations of cholesterol do not begin to exhaust the very complex secondary and higher-order degradations that occur. The most polar cholesterol oxidation products found in once pure air-aged cholesterol are the steroid acids 3β-hydroxychol-5-enic acid, 3β-hydroxy-22,23-bisnorcholenic acid, and 3β-hydroxyandrost-5-ene-17β-carboxylic acid, but a host of other very polar unidentified products are formed, yet to be examined.[38]

There are three kinds of reactions undergone by the initially formed 7-hydroperoxides, **28** and **29:** (1) as previously noted, in the presence of cholesterol the 7-hydroperoxides epoxidize the Δ^5-double bond by donation of the terminal peroxide oxygen atom, forming the corresponding 3β,7-alcohols as co-product; (2) the 7α-hydroperoxide **28** is epimerized spontaneously to the 7β-hydroperoxide **29;** (3) the 7-hydroperoxides are transformed to nonperoxidic products under a variety of conditions.

Epoxidations of the Δ^5-double bond of cholesterol has already been discussed, as also has 5,6-epoxide hydration. Moreover, those oxysterols retaining the Δ^5-double bond may be epoxidized and the resultant 5,6-epoxides hydrated in the same fashion, as 5,6α-epoxy-5α-cholestane-3β,7α-diol and 5α-cholestane-3β,5,6β,7α-tetraol are cholesterol autoxidation products,[59] and mixed 5,6ξ-epoxy-5ξ-sitostane-3β,7ξ-diols are formed from sitosterol (24-ethyl-(24R)-cholest-5-en-3β-ol) 3β-stearate.[60] The 3β,5α,6β-triol, **15,** may be further dehydrogenated to 3β,5-dihydroxy-5α-cholestan-6-one.

Epimerization of the quasiaxial 7α-hydroperoxide **28** to the quasiequatorial 7β-hydroperoxide **29** occurs spontaneously, but the reverse epimerization has not been observed. The rearrangement involves the epimeric 7-peroxyl radicals **27,** as indicated in Equations 1 and 2, with dissociation of dioxygen and recombination of the C7 radical **27** with dioxygen.[43,56,61]

$$7\alpha\text{-OOH } \mathbf{28} \rightarrow 7\alpha\text{-OO}\cdot \ \mathbf{27} \rightarrow \text{C7}\cdot \ \mathbf{26} + O_2 \tag{1}$$

$$\text{C7}\cdot \ \mathbf{26} + O_2 \rightarrow 7\beta\text{-OO}\cdot \ \mathbf{27} \rightarrow 7\beta\text{-OOH } \mathbf{29} \tag{2}$$

It is the thermal decomposition of the initially formed sterol hydroperoxides that generates the very complex spate of oxysterols encountered in autoxidized cholesterol The 7-hydroperoxides **28** and **29** are unstable under conditions involving heat and reducing agents. Reducing agents transform the 7-hydroperoxides to the corresponding 3β,7-diols (3β,7α-diol **2** from 7α-hydroperoxide **28,** 3β,7β-diol **36** from 7β-hydroperoxide **29**), and the thermal decomposition of each 7-hydroperoxide likewise yields the corresponding 3β,7-diol together with the 7-ketone, **35** (Figure 6).

The thermolysis of the 7-hydroperoxides may be formulated in the same manner of other organic hydroperoxides (Equation 3), with

$$\text{R–O–O–H } \mathbf{28/29} \rightarrow \text{R–O}\cdot \ \mathbf{33/34} + \text{HO}\cdot \tag{3}$$

homolytic cleavage of the peroxide bond yielding the cholesterol 7-oxyl radicals, **33** and **34**. Peroxide bond homolysis putatively generates HO· that also oxidizes cholesterol to the 3β,7-diols **2** and **36** and 7-ketone **35** *inter alia* (cf. Section III.B.6).

It is likely that both 7α- and 7β-oxyl radicals **33** and **34** exist, but the point has not been directly established. Stabilization of either 7-oxyl radical by hydrogen abstraction gives the stable product 3β,7-diol **2** or **36**. Alternatively, stabilization by β-scission of the C7/H7 bond yields the 7-ketone **35**. Both processes occur together, generating the classic pattern of oxysterols 3β,7-diols **2** and **36** and 7-ketone **35** (with 7-hydroperoxides **28** and **29**) evincing radical oxidation of cholesterol, whether the reaction occurs in solid, solution, or dispersed states.

In addition to these reactions of the 7-oxyl radicals **33** and **34**, thermal elimination reactions of the stable oxysterols, or of radicals derived from them, occur. The most commonly encountered is the dehydration of the 7-ketone **35** to cholesta-3,5-dien-7-one (**37**) (Figure 6) and that of 7α-hydroxycholest-4-en-3-one to the $\Delta^{4,6}$-3-ketone, **42** (Figure 7) under alkaline conditions and by heat. These oxysterols are readily recognized by their increased chromatographic mobility (less polar than cholesterol) and characteristic ultraviolet light absorption.

Thermal eliminations of the 7-hydroperoxides **28** and **29** also occur, with formation of cholesta-2,4,6-triene.[61] Thermal decomposition of the 7-hydroperoxides is accompanied by interconversions of the 7-hydroperoxides, **28** and **29**; 3β,7-diols, **2** and **36**; or 7-oxyl radicals, **33** and **34**. Less vigorous conditions acting on other sterol hydroperoxides yield sterol olefins; for example, the epimeric 3β-hydroxy-5α-cholest-7-ene-6-hydroperoxides (formed by radiation-induced autoxidation of 5α-cholest-7-en-3β-ol) yield cholesta-5,7-dien-3β-ol,[62] and among the autoxidation products of sitosterol 3β-stearate is 24-ethylcholesta-5,24-dien-3β-ol. The oxygen-dependent dehydrogenation could occur via formation of a 24- or 25-hydroperoxide with subsequent elimination of H_2O_2 to yield the olefin.[63,64] Formation of desmosterol (cholesta-5,24-dien-3β-ol) as a decomposition product of either cholesterol 24- or 25-hydroperoxide, though conceptionally possible in the same manner, has not been observed.

In addition to these subsequent reactions of initially formed cholesterol autoxidation products, each initially formed oxysterol is subject to further autoxidation. Formation of 5α-cholestane-3β,5,6β,7α-tetraol via epoxidation of the 3β,7α-diol **2** has already been noted to be an oxysterol produced by two separate oxidations of cholesterol. Furthermore, autoxidation of the Δ^5-3-ketone, **18**, yields the epimeric 6α- and 6β-hydroperoxides, **20** and **21**, whose subsequent decomposition in turn yields the corresponding 6α- and 6β-hydroxy-Δ^4-

3-ketones **22** and **23** via the postulated 6-oxyl radical and hydrogen abstraction, and the Δ^4-3-ketone **24** by β-scission.[53] Thermal decomposition of the epimeric 6-alcohols **22** and **23** additionally yields 5α-cholestane-3,6-dione (**25**).

Multiple attack of dioxygen in the sterol side chain also may occur. The epimeric 3β-hydroxy-(20RS)-pregn-5-ene-20-hydroperoxides recovered from air-oxidized cholesterol suggest a double attack of dioxygen, initially accompanied by C20/C22 bond scission with subsequent additional radical attack and formation of the degraded C_{21}-20-hydroperoxides.[65] Additionally, a double oxidative attack of dioxygen on cholesterol is implied by the reported recovery of 3β,20-dihydroxy-(20RS)-cholest-5-ene-25-hydroperoxide from autoxidized cholesterol.[66]

These subsequent transformations of initially formed oxysterols have dealt mainly with A- and B-ring alterations of the sterol nucleus, as these are the most prominent ones encountered. As it appears that no nuclear carbon-carbon bonds are cleaved in cholesterol autoxidations (no secosterols have ever been discovered among cholesterol autoxidation products), there remains the subsequent reactions of the oxysterols bearing side-chain hydroperoxide features. The same general alteration processes described for the 7-hydroperoxides **28** and **29** also occur with the side-chain hydroperoxides, thus peroxide bond scission and hydrogen abstraction and β-scission reactions of the resultant oxyl radical. However, in the case of the side-chain hydroperoxides, the β-scission reactions include carbon-carbon cleavages yielding degraded oxysterols as well as carbon-hydrogen scissions yielding ketones.

The prominent autoxidation at the tertiary C25 site yielding the 25-hydroperoxide **32** is followed by subsequent hydrogen abstraction by the postulated 25-oxyl radical, yielding the 3β,25-diol, **3**. Also formed are the degraded oxysterols chol-5-en-3β-ol and 3β-hydroxy-27-norcholest-5-en-25-one as β-scission products, in which the C24/C25 and C25/C27 bonds, respectively, are cleaved.

The thermal degradation of the 25-hydroperoxide **32** also generates another degraded product pregn-5-en-3β-ol in which the C20/C22 bond is broken. The transformation may be formulated as involving homolysis of the 25-hydroperoxide **32** peroxide bond, with the 25-oxyl radical thereby formed abstracting the tertiary C20 hydrogen to yield 3β,25-dihydroxycholest-5-en-20-yl radical that then yields pregn-5-en-3β-ol by β-scission of the C20/C22 bond and 2-methylpent-4-en-2-ol identified as a volatile oxidation product of cholesterol (*vide infra*).

Carbon-carbon bond cleavages also occur in the decomposition of other cholesterol side-chain hydroperoxides. The isomeric tertiary hydroperoxides 3β-hydroxy-(20S)- and (20R)-cholest-5-ene-20-hydroperoxides both formed from cholesterol decompose by the same processes, homolysis of the 20-hydroperoxide peroxide bond yielding the corresponding 20-oxyl radicals, stabilized by hydrogen abstraction to give the (20S)-3β,20-diol **5** and (20R)-cholest-5-ene-3β,20-diol. Additionally, 3β-hydroxypregn-5-en-20-one (pregnenolone) is formed by β-scission of the C20/C22 carbon-carbon bond of the 20-oxyl radical.

Yet other complexity also accompanies the decomposition of the 20-hydroperoxides. The C_{19}-steroids androst-5-ene-3β,17β-diol, androst-5-en-3β-ol, and androsta-5,16-dien-3β-ol are formed, by β-scission of the C17/C20 bond of the 20-oxyl radical. The putative intermediate 3β-hydroxyandrost-5-en-17-yl radical must then undergo reaction with HO· formerly present from homolysis of the 20-hydroperoxide peroxide bond to yield androst-5-ene-3β,17β-diol. Disproportionation of the 17β-yl radical then would yield androst-5-en-3β-ol and androsta-5,16-dien-3β-ol.

The epimeric cholesterol 24-hydroperoxides decompose via peroxide bond homolysis and β-scission of the 24-oxyl radical only, with no hydrogen abstraction processes detected. Oxysterol products include 3β-hydroxycholest-5-en-24-one from cleavage of the C24/H24 bond, 3β-hydroxychol-5-en-24-al from C24/C25 bond scission, and 24-norchol-5-en-3β-ol from cleavage of the C23/C24 bond. Further oxidation of the 24-aldehyde may be the process by which 3β-hydroxychol-5-enic acid, found as an autoxidation product of cholesterol, is formed.

The cholesterol 26-hydroperoxides 3β-hydroxy-(25R)- and (25S)- cholest-5-ene-26-hydroperoxides decompose by hydrogen abstraction yielding the corresponding 26-alcohols (25R)- and (25S)-cholest-5-ene-3β,26-diols and by β-scission of the C26/(C27)/H26(27) bond, yielding the 26-aldehydes 3β-hydroxy-(25RS)-cholest-5-en-26-als. Scission of the C25/C26 bond further yields 27-norcholest-5-en-3β-ol and (25ξ)-27-norcholest-5-ene-3β,25-diol.

These indicated bond cleavages are confirmed by the formation of volatile degradation products in air-aged cholesterol. Assuming scission of only one carbon-carbon bond predominates, the C_3-compound acetone formed implies cleavage of the C24/C25 bond, the C_4-compounds 2-methylprop-1-ene, 2-methylpropan-1-ol, 2-methylpropan-2-ol, and 2-methylpropionic acid that of the C23/C24 bond. The C_5-compounds 2-methylbutan-2-ol and 3-methylbutan-2-one arise from C22/C23 bond scission; the C_6-compounds 2-methypent-1-ene, 2-methylpentan-2-ol, and 2-methylpent-4-en-2-ol establish C20/C22 bond cleavage. Butane-2-one, pentan-2-one, acetic acid, and ethanol, also present among volatiles, require breakage of at least two carbon-carbon bonds.

The bond cleavages inferred by these volatiles are obviously not simple reactions but more complex ones in which radical transfer processes or multiple autoxidative attacks must occur. The four C_4 to C_6 secondary alcohols 2-methypropan-2-ol, 2-methylbutan-2-ol, 2-methylpentan-2-ol, and 2-methylpent-4-en-2-ol as examples appear to derive from the 25-hydroperoxide **32** but involve cleavage of the C23/C24, C22/C23, and C20/C22 bonds, with formation of degraded C_{23}-, C_{22}-, and C_{21}-oxysterols. The established cholesterol autoxidation products C_{23}-oxysterol 24-norchol-5-en-3β-ol derived from the cholesterol 24-hydroperoxide, the C_{22}-oxysterol 3β-hydroxy-22,23-bisnorchol-5-enic acid isolated from air-aged cholesterol, and the C_{21} product pregnenolone support this formulation in detail.

The isomeric 3β-hydroxy-(20R)- and (20S)-pregn-5-ene-20-hydroperoxides isolated from air-aged cholesterol[65] represent scission of the C20/C22 bond with concommitant radical formation at C20 and subsequent reaction with $^3O_2$. The indicated bond cleavage can arise via β-scission of a 20- or 22-oxyl radical derived from the corresponding parent 20- or 22-hydroperoxide.

Clearly many complex processes occur in the subsequent reactions of initially formed cholesterol hydroperoxides. More exact studies of products formed and radical transfer and bond cleavage reactions have not been conducted.

B. OTHER ACTIVE OXYGEN SPECIES

Remarks to this point pertain to cholesterol oxidations in which the molecular oxygen ($^3O_2$) of the air is the oxidant. However, the matter is a bit more complicated, as other species of dioxygen potentially present in inspired air or generated *in vivo* may participate in the oxidation of cholesterol. Among these active oxygen species are $^1O_2$, $O_2^{\cdot -}$, H_2O_2, $O_2^{\cdot +}$, O_3, and HO·, as well as the nitrogen oxides (NO_x). A relatively satisfactory description of the processes and oxidation products formed from cholesterol is now available, and the present treatment attempts to provide a terse appreciation of the matter. Greater details are variously presented in a series of items published over the past decade.[1,2]

With the exceptions of $O_2^{\cdot -}$ that does not oxidize cholesterol and $O_2^{\cdot +}$, not likely to be present in biological systems, these active oxygen species oxidize cholesterol under conditions likely to be encountered in biological systems. In the case of $O_2^{\cdot -}$, its disproportionation may generated H_2O_2 from which HO· may form, both species oxidizing cholesterol Yet other active oxygen species formed with transition metal ions might also oxidize cholesterol, but such reactions have not been examined. Moreover, $O_2^{\cdot -}$ may contribute to oxysterol formation

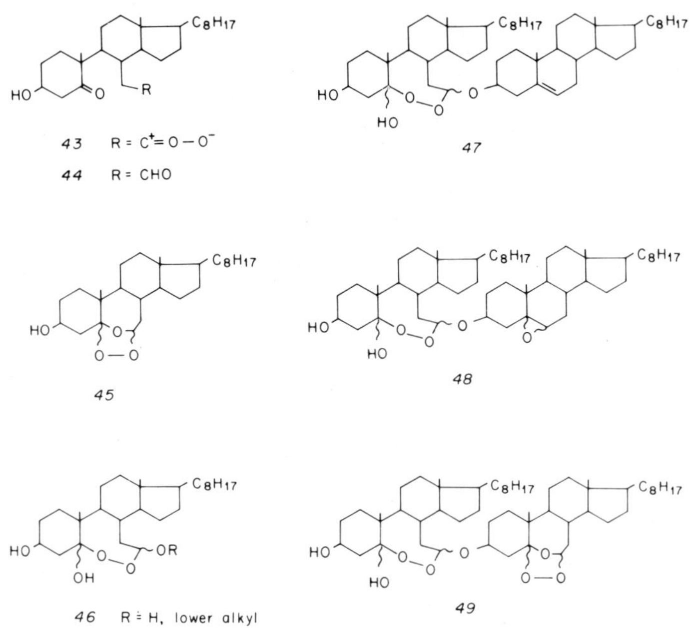

FIGURE 8. Oxysterol products of cholesterol ozonization.

in yet another way, as O_2^- reacts with sterol hydroperoxides, apparently generating the 7α-oxyl radical **33** from the 7α-hydroperoxide **28**, 3-β-hydroxy-5α-cholest-6-ene-5-oxyl radical (**40**) from the 5α-hydroperoxide **38**.[67]

1. Ozone

Ozone present at low levels in the atmosphere is a powerful oxidant that generates another complex set of oxysterols from cholesterol. Two separate processes appear to operate: the major one involving classic 1,2,4-trioxolane ozonide formation in which peroxidic sterols are formed with scission of the C5/C6 double bond, and epoxidation of the Δ^5-double bond in which are formed the isomeric 5,6-epoxides **13** and **14**. In nonparticipating solvents the attack of O_3 on the Δ^5-double bond proceeds putatively via the Criegee zwitterionic intermediate 3β-hydroxy-5-oxo-5,6-secocholestane-6-carbonyl oxide (**43**) to the isomeric cholesterol ozonides 5,6ξ-epidioxy-5ξ-B-homo-6-oxacholestan-3β-ol (**45**) (Figure 8). Protic solvents (water, alcohols) participate in the reaction of the carbonyl oxide, **43**, yielding 6ξ-hydroxy (or alkoxy)-5,6ξ-epidioxy-5ξ-5,6-secocholestane-3β,5ξ-diol (**46**), accompanied by the isomeric 5,6-epoxides, **13** and **14**. Decomposition of the 5,6-epidioxides, **46**, yields 3β-hydroxy-5-oxo-5,6-secocholestan-6-al (**38**).[68,69]

In more concentrated solutions of cholesterol the 3β-hydroxyl group participates in the reaction as a protic reactant, yielding the sterol dimer 6ξ-cholest-5′-en-3′β-yloxy-5ξ-5ξ,6ξ-epidioxy-5ξ-5,6-secocholestane-3β,5-diol (**47**). Any 5,6-epoxides **13** and **14** or ozonides **39** formed also react as alcohols with carbonyl oxide **43**, yielding isomeric dimer epoxides 5,6ξ-epidioxy-6ξ-5′,6′-epoxycholestan-3′-β-yloxy-5ξ-5,6-secocholestane-3β,5ξ-diols (**48**) and isomeric dimer ozonides 5,6ξ-epidioxy-6ξ-5′,7′-epidioxy-B′-homo-6′-oxacholestan-3′β-yloxy-5ξ-5,6-secocholestane-3β,5-ols (**48**), respectively. Moreover, the Δ^5-double bond of the dimer **47** is subject to ozonization, generating the same dimeric epoxides **48** and ozonides **49** but also trimeric, tetrameric, and higher oligomeric ozonide species.

As O_3 is generated from $^3O_2$ and the two species exist together, some autoxidation might accompany cholesterol ozonization, particularly in that radical character has been advanced for Criegee carbonyl oxides such as oxysterol **43** and radical oxidations noted in some ozonizations. However, the absence of oxysterols **2**, **13** to **15**, **28**, **29**, **35**, and **36** derived by cholesterol autoxidation among ozonization products **44** to **49**, establishes that cholesterol is more sensitive to oxidation by O_3 than by $^3O_2$ and that autoxidation does not accompany ozonization. The great complexity of $^3O_2$ and O_3 oxidations of cholesterol contrast with the simpler reactions of other active oxygen species.

Both $^3O_2$ and O_3 exist in the atmosphere, and the two oxygen species together account for the great diversity of oxysterols identified as derived from. Oxysterols derived by the action of O_3 have not been sought in human tissues, but defined ozonides of olefinic pentacyclic triterpenoids have been isolated from plant materials.[70,71] These novel and provocative reports appear not to have been exploited further.

2. Singlet Molecular Oxygen

Electronically excited (singlet) dioxygen reacts with cholesterol via the "ene" reaction mechanism not involving initial radical generation in the substrate. Rather, the dioxygen species is the reactant that is activated by electronic excitation. Formation of the major product 5α-hydroperoxide **38** (Figure 7) involves allylic shift of the B-ring double bond characteristic of $^1O_2$ reactions with olefins. Traces of the epimeric 3β-hydroxycholes-4-ene-6α- and 6β-hydroperoxides also form, the allylic shift of the Δ^5-double bond involving the A-ring in these examples. These Δ^6-5α- and epimeric Δ^4-6-hydroperoxide products are thus entirely different from the Δ^5-7-hydroperoxides formed by autoxidation. Indeed, the 5α-hydroperoxide **38** has become an accepted indicator of past action of $^1O_2$ in systems containing cholesterol.

The 5α-hydroperoxide **38** is readily rearranged allylically to the isomeric 7α-hydroperoxide **28**. The transformation appears to involve the corresponding 3β-hydroxy-5α-cholest-6-ene-5-peroxyl radical (**39**) and 7α-peroxyl radical **33**, proceeding by a sigmatropic [2,3]-rearrangement in which the peroxide oxygens of 5α-hydroperoxide **38** are retained in 7α-hydroperoxide **28**. In distinction, epimerization of the 7α-hydroperoxide **28** to the 7β-hydroperoxide **29** involves oxygen dissociation from the sterol.[42] The different mechanisms of isomerization are also suggested by the differential metabolism of the 5α- and 7α-hydroperoxides, where cultures of *Salmonella typhimurium* rearrange the 5α-hydroperoxide **38** to the 7α-hydroperoxide **28** but do not epimerize the 7α-hydroperoxide.[72]

The 5α-hydroperoxide **38** is thermally decomposed to the $\Delta^{4,6}$-3-one **42** as chief distinctive product; also formed are epimeric 3β,7-diols, **2** and **36**; 7-ketones, **35** and **37**; and elimination products cholesta-2,4,6-triene and cholesta-3,5,7-triene.[61] The decomposition of 5α-hydroperoxide **38** to 3β,7α-diol **2** and 7-ketone **35** appears to involve formation of 3β-hydroxy-5α-cholest-6-ene-5-oxyl radical (**40**), rearrangement of which yields the 7α-oxyl radical **33**, from which derive the 7-oxygenated products **2** and **35**. Moreover, the 5α-hydroperoxide **38** reacts with $O_2^{\bar{}}$ to give 5α-cholest-6-ene-3β,5α-diol (**41**) and 7-ketone, **35**. The 3β,5α-diol **41** would derive from the 5α-oxyl radical **40** by hydrogen abstraction, the 7-ketone **35** from rearrangement of the 5α-oxyl radical **40** to the 7α-oxyl radical **33** and subsequent β-scission.[67]

Thermal decomposition of the trace $^1O_2$ products epimeric 3β-hydroxycholest-4-ene-6-hydroperoxides yields the corresponding epimeric 6-alcohols cholest-4-ene-3β,6α- and -3β,6β-diols and 6-ketone 3β-hydroxycholest-4-en-6-one.

As $^1O_2$ is derived from and relaxes to $^3O_2$, both dioxygen species may react with cholesterol under certain conditions, the epimeric 7-hydroperoxides **28** and **29** being formed from $^3O_2$ attack, the 5α-hydroperoxide **38** from $^1O_2$ attack. Identification of the 5α-hydroperoxide **38** or its unique alteration products 3β,5α-diol **41** and $\Delta^{4,6}$-3-ketone **42** infer participation of $^1O_2$ in oxidations,[73] but should the rearrangement of the 5α-hydroperoxide **38** occur, evidence of past $^1O_2$ action would be lost. Under those conditions where the integrity of the 5α-hydroperoxide **38** is not preserved, there may be gained the misleading impression that cholesterol autoxidation rather than photosensitized oxidation occurred.

Depending on conditions the oxidation of cholesterol by $^1O_2$ and/or $^3O_2$ may occur. Moreover, oxidation of cholesterol by the more reactive $^1O_2$ species, yielding the 5α-hydroperoxide **38** whose thermal decomposition (peroxide bond homolysis) may initiate radical autoxidation of cholesterol by $^3O_2$. Product analysis over the course of an experiment suggests this formulation is so.[51,74,75] Recovery of both 5α-hydroperoxide **38** and 7α-hydroperoxide **28** in experiments where $^1O_2$ oxidations and radical autoxidations might both occur or where only $^1O_2$ oxidations occur but with partial isomerization of the 5α-hydroperoxide **38** to the 7α-hydroperoxide **28** may complicate interpretations.[76]

The oxidation of cholesterol 3β-esters by $^1O_2$ follows another course, yielding not 5α-hydroperoxide **38** 3β-esters but the 7α-hydroperoxide **28** 3β-esters.[43]

The Δ^5-3-ketone **18** also undergoes different reactions with the two dioxygen species, forming the epimeric 6-hydroperoxy-Δ^4-3-ketones **20** and **21** with $^3O_2$, but 5-hydroperoxy-5α-cholest-6-en-3-one with $^1O_2$. Elimination of the 5α-hydroperoxyl function yields the $\Delta^{4,6}$-3-ketone **42**, the same oxysterol obtained in the thermal decomposition of the 5α-hydroperoxide **38** from $^1O_2$ oxidation of cholesterol.

3. Hydrogen Peroxide

Hydrogen peroxide transforms cholesterol to the isomeric 5,6-epoxides **13** and **14**, with the 5β-6β-epoxide **14** predominating. Accordingly, the reaction has the same outcome as autoxidation where the initially formed 7-hydroperoxides **28** and **29** epoxidize cholesterol in the same fashion. Lipid peroxidations of cholesterol *in vitro* yield the same 5,6-epoxides but with somewhat diminished ratios of 5β,6β-epoxide **14** to 5α-6α-epoxide **13**. Yet other oxygen species O_3 and HO· epoxidize cholesterol to mixtures of 5,6-epoxides; indeed, epoxidation of cholesterol is that one reaction of cholesterol that occurs most broadly with all manner of oxygen species.[10,11]

Systems where H_2O_2 disproportionates, cholesterol may be oxidized by the dioxygen species formed, whether $^3O_2$ or $^1O_2$, as well as by HO· possibly formed from H_2O_2 by peroxide bond homolysis (catalyzed by transition metal ion). Hydrogen peroxide is thus the parent of active oxygen species as well as active itself in cholesterol oxidations.[73]

Also among oxysterols formed in aqueous systems containing H_2O_2 is 5α-cholestane-3β,6β-diol, a formal product of hydration of the cholesterol Δ^5-double bond.[73] Under other conditions in photolyzed aqueous systems, addition of water to the Δ^5-double bond occurs, yielding the isomeric 5β-cholestane-3β,5-diol, *inter alia.*[77] Neither 5α-cholestane-3β,6β-diol nor 5β-cholestane-3β,5-diol are among the oxysterols found in tissues.

4. Alkyl Peroxyl and Oxyl Radicals

Cholesterol hydroperoxide formation involves generation of sterol peroxyl radicals; cholesterol hydroperoxide decomposition involving peroxide bond homolysis involves sterol oxyl radicals. Both oxygen radicals are implicated in the radical chain reactions of cholesterol autoxidation, via hydrogen abstraction reactions. Indeed, it is by these very means that the radical chains are posited to proceed, with the cholesterol carbon-centered radicals themselves being formal oxidation products of cholesterol.

Demonstration of the oxidation of cholesterol by peroxyl radicals **27**, **31**, etc. (Figure 5) or 7-oxyl radicals **33** and **34** (Figure 6) is compromised by the limitations of generating the single radical species uniquely. In systems containing transition metal ions M^{n+} and $M^{(n+1)+}$ both radicals may exist, and interconversions of peroxyl and oxyl radicals may occur, as in Equations 4 to 6.

$$Mn^{n+} + ROOH \rightarrow M^{(n+1)+} + RO\cdot + HO^- \qquad (4)$$

$$M^{(n+1)+} + ROOH \rightarrow M^{n+} + ROO\cdot + H^+ \quad (5)$$

$$ROOH + R'O\cdot \rightarrow ROO\cdot + R'OH \quad (6)$$

The direct oxidation of cholesterol by specific peroxyl or oxyl radicals generated by these means has not been described, but oxidation of cholesterol 3β-methyl ether and 3β-acetate by these species in model chemical systems has. Alkyl peroxyl radicals appear to epoxidize the 3β-methyl ether; alkyl oxyl radicals appear to abstract the C7 hydrogen, yielding 7-alkylperoxide derivatives as well as the 3β,7-diols **2** and **36** and 7-ketone **35**.[78-82]

In other model aqueous systems in which the trichloromethyl peroxyl radical $CCl_3OO\cdot$ is formed from CCl_4 subjected to γ-radiation in the presence of air, 5α-6α- and 5β,6β-epoxides **13** and **14** result.[83,84] However, in model nonaqueous systems in which $CCl_3OO\cdot$ is formed from reaction of CCl_4 with KO_2 the 3β,7-diols **2** and **36** and 7-ketone **35** are formed.[85] Because of the limitations of these model systems in creating specific peroxyl and oxyl radicals only, these results may not be definitive. In any event, organic peroxyl and oxyl radicals do appear to oxidize cholesterol.

5. Dioxygen Cation

The dioxygen cation O_2^+, though an unlikely oxygen species encountered in biological systems, is formed *inter alia* by energetic electric discharge and oxidizes cholesterol to many products. Among oxysterols identified are the 3β,7-diols **2** and **36**, the 7-ketone **35**, and the 5,6-epoxides **13** and **14**. Sterol hydroperoxides were not detected but would not have survived the experimental conditions.

6. Hydroxyl Radical

The highly reactive HO· reacts with cholesterol in organic or aqueous systems to yield the epimeric 3β,7-diols **2** and **36** and 7-ketone **35** by C7 hydrogen abstraction and the isomeric 5,6-epoxides **13** and **14** and 3β,5α,6β-triol **15** by radical addition. In the absence of air no sterol hydroperoxides are formed, but with air in the system some radical autoxidation may occur.[86]

C. OTHER OXIDANTS

The defined active oxygen species that may be generated under controlled circumstances are not the only agents capable of oxidizng cholesterol to form oxysterols that may be encountered in biological systems. The full extent of the possibilities has yet to be recognized, but several cases of interest have been described. Among such oxidizing agents are the hypohalous acids, nitrogen oxides, and certain xenobiotic substances.

1. Hypohalous Acids

The reaction of cholesterol with hypochlorous (HOCl) or hypobromous (HOBr) acids to form chlorohydrin or bromohydrin derivatives, respectively, must also be considered as means of formation of oxysterols *in vivo*. Reaction of cholesterol with aqueous HOCl yields chlorohydrins 5-chloro-5α-cholestane-3β,6β-diol (**17**), 6α-chloro-5β-cholestane-3β,5-diol, and 6β-chloro-5α-cholestane-3β,5-diol (**16**) (Figure 3), *inter alia*.[87] In pH 4.65 aqueous systems the common oxysterols 3β,7-diols **2** and **36**, 7-ketone **35**, and 5,6-epoxides **13/14** form. It also appears that human polymorphonuclear leukocytes generating HOCl oxidize cholesterol to chlorohydrins *in vitro*.[18,19] Moreover, the apparent generation of HOBr by human blood eosinophiles *in vitro*[88] suggests the unexplored possibility that cholesterol bromohydrins form analogously.

The 6β-chloro derivative **16** is also formed from the 5α,6α-epoxide **13**, the 5α-chloro derivative **17** from the 5β,6β-epoxide **14** in aqueous HCl. In base, chlorohydrins **16** and **17** are reconverted to the corresponding parent 5,6-epoxide (Figure 3). Thereby dietary 5,6-epoxides might be transformed to derivative chlorohydrins in the acid stomach, with reconversion of chlorohydrin to 5,6-epoxide in the alkaline small intestine.[89] Cholesterol halohydrins have not been sought in tissues nor have biological activities been considered for these oxysterols.

Oxidation of cholesterol by *N*-halogenated compounds formed under the action of hypohalous acids has not been investigated, but one may presume that oxysterols result.

2. Nitrogen Oxides

The reaction of cholesterol with NO_2 potentially present in ambient air occurs in aqueous systems, yielding cholesterol 3β-nitrate. In organic solvents and in mouse skin exposed to NO_2 cholesterol 3β-nitrite (and not 3β-nitrate) was formed. A minor product formed in organic solvent was 6α-nitro-5α-cholestane-3β,5-diol 5α-nitrite, readily hydrolyzed to 6α-nitro-5α-cholestane-3β,5-diol.[90,91]

The potential of NO_2 in inspired air to provoke *in vivo* oxidation has been demonstrated in rats, where the isomeric 5,6-epoxides **13** and **14** were formed, in the **13**:**14** ratio of about 8:1.[92] As the formation *in vitro* of the radical species nitric oxide (NO·) by endothelial cells, macrophages, and neutrophiles has been demonstrated, there may be an endogenous source of NO_2 formed from NO· that could oxidize cholesterol, a matter not yet investigated.

3. Xenobiotic Substances

The ingestion of xenobiotic substances not necessarily considered as oxidants may cause the *in vivo* oxidation of cholesterol to odd products not commonly included among the oxysterols. Nonetheless, thoughtful examination of oxysterol biological effects should now include such possible products, as there is evidence that ingested xenobiotic substances, including exotic foods, drugs, and environmental chemicals induce special oxidizing conditions *in vivo*. Thus, tissues *ex vivo* and *in vitro* exposed to air and unprotected by antioxidants soon contain oxysterols derived from cholesterol autoxidation. The insidious nature of autoxidation is best exemplified in the increase in amounts of oxysterols in human blood at various times following collection (cf. Section III.A).

Administration of carbon tetrachloride (CCl_4), known to induce endogenous lipid peroxidations, to rats led to formation of a trichloromethyl adduct of cholesterol, perhaps a 7-trichloromethylcholest-5-en-3β-ol.[93] Other *in vivo* evidence of interaction of cholesterol with xenobiotic substances includes formation of xenobiotic cholesterol esters from ingested xenobiotic carboxylic acids.[94] These examples establish cholesterol as an endogenous component capable of sequestration of xenobiotic agents potentially harmful to the living system.

Xenobiotic agents such as nitrosobenzene also oxidize cholesterol at the Δ^5-double bond, forming a stable adduct aminoxyl (nitroxyl) radical.[95] Such reactions and products necessarily have not been sought in living systems.

These results suggest that cholesterol is oxidized by a variety of reactive species not heretofore considered as contributing to tissue oxysterols. These presently meager discoveries nonetheless pose the speculation that cholesterol acts as a protective agent against such xenobiotic agent insults. Were cholesterol to serve as antioxidant or interceptor of oxidants and toxic agents *in vivo*, the presence of the common oxysterols 3β,7-diols, **2** and **36**; 7-ketone, **35**; 5,6-epoxides, **13** and **14**; etc., evincing radical oxidations of cholesterol in living systems, would have additional rationale. Some experimental evidence of the antioxidant action of cholesterol is erythrocytes has been recorded.[96,97]

Other dietary manipulations also provoke *in vivo* lipid peroxidations of cholesterol. Rabbits fed cholesterol (1% in diet, 2 months) developed marked elevations of plasma and liver levels of 5,6-epoxides **13** and **14**,[98] and mice fed cholesterol (3%, 1 or 2 days) showed elevated liver levels of 3β,7-diols **2** and **36** and 7-ketones **35** (as well as of the 3β,24-diol **7** and 3β,26-diol **4** implicated in hepatic bile acid biosynthesis).[99]

IV. SUMMARY

Enzymic and nonenzymic reactions are implicated in the formation of the numerous oxysterols from cholesterol *in vivo*, and as the oxysterols exhibit impressive biological activities it is important to distinguish among the several processes, thereby to approach balanced judgement anent physiological relevance. The enzyme reactions clearly must provide oxysterol metabolites for physiological matters, but the formation of some oxysterols by autoxidation, lipid peroxidation, or dioxygenase actions occurs without presently recognized reason or purpose.

We may summarize aspects of possible physiological relevance of oxysterols found in tissues as follows:

1. Oxysterols are formed enzymically for physiological purposes, the case of the 3β,7α-diol **2** as bile acid precursor being a paramount example.
2. Oxysterols may serve as regulatory agents implicated in *de novo* sterol biosynthesis and other *in vivo* events, perhaps as moderators of plasma membrane permeability, function, and stability and as inhibitors of 3-hydroxy-3-methylglutaryl Coenzyme A reductase.
3. Oxysterols may be formed adventitiously *in vivo* where appropriate oxidizing conditions exist, possibly escaping antioxidant protections; the oxysterols being toxic thereby pose health risks.
4. Oxysterols may merely represent the scavenging activity of cholesterol itself as an antioxidant and as an interceptor of other toxic species, with esterification and excretion providing detoxification protection.
5. In some cases oxysterols may merely evince imperfect analysis or isolation methods, being formed during sample collection and processing.

There are advocates of each viewpoint, and clearly combinations of these diverse possibilities are feasible. Aspects of these concepts will be discussed in the ensuing chapters of this monograph, with the hope of elucidating some of the yet unsettled features of this provocative topic.

REFERENCES

1. **Smith, L. L.,** *Cholesterol Autoxidation,* Plenum Press, New York, 1981.
2. **Smith, L. L.,** Cholesterol autoxidation 1981-1986, *Chem. Phys. Lipids,* 44, 87, 1987.
3. **Smith, L. L.,** Cholesterol oxidation, in *Membrane Lipid Oxidations,* Vigo-Pelfrey, C., Ed., CRC Press, Boca Raton, FL, 1, 129, 1990.
4. **Maerker, G.,** Cholesterol autoxidation — current status, *J. Am. Oil Chem. Soc.,* 64, 388, 1987.
5. **Smith, L. L. and Johnson, B. H.,** Biological activities of oxysterols, *Free Radicals Biol. Med.,* 7, 285, 1989.
6. **Beck, J. P. and Crastes de Paulet, A., Ed.,** *Activités Biologiques des Oxystérols,* Editions INSERM, Paris, 1987.
7. **Velisek, J., Davidek, J., and Weber, N.,** Formation of $\Delta^{3,5}$-diene and 3-chloro-Δ^{5}-ene analogues of sterols in protein hydrolysates. *J. Agric. Food Chem.,* 34, 660, 1986.
8. **Seher, A.,** Sterols in the chemistry and technology of edible fats and oils, in *Lipids. Vol. 2. Technology,* Paoletti, R., Jacini, G., and Porcellati, G., Eds., Raven Press, New York, 1976, 293.

9. **Emmons, G. T., Pyrek, J. S., Dam, R., Martin, M., Kudo, K., and Schroepfer, G. J.,** 5α-Cholest-8(14)-en-3β-ol-15-one, a potent regulator of cholesterol metabolism: occurrence in rat skin, *J. Lipid Res.*, 29, 1039, 1988.
10. **Gumulka, J., Pyrek, J. S., and Smith, L. L.,** Interception of oxygen species in aqueous media by cholesterol: formation of cholesterol epoxides and secosterols, *Lipids,* 17, 197, 1982.
11. **Smith, L. L. and Jaworski, J.,** Cholesterol epoxidation by defined species of oxygen, in *Oxygen Radicals in Biology and Medicine,* Simic, M. G., Taylor, K. A., Ward, J. F., and von Sonntag, C., Eds., Plenum Press, New York, 1988, 313.
12. **Nelson, J. A., Steckbeck, S. R., and Spencer, T. A.,** Biosynthesis of 24,25-epoxycholesterol from squalene 2,3;22,23-dioxide, *J. Biol. Chem.*, 256, 1067, 1981.
13. **Steckbeck, S. R., Nelson, J. A., and Spencer, T. A.,** Enzymic reduction of an epoxide to an alcohol, *J. Am. Chem. Soc.*, 104, 893, 1982.
14. **Spencer, T. A., Gayen, A. K., Phirwa, S., Nelson, J. A., Taylor, F. R., Kandutsch, A. A., and Erickson, S. K.,** 24(S),25-Epoxycholesterol. Evidence consistent with a role in the regulation of hepatic cholesterogenesis, *J. Biol. Chem.*, 260, 13391, 1985.
15. **Taylor, F. R., Kandutsch, A. A., Gayen, A. K., Nelson, J. A., Steckbeck Nelson, S., Phirwa, S., and Spencer, T. A.,** 24,25-Epoxysterol metabolism in cultured mammalian cells and repression of 3-hydroxy-3-methylglutaryl-CoA reductase, *J. Biol. Chem.*, 261, 15039, 1986.
16. **Kabore, S. A., Combaut, G., Vidal, J.-P., Codomier, L., Passet, J., Girard, J.-P., and Rossi, J.-C.,** Sterols of the red alga, *Rissoella verruculosa, Phytochemistry,* 22, 1239, 1983.
17. **Teng, J. I. and Smith, L. L.,** Sterol metabolism. XXXVI. On the oxidation of cholesterol by dioxygenases, *Bioorg. Chem.*, 5, 99, 1976.
18. **Foote, C. S., Abakerli, R. B., Clough, R. L., and Shook, F. C.,** On the question of singlet oxygen production in leucocytes, macrophages and the dismutation of superoxide anion, in *Biological and Clinical Aspects of Superoxide and Superoxide Dismutase,* Bannister, W. H. and Bannister, J. V., Eds., Elsevier/North Holland, New York, 1980, 222.
19. **Foote, C. S., Abakerli, R. B., Clough, R. L., and Lehrer, R. I.,** On the question of singlet oxygen production in polymorphonuclear leucocytes, in *Bioluminescence and Chemiluminescence, Basic Chemistry and Analytical Applications,* DeLuca, M. A. and McElroy, W. D., Eds., Academic Press, New York, 1981, 81.
20. **Smith, L. L., Teng, J. I., Lin, Y. Y., Seitz, P. K., and McGehee, M. J.,** Sterol Metabolism. XLVII. Oxidized cholestrerol esters in human tissues, *J. Steroid Biochem.*, 14, 889, 1981.
21. **Smith, L. L., Teng, J. I., Lin, Y. Y., Seitz, P. K., and McGehee, M. J.,** Lipid peroxidations of cholesterol, in *Lipid Peroxides in Biology and Medicine,* Yagi, K., Ed., Academic Press, New York, 1982, 89.
22. **Brooks, C. J. W., Cole, W. J., MacLachlan, J., and Lawrie, T. D. V.,** Some aspects of the analysis of minor oxygenated sterols in serum and in serum lipoprotein fractions, *J. Am. Oil Chem. Soc.*, 62, 622, 1985.
23. **Brooks, C. J. W., McKenna, R. M., Cole, W. J., MacLachlan, J., and Lawrie, T. D. V.,** 'Profile' analysis of oxygenated sterols in plasma and serum, *Biochem. Soc. Trans.*, 11, 700, 1983.
24. **Björkhem, I.,** Assay of unesterified 7-oxocholesterol in human serum by isotope dilution-mass spectrometry, *Anal. Biochem.*, 154, 497, 1986.
25. **Björkhem, I., Reihnér, E., Angelin, B., Ewerth, S., Åkerland, J.-E., and Einarsson, K.,** On the possible use of the serum level of 7α-hydroxycholesterol as a marker for increased activity of the cholesterol 7α-hydroxylase in humans, *J. Lipid Res.*, 28, 889, 1987.
26. **Koopman, B. J., van der Molen, J. C., and Wolthers, B. G.,** Determination of some hydroxycholesterols in human serum samples, *J. Chromatogr.*, 416, 1, 1987.
27. **Kudo, K., Emmons, G. T., Casserly, E. W., Via, D. P., Smith, L. C., Pyrek, J. S., and Schroepfer, G. J.,** Inhibitors of sterol synthesis. Chromatography of acetate derivatives of oxygenated sterols, *J. Lipid Res.*, 30, 1097, 1989.
28. **van Doormaal, J. J., Smit, N., Koopman, B. J., van der Molen, J. C., Wolthers, B. G., and Doorenbos, H.,** Hydroxycholesterols in serum from hypercholesterolaemic patients with and without bile acid sequestrant therapy, *Clin. Chim. Acta,* 181, 273, 1989.
29. **Addis, P. B., Emanuel, H. A., Bergmann, S. D., and Zavoral, J. H.,** Capillary GC quantification of cholesterol oxidation products in plasma lipoproteins of fasted humans, *Free Radical Biol. Med.*, 7, 179, 1989.
30. **Yamamoto, Y., Brodsky, M. H., Baker, J. C., and Ames, B. N.,** Detection and characterization of lipid hydroperoxides at picomole levels by high-performance liquid chromatography, *Anal. Biochem.*, 160, 7, 1987.
31. **Johnson, M. A. and Croteau, R.,** Biosynthesis of ascaridole: iodide peroxidase-catalyzed synthesis of a monoterpene endoperoxide in soluble extracts of Chenopodium ambrosioides fruit, *Arch. Biochem. Biophys.*, 235, 254, 1984.

32. **Guyot, M., Davoust, D., and Belaud, C.,** Hydroperoxy-24 vinyl-24 cholesterol, nouvel hydroperoxyde naturel isole de deux tunicates: *Phallusia mamillata* et *Ciona intestinales*, *Tetrahedron Lett.*, 23, 1905, 1982.
33. **Guyot, M., Morel, E., and Belaud, C.,** A cytotoxic hydroperoxide of marine origin and related compounds: synthesis and biological study, *J. Chem. Res. (S)*, 188, 1983.
34. **Herz, W., Watanabe, K., Kulanthaivel, P., and Blount, J. F.,** Cycloartanes from *Lindheimera texana*, *Phytochemistry*, 24, 2645, 1985.
35. **Smith, L. L., Kulig, M. J., and Teng, J. I.,** Sterol metabolism. XL. On the failure of superoxide radical anion to react with cholesterol, *Chem. Phys. Lipids*, 20, 211, 1977.
36. **Fischer, K.-H., Laskawy, G., and Grosch, W.,** Quantitative Analyse von Autoxidationsprodukten des Cholesterols in tierschen Lebensmitteln, *Z. Lebensm. Unters. Forsch.*, 181, 14, 1985.
37. **Sanghvi, A., Grassi, E., Bartman, C., Lester, R., Galli Kienle, M., and Galli, G.,** Measurement of cholesterol 7α-hydroxylase activity with selected ion monitoring, *J. Lipid Res.*, 22, 720, 1981.
38. **Ansari, G. A. S., Walker, R. D., Smart, V. B., and Smith, L. L.,** Further investigations of mutagenic cholesterol preparations, *Food Chem. Toxicol.*, 20, 35, 1982.
39. **Csiky, I.,** Trace enrichment and separation of cholesterol oxidation products by adsorption high-performance liquid chromatography, *J. Chromatogr.*, 241, 381, 1982.
40. **Dumaswala, R., Setchell, K. D. R., Zimmer-Nechemias, L., Iida, T., Goto, J., and Nambara, T.,** Identification of 3α,4β,7α-trihydroxy-5β-cholestanoic acid in human bile: reflection of a new pathway in bile acid metabolism in humans, *J. Lipid Res.*, 30, 847, 1989.
41. **Björkhem, I., Skrede, S., Buchmann, M. S., East, C., and Grundy, S.,** Accumulation of 7α-hydroxy-4-cholesten-3-one and cholesta-4,6-dien-3-one in patients with cerebrotendinous xanthomatosis; effect of treatment with chenodeoxycholic acid, *Hepatology*, 7, 266, 1987.
42. **Sevilla, C. L., Becker, D., and Sevilla, M. D.,** An electron spin resonance investigation of radical intermediates in cholesterol and related compounds: relation to solid-state autoxidation, *J. Phys. Chem.*, 90, 2963, 1986.
43. **Beckwith, A. L. J., Davies, A. G., Davison, I. G. E., Maccoll, A., and Mruzek, M. H.,** The mechanisms of the rearrangements of allylic hydroperoxides: 5α-hydroperoxy-3β-hydroxycholest-6-ene and 7α-hydroperoxy-3β-hydroxycholest-5-ene, *J. Chem. Soc. Perkin Trans. II*, 815, 1989.
44. **Akihisa, T., Matsumoto, T., Sakamaki, H., Take, M., and Ichinohe, Y.,** Platinum-catalyzed oxidation of cholesterol, *Bull. Chem. Soc. Jpn.*, 59, 680, 1986.
45. **Weenen, H. and Porter, N. A.,** Autoxidation of model membrane systems: cooxidation of polyunsaturated lecithins with steroids, fatty acids, and α-tocopherol, *J. Am. Chem. Soc.*, 104, 5216, 1982.
46. **Muto, T., Tanaka, J., Miura, T., and Kimura, M.,** Iron-catalyzed autoxidation of liposomal cholesterol, *Chem. Pharm. Bull.*, 31, 1561, 1983.
47. **Muto, T., Tanaka, J., Miura, T., and Kimura, M.,** Autoxidation of cholesterol in the presence of unsaturated fatty acid and iron-catalyst, *J. Pharm. Dyn.*, 6, s-15, 1983.
48. **Terao, J., Sugino, K., and Matsushita, S.,** Fe^{2+} and ascorbic acid induced oxidation of cholesterol in phosphatidylcholine liposomes and its inhibition by α-tocopherol, *J. Nutr. Sci. Vitaminol.*, 31, 499, 1985.
49. **Lijana, R. C., McCracken, M. S., and Rudolph, C. J.,** The oxidation of cholesterol in vesicles, *Biochim. Biophys. Acta*, 879, 247, 1986.
50. **Sevanian, A. and McLeod, L. L.,** Cholesterol autoxidation in phospholipid membrane bilayers, *Lipids*, 22, 627, 1987.
51. **Girotti, A. W., Bachowski, G. J., and Jordan, J. E.,** Lipid peroxidation in erythrocyte membranes: cholesterol product analysis in photosensitized and xanthine oxidase-catalyzed reactions, *Lipids*, 22, 401, 1987.
52. **Wu, G.-S.,Stein, R. A., Mead, J. F., and Mustafa, M. A.,** Formation of oxygenated cholesterol products in liposomes and in vivo peroxidation, in *Lipid Peroxidation in Biological Systems*, Sevanian, A., Ed., American Oil Chemists Society, Chicago, IL, 1988, 129.
53. **Ansari, G. A. S. and Smith, L. L.,** Sterol metabolism. XLI. Cholesterol A-ring autoxidations, *Chem. Phys. Lipids*, 22, 55, 1978.
54. **de la Mare, P. B. and Wilson, R. D.,** Autoxidation of cholest-5-en-3-one, and its accompanying isomerization, in acetic acid, *J. Chem. Soc. Perkin Trans. II*, 157, 1977.
55. **Maerker, G. and Bunick, F. J.,** Cholesterol oxides. II. Measurement of the 5,6-epoxides during cholesterol oxidation in aqueous dispersions, *J. Am. Oil Chem. Soc.*, 63, 771, 1986.
56. **Teng, J. I., Kulig, M. J., Smith, L. L., Kan, G., and van Lier, J. E.,** Sterol metabolism. XX. Cholesterol 7β-hydroperoxide, *J. Org. Chem.*, 38, 119, 1973.
57. **Korahani, V., Bascoul, J., and Crastes de Paulet, A.,** Autoxidation of cholesterol fatty acid esters in solid state and aqueous dispersion, *Lipids*, 17, 703, 1982.
58. **Bascoul, J., Domergue, N., and Crastes de Paulet, A.,** Preparation of autoxidized cholesterol mixtures from [$4\text{-}^{14}C$] cholesterol and cholesteryl fatty acid esters, *J. Steroid Biochem.*, 19, 1779, 1983.

59. **Kawata, M., Tohma, M., Sawaya, T., and Kimura, M.,** A new product of cholesterol by metal-free autoxidation in aqueous dispersion, *Chem. Pharm. Bull.,* 24, 3109, 1976.
60. **Yanishlieva-Maslarova, N., Schiller, H., and Seher, A.,** Die Autoxidation von Sitosterin. III. Sitosterylstearate, *Fette Seifen Anstrichm.,* 84, 308, 1982.
61. **Teng, J. I., Kulig, M. J., and Smith, L. L.,** Sterol metabolism. XXII. Gas chromatographic differentiation among cholesterol hydroperoxides, *J. Chromatogr.,* 75, 108, 1973.
62. **Kulig, M. J. and Smith, L. L.,** Sterol metabolism. XXXII. Radiation-induced oxidation of isomeric cholesten-3β-ols, *J. Org. Chem.,* 39, 3398, 1974.
63. **Yanishlieva, N. and Marinova, E.,** Autoxidation of sitosterol. I. Kinetic studies on free and esterified sitosterol, *Riv. Ital. Sostanze Grasse,* 57, 477, 1980.
64. **Yanishlieva, N., Schiller, H., and Marinova, E.,** Autoxidation of sitosterol. II. Main products formed at ambient and high temperature treatment with oxygen, *Riv. Ital. Sostanze Grasse,* 57, 572, 1980.
65. **van Lier, J. A., Kan, G., and Langlois, R.,** The epimeric 20-hydroperoxy-5-pregnen-3β-ols; thermal and enzymatic decomposition, *Steroids,* 21, 521, 1973.
66. **Shih, M. and Tipton, C. L.,** Isolation and characterization of calmodulin-inhibitory cholesterol hydroperoxides, *FASEB J.,* 2, A1580, 1988.
67. **Made Gowda, N. M. and Smith, L. L.,** On reaction of sterol hydroperoxides with superoxide, *J. Steroid Biochem.,* 20, 917, 1984.
68. **Gumulka, J. and Smith, L. L.,** Ozonization of cholesterol, *J. Am. Chem. Soc.,* 105, 1972, 1983.
69. **Jaworski, K. and Smith, L. L.,** Ozonization of cholesterol in nonparticipating solvents, *J. Org. Chem.,* 53, 545, 1988.
70. **Ageta, H., Shiojima, K., Kamaya, R., and Masuda, K.,** Fern constituent: naturally occurring adian-5-ene ozonide in the leaves of *Adiantum monochlamys* and *Oleandra wallichi, Tetrahedron Lett.,* 899, 1978.
71. **Itokawa, H., Tachi, Y., Kamano, Y., and Iitaka, Y.,** Structure of gilvanol, a new triterpene isolated from *Quercus gilva* Blume, *Chem. Pharm. Bull.,* 26, 331, 1978.
72. **Smith, L. L., Made Gowda, N. M., and Teng, J. I.,** Sterol hydroperoxide metabolism by *Salmonella typhimurium, J. Steroid Biochem.,* 26, 259, 1987.
73. **Smith, L. L., Kulig, M. J., Miiller, D., and Ansari, G. A. S.,** Oxidation of cholesterol by dioxygen species, *J. Am. Chem. Soc.,* 100, 6206, 1978.
74. **Bachowski, G. T., Thomas, J. P., and Girotti, A. W.,** Ascorbate-enhanced lipid peroxidation in photooxidized cell membranes; cholesterol product analysis as a probe of reaction mechanism, *Lipids,* 23, 580, 1988.
75. **Kalyanaraman, B., Feix, J. B., Sieber, F., Thomas, J. P., and Girotti, A. W.,** Photodynamic action of merocyanine 540 on artificial and natural cell membranes: involvement of singlet molecular oxygen, *Proc. Natl. Acad. Sci. U.S.A.,* 84, 2999, 1987.
76. **Galaris, D., Mira, D., Sevanian, A., Cadenas, E., and Hochstein, P.,** Co-oxidation of salicylate and cholesterol during the oxidation of metmyoglobin by H_2O_2, *Arch. Biochem. Biophys.,* 262, 221, 1988.
77. **Waters, J. A. and Witkop, B.,** Photosensitized hydration of cholesterol, *J. Org. Chem.,* 34, 3774, 1966.
78. **Kimura, K. and Muto, T.,** On the reaction of cholesteryl acetate with *tert*-butyl hydroperoxide in the presence of tris(acetylacetonato)iron (III), *Chem. Pharm. Bull.,* 27, 109, 1979.
79. **Kimura, K. and Muto, T.,** The reactions of cholesteryl acetate with *tert*-butyl hydroperoxide and molybdenum complexes, *Chem. Pharm. Bull.,* 29, 35, 1981.
80. **Kimura, K. and Muto, T.,** Further studies on the oxy-functionalization of cholesteryl acetate with *tert*-butyl hydroperoxide and tris(acetylacetonato)iron(III), *Chem. Pharm. Bull.,* 29, 1862, 1981.
81. **Muto, T., Tanaka, J., Miura, T., and Kimura, K.,** Iron-catalyzed autoxidation of cholesterol in the presence of unsaturated long-chain fatty acid, *Chem. Pharm. Bull.,* 30, 3172, 1982.
82. **Muto, T., Urano, C., Hayashi, T., Miura, T., and Kimura, K.,** On the mode of oxygenation with the ferric perchlorate-hydrogen peroxide system, *Chem. Pharm. Bull.,* 31, 1166, 1983.
83. **van Lier, J. E. and Langlois, R.,** Chlorinated hydrocarbon mediated cholesterol degradation, *J. Am. Oil Chem. Soc.,* 62, 621, 1985.
84. **van Lier, J. E.,** Degradation of cholesterol. Hydroperoxy intermediates and peroxy radical mediated epoxidation, in *Activités biologiques des oxystérols,* Beck, J. P. and Crastes de Paulet, A., Eds., Editions INSERM, Paris, 1988, 15.
85. **Kaphalia, B. S. and Ansari, G. A. D.,** Trichloromethyl peroxyl radical ($Cl_3COO\cdot$) mediated oxidation of cholesterol, *Chemosphere,* 20, 5, 1990.
86. **Ansari, G. A. S. and Smith, L. L.,** The oxidation of cholesterol by hydroxyl radical, *Photochem. Photobiol.,* 30, 147, 1979.
87. **Lindgren, B. O.,** Reactions of sterols with bleaching agents: reactions of cholesterol, *Sven. Papperstidn.,* 70, 532, 1967.
88. **Weiss, S. J., Test, S. T., Eckmann, C. M., Roos, D., and Regiani, S.,** Brominating oxidants generated by human eosinophiles, *Science,* 234, 200, 1986.

89. **Maerker, G., Nungesser, E. H., and Bunick, F. J.,** Reaction of cholesterol 5,6-epoxides with simulated gastric juice, *Lipids,* 23, 761, 1988.
90. **Kobayashi, T. and Kubota, K.,** The reaction of nitrogen dioxide with lung surface components: the reaction with cholesterol, *Chemosphere,* 9, 777, 1980.
91. **Mirvish, S. S., Babcock, D. M., Deshpande, A. D., and Nagel, D. L.,** Identification of cholesterol as a mouse skin lipid that reacts with nitrogen dioxide to yield nitrosating agent and of cholesteryl nitrite as the nitrosating agent produced in a chemical system from cholesterol, *Cancer Lett.,* 31, 97, 1986.
92. **Sevanian, A., Mead, J. F., and Stein, R. A.,** Epoxides as products of lipid autoxidation in rat lungs, *Lipids,* 14, 634, 1979.
93. **Ansari, G. A. S., Moslen, M. T., and Reynolds, E. S.,** Evidence for *in vivo* covalent binding of $\cdot CCl_3$ derived from CCl_4 to cholesterol of rat liver, *Biochem. Pharmacol.,* 31, 3509, 1982.
94. **Fears, R.,** Lipophilic xenobiotic conjugates: the pharmacological and toxicological consequences of the participation of drugs and other foreign compounds as substrates in lipid biosynthesis, *Progr. Lipid Res.,* 24, 177, 1985.
95. **Sridhar, R. and Floyd, R. A.,** An electron paramagnetic resonance study of the reaction of nitrosobenzene with cholesterol, *Can. J. Chem.,* 60, 1574, 1982.
96. **Jain, S. K.and Shohet, S. B.,** Apparent role of cholesterol as an erythrocyte membrane antioxidant, *Clin. Res.,* 29, 336A, 1981.
97. **Bereza, U. L., Brewer, G. J., and Hill, G. M.,** Effect of dietary cholesterol on erythrocyte peroxidant stress in vitro and in vivo, *Biochim. Biophys. Acta,* 835, 434, 1985.
98. **Watabe, T., Isobe, M., and Kanai, M.,** Cholesterol diet increases plasma and liver concentrations of cholesterol epoxides and cholestanetriol, *J. Pharm. Dyn.,* 3, 553, 1980.
99. **Saucier, S. E., Kandutsch, A. A., Gayen, A. K., Swahn, D. K., and Spencer, T. A.,** Oxysterol regulators of 3-hydroxy-3-methylglutaryl-CoA reductase in liver. Effect of dietary cholesterol, *J. Biol. Chem.,* 264, 6863, 1989.

Chapter 3

METHODS OF ANALYSIS OF CHOLESTEROL OXIDES

Peter S. W. Park and Paul B. Addis

TABLE OF CONTENTS

I. INTRODUCTION

Cholesterol is susceptible to oxidation under a variety of conditions. As an unsaturated lipid, cholesterol in contact with air is autoxidized forming cholesterol hydroperoxides, from which over 30 secondary oxidation products have been reported.[1] Cholesterol in cell membranes can also be oxidized via either enzymatic or free radical-mediated peroxidation processes.[2-5] Whether it originates from autoxidative or enzymic processes, cholesterol oxidation has received much scientific attention due to its undesirable implications in human health.[6-11] Because of the tendency of cholesterol to undergo spontaneous air oxidation, human diets can be a major source of cholesterol oxide intake. It was not until recently, however, that cholesterol oxidation in foods has been more reliably analyzed. This delay may be accounted for by the technical difficulties associated with cholesterol oxide analysis. Cholesterol oxides share very similar chemical structures, which requires that the analytical method possess a high resolving power. The presence of cholesterol oxides at trace levels also demands that the methods be capable of sensitive detection. This trace analysis is often complicated by interference by a diversity of compounds in food or biological specimens. For cholesterol oxides analysis, therefore, it is obvious that a dependable, reliable, and sensitive analytical method is highly desired. Instability of some cholesterol oxides also requires that samples be handled with care during sample preparation work-ups.

Depending upon the sample content, an analytical method may be viewed as consisting of a series of stepwise procedures to accomplish the following goals:

1. Extraction of total lipids
2. Enrichment of cholesterol oxides
3. Separation and detection of cholesterol oxides
4. Confirmation of structural identity

The purpose of this chapter is to review the analytical techniques that have been used for the analysis of cholesterol oxides, and in particular how these analytical methods can be applied to identify and quantify cholesterol oxides in food and biological specimens. Much of the discussion will be devoted to chromatography and mass spectrometry.

II. EXTRACTION OF TOTAL LIPIDS

Sterols are a minor fraction in food and biological specimens which are a complex mixture. Taking beef tallow for instance, the cholesterol content is 600 to 1200 mg/g,[12-14] the rest being mostly triglycerides. In biological tissues cholesterol is by far less, about 0.6 mg/g of bovine muscle[15] and about 1.4 mg/g of fresh mouse liver.[16] Therefore, the preparation of sterol samples as cleanly as possible is obviously a crucial factor for the successful analysis of cholesterol oxides in food and biological samples. The first part of this requirement can be achieved by extracting total lipids from samples with organic solvents in which cholesterol and its oxides are co-extracted.

For quantitative measurement of cholesterol, total lipids were extracted from freeze-dried, ready-to-serve foods using petroleum ether in a Soxhlet apparatus for 3 to 6 h.[17] The main advantage of this extraction procedure was to avoid the troublesome emulsion often formed in mixed solvent systems. It should be noted, however, that the strength of the extracting solvent is important to ensure good extraction yield of cholesterol oxides, whose polarities range widely from very nonpolar to polar. The efficiency of various organic solvents were compared for the extraction of lipids and cholesterol from the short-necked clam and top shell.[18] Among the solvents tested, including chloroform/methanol (2/1, v/v), hot benzene, acetone, and diethyl ether, chloroform/methanol (2/1, v/v) was shown to be the most suitable solvent system to extract both lipids and cholesterol thoroughly.

For the extraction of total lipids in food and biological samples, the most popular method is the procedure of Folch et al.[19] in which samples are homogenized in a mixture of chloroform/methanol (2/1, v/v) as the extracting solvents. The ratio of sample to extracting solvent mixture is usually 1:20 (w/v). The homogenates are filtered and the filtrates containing lipid extracts are washed with distilled water or salt solution to remove nonlipid material. After removal of moisture from the lipid extracts by anhydrous sodium sulfate treatment, solvents are removed by evaporation under a nitrogen stream. This extraction procedure has frequently been used to extract total lipids for subsequent analysis of cholesterol oxides from foods, serum, and tissue samples.[16,20-24] The procedure of Bligh and Dyer,[25] whose chemical primciple is basically the same as that of Folch et al.[19] was also employed for the extraction of lipids from irradiated beef.[26]

Other solvents may also be used, depending on the nature of the sample matrix containing cholesterol oxides. Acetone was used to extract lipids containing cholesterol epoxides from egg yolk solids.[27,28] Acetonitrile was preferred to the conventional choloroform/methanol mixture for the extraction of total lipids from plasma and foods for determination of 25-hydroxycholesterol.[29] Hexane/2-propanol (3/2, v/v) was used to extract lipids from various Swedish foods.[30-32] After washing the extracts with 0.47 *M* Na_2SO_4 solution, the organic solvent layer was collected for cholesterol oxide quantitation. In another method, after dispersing food samples in ethanol, dichloromethane was used to extract lipids.[33]

Hexane was also used to extract lipids for the analysis of cholesterol oxides in butterfat,[34] nonfat dry milk,[35] and heated butter.[36] Some cholesterol oxides, however, such as 25-hydroxycholesterol and 5α-cholestane-3β,5,6β-triol, are not freely soluble in nonpolar solvents such as hexane or petroleum ether. Consequently, the extraction of such oxides into total lipids using nonpolar solvents is not likely to be effective, apart from limited capability in crossing the aqueous or phospholipid layer of cell membrane prior to solubilizing lipids. When the samples are already in the form of lipid extracts, the total lipid extraction step may be omitted and the next step can be directly implemented as is the case with beef or port fats[12-14,37] and milk fats.[33,38,39]

III. ENRICHMENT OF CHOLESTEROL OXIDES

The total lipid extracts are a crude preparation. The major portion of the extracts is triglycerides and/or phospholipids and sterols are a very minor component. Accordingly, cholesterol oxides will be at trace levels, if present at all. The trace nature of cholesterol oxides, therefore, requires enrichment by further isolating cholesterol oxides from the bulk of accompanying lipids in the total lipid extracts. After this enrichment, cholesterol oxides can be analyzed with less interference and concentrated to above the detection level. This enrichment is commonly implemented by either saponification or chromatographic means.

A. SAPONIFICATION

The basic principle of saponification is to hydrolyze the ester bonds of triglycerides and phospholipids in alkaline media such as methanolic or ethanolic KOH, converting them to water-soluble derivatives. After the addition of water to the saponified mixture, cholesterol oxides are extracted with suitable organic solvents into the nonsaponifiable fraction. It is usual to carry out saponification on the total lipid extracts. With some samples, such as human serum[40,41] and human breast fluid,[42] it has also been directly applied, omitting the total lipid extraction step. This direct saponification on such samples, however, was followed by subsequent supplemental purification of nonsaponifiable extracts by thin-layer chromatography (TLC) or high-performance liquid chromatography (HPLC),[40-42] whereas nonsaponifiables from the total lipid extracts are ready for chromatographic resolution. To facilitate ester bond hydrolysis, the usual practice is to reflux the reaction mixture. Unfortunately, despite the damaging effect of the conventional hot saponification on 7-ketocholesterol integrity, it was often used for the determination of cholesterol oxides without proper precautions against artifact formation. Two cholesterol oxides — 7-ketocholesterol and cholesterol α-epoxide — deserve special attention due to controversy over stability as related to saponification.

The instability of 7-ketocholesterol in hot alkaline media has long been known and cholesta-3,5-dien-7-one was suggested as the major reaction product.[43] Chicoye et al.[44,45] found a large amount of cholesta-3,5-dien-7-one with numerous other degradation products from the TLC analysis of 7-ketocholesteryl acetate treated with a hot alkaline solution. The conversion of 7-ketocholesterol to cholesta-3,5-dien-7-one was recently reaffirmed from the mass spectrometric analysis of an unknown compound which newly emerged during hot saponification of 7-ketocholesterol.[46] Although cholesta-3,5-dien-7-one is among the established cholesterol oxides, it is rarely reported, whereas 7-ketocholesterol is among the more common products from cholesterol autoxidation. It is our experience that 7-ketocholesterol was the most predominant species among cholesterol oxides found in various food products; in contrast cholesta-3,5-dien-7-one was seldom noticed.[13,14,47,48] Therefore, when hot saponification is used, the detection of cholesta-3,5-dien-7-one should be suspected as an artifact arising from the thermal dehydration of 7-ketocholesterol rather than as originally present in the sample.

Cholesta-3,5-dien-7-one was identified as a minor sterol in the egg yolk of the developing embryo.[49] Although the samples were saponified in the presence of added antioxidant (pyrogallol), it seems reasonable to suspect this dienone to be an artifact due to the use of hot saponification. Anhydrous milk fat (AMF) was saponified without a solvent by merely grinding AMF in a mortar with aqueous KOH and Celite 545.[35] Nonsaponifiables were then extracted with hexane for 24 h in a Soxhlet thimble to avoid the troublesome emulsion during the solvent extraction of nonsaponifiables. The nonsaponifiables must have been heated, however, to implement the Soxhlet extraction, which would exert the same effect as hot saponification. Accordingly, the finding of this dienone in AMF should also be suspected to be an artifact derived from 7-ketocholesterol. The detection of this dienone compound, and the absence of 7-ketocholesterol in heated tallow[12] may also be ascribed to the employment of hot saponification during sample preparation.

Recently, Bascoul et al.[50] reported a dienone amount twice that of 7-ketocholesterol after refluxing heated beef fats in alkaline solution for 2 h. This finding, however, conflicts with reports by others. Chicoye et al.[44] performed hot saponification deliberately to destroy 7-ketocholesterol which interfered with cholesterol α-epoxide during TLC. Park and Addis[13,14] found 7-ketocholesterol was the most predominant oxide species in heated fats and that no measurable dienone was present after saponification overnight at room temperature.

For the successful recovery of 7-ketocholesterol, the merit of cold saponification was recognized early by Bergstrom and Wintersteiner.[43] The cold saponification was adopted by Williams and Pearson[37] to detect 7-ketocholesterol in pork fat, although its identification was only speculative. Chicoye et al.[51] demonstrated that saponification at ambient temperature was successful and effective in recovering 7-ketocholesterol formed from cholesterol-containing foods. Cold saponification was also used in the analysis of cholesterol oxides in human atherosclerotic aorta, liver, and plasma samples.[21] When 7-ketocholesterol added to beef fats was saponified overnight at room temperature, the recovery was found to be about 96%, similar to recoveries of other common cholesterol oxides tested.[13]

A contradictory result was found in a recent study suggesting that cold saponification was not better than hot saponification, leaving only 11 to 40% of the original 7-ketocholesterol content.[52] This claim, however, is hard to accept in view of the above-mentioned disucssion. Moreover, another recent study[33] reported almost 100% recovery of 7-ketocholesterol from lipid extracts that were stirred at 40°C for 10 min in alcoholic KOH and then left overnight at room temperature. We feel that low recovery can result even from cold saponification depending upon how it is implemented (details of which are seldom available from published reports). The lipid extracts are not easily dissolved in polar alcoholic solution. If a little heat is applied to hastily disperse lipid extracts in alcoholic solution, this would lead to partial degradation of 7-ketocholesterol according to our own experience (data unpublished). The contradictory result[52] may be attributable to failure to follow all the precautions in handling 7-ketocholesterol in alkaline solution.

Another controversial statement about saponification was made regarding the recovery of cholesterol α-epoxide. Its recovery was claimed to be only 25% following a hot saponification procedure.[53] The poor recovery was attributed to the nucleophilicity of α-epoxide causing opening of the highly strained 3-membered ring structure during hot saponification. Consequently, the definite avoidance of saponification was proposed for any quantitation procedure for cholesterol epoxides. It is doubtful, however, whether α-epoxide is likely to undergo ring opening in alkaline media as much as in acidic media. Ring opening will be even more unlikely when no heat is applied, such as in cold saponification. The resistance of α-epoxide against ring opening in alkaline media is indirectly supported by the results of Chicoye et al.[51] which indicated that β-epoxide was not structurally altered when irradiated egg solids or a dispersion of cholesterol as a source of cholesterol oxides were treated with hot KOH solution at 81°C for 2 h. Gruenke et al.[42] was concerned about the possible formation of 5α-cholestane-3β,5,6β-triol from the chemical hydrolysis of epoxides during the processing when they subjected human breast fluids to saponification at 55°C for 15 min. Based on combined gas chromatography-mass spectrometry (GC-MS) analysis, they found no evidence of the presence of 5α-cholestane-3β,5,6,β-triol, indicating that the ring structure of the epoxides remained stable during saponification.

As well as effective extraction of total lipids, one should evaluate the adequacy of the extracting solvent for the successful recovery of cholesterol oxides into nonsaponifiable extracts. For the extraction of neutral sterols, nonpolar solvents such as hexane or petroleum ether are commonly used.[15,54,55] They also can be used for nonpolar cholesterol oxides such as cholesta-3,5-dien-7-one.[35,49] The more commonly reported cholesterol oxides, however, are more polar than cholesterol and some of these are not freely soluble in hexane or petroleum ether. The alleged poor recovery of α-epoxide[53] following hot saponification, therefore, might be better accounted for by the incomplete extraction of α-epoxide due to the poor choice of extracting solvent, i.e., petroleum ether, rather than by its structural alteration due to ring opening. For the extraction of cholesterol oxides, the efficiency of 4 organic solvents including chloroform, ethyl ether, isopropyl ether, and petroleum ether was evaluated.[46] Ethyl ether was the best solvent for extracting cholesterol oxides ranging from 25-hydroxycholesterol to 5α-cholestane-3β,5,6β-triol, which represented nonpolar and the most polar oxides. In contrast, petroleum ether was the worst solvent for these sterols.

In the course of studying the identification and quantitative estimation of cholesterol oxides formed in heated fats, Park and Addis[13] examined recoveries of the more common cholesterol oxides subjected to cold saponification. Except for 5α-cholestane-3β,5,6β-triol, recoveries of oxides, including cholesterol α-epoxide were no lower than 80%. After the use of 5α-cholestane as an internal standard, recoveries of oxides were compensated and did not require any additional correction factors except for 5α-cholestane-3β,5,6β-triol. Pie et al.[33] ended up with more than 100% instead of lower recoveries for both epimeric epoxides after cold saponification, which further supports the use of saponification to enrich cholesterol oxides, including cholesterol epoxides.

B. COLUMN CHROMATOGRAPHY

Column chromatography is another useful means of achieving further enrichment of sterols from total lipid extracts. The interaction between lipid components and the adsorbent

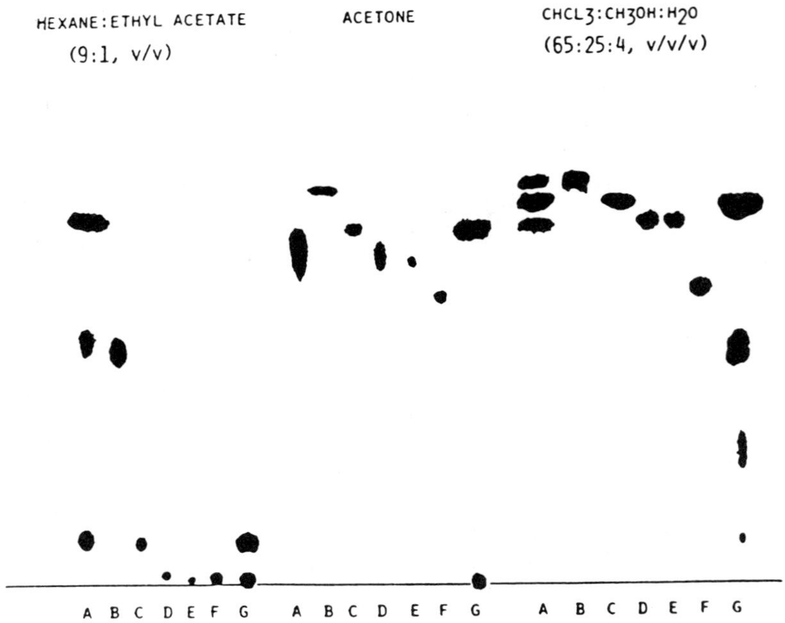

FIGURE 1. TLC of some representative lipid components in different irrigating solvents: (A) nonpolar lipid mix; (B) triolein; (C) cholesterol; (D) 25-hydroxycholesterol; (E) 7-ketopregnenolone; (F) 5α-cholestane-3β,5,6β-triol; (G) polar lipid mix. (From Park, S. W. and Addis, P. B., *J. Food Sci.*, 50, 1437, 1985. With permission.)

in the column for a given solvent or solvent system is the key factor determining the efficiency of isolating cholesterol oxides or eliminating other interfering substances. In adsorption column chromatography with silicic acid, elution starting with a nonpolar solvent and progressing toward polar solvents will release lipid components from the column in increasing order of polarity, i.e., esterified cholesterols, triglycerides, free sterols, and phospholipids. The earlier use of this column chromatography was for preliminary purification of the nonsaponifiables rather than an original means of isolating cholesterol oxides from samples.[35,37,49] Kaneda et al.[18] evaluated the effectiveness of further column chromatographic purification of the nonsaponifiables for the quantitation of cholesterol in foods. Results indicated that the use of florisil column chromatography as a further pretreatment contributed to improved separation by gas chromatography (GC) but decreased recovery of cholesterol. With the currently available HPLC or capillary GC as a final analytical tool, the need for preliminary purification of the nonsaponifiables has been decreased due to the excellent resolving power of HPLC or GC.

Although silicic acid column chromatography has long been used for the separation of different lipid classes,[56] only a few researchers have tried this method as a purification procedure in place of saponification prior to cholesterol oxide analysis. Tsai et al.[53] devised silicic acid column chromatography procedures to prepare analytes prior to HPLC determination of epimeric epoxides. Lipid extracts from egg yolk were charged in a 5-ml disposable glass pipet packed with 1 g silicic acid. After washing with 30 ml 10% ethyl acetate in hexane, eluates with 20 ml hexane/ethyl acetate (5/3, v/v) were collected as cholesterol epoxide-containing fractions.

Park and Addis[22] devised a similar procedure to analyze most of the more common cholesterol oxides. After irrigating different lipid classes on silica gel plate with different developing solvents, they found cholesterol and its oxides rarely moved on silica gel plates using 10% and 20% ethyl acetate in hexane, whereas esterified cholesterol and triglycerides migrated rapidly (Figure 1). In the meantime, with acetone as the developing solvent, sterol fractions ranging from cholesterol to 5α-cholestane-3β,5,6β-triol migrated almost with the same mobilities but phospholipids remained at the origin due to their insolubility in acetone. Based on those observations, they optimized column chromatographic purification procedures: two consecutive washings with 28 ml hexane/ethyl acetate (9/1) and 20 ml hexane/ethyl acetate (8/2, v/v) through 4-g silica gel columns (16 × 140 mm) removed triglycerides and most of the cholesterol. Subsequent elution with 45 ml acetone released the cholesterol oxide fraction free from phospholipid contamination.[22] This purification procedure was later improved by De Vore[57] who used commercially available disposable columns for the analysis of 7-ketocholesterol. With 2.8-ml-volume silica columns, solvent consumption was reduced to 2 ml each for each washing and elution.[57]

Morgan and Armstrong[58,59] enriched epimeric epoxides from egg yolk lipid extracts by passage through silica Sep-pak cartridges. When the lipids suspended in 1 ml toluene/ethyl acetate (3/2, v/v) were passed through 3-ml Diol columns, this resulted in 99% recoveries of cholesterol oxides from liposomes.[5] It is noteworthy that cholesterol oxides in the fraction collected through column chromatography of total lipid extracts represent the oxidation of only unesterified cholesterol, because cholesteryl esters are nonpolar and were removed along with triglycerides. Not only isolation but also prefractionation of cholesterol oxides were also reported;[52] prefractionation, however, has no obvious advantages. Missler et al.[60] purified lipid extracts by silica gel liquid chromatography and then by preparative HPLC prior to GC quantitation. As part of rigorous sample purification procedures, 25-hydroxycholesterol recovered from cold saponification was further purified through C18 Sep-Pak cartridges followed by C18 reversed-phase HPLC.[29] Koopman et al.[40] carried out purification of nonsaponifiable extracts from serum samples through reversed-phase HPLC, in which the first 5-ml fraction following injection was collected for subsequent GC analysis.

Taking advantage of the difference in molecular size is another approach to isolating sterols from triglycerides or phospholipids. For the isolation of steroids from lipid extracts of bovine corpus luteum, gel chromatography with hydrophobic Sephadex was demonstrated to be successful.[61] Eisman et al.[62] subjected lipid extracts of human plasma to two successive gel chromatographies with Sephadex LH-20 prior to the quantitation of 25-hydroxyvitamins, D_2 and D_3, by HPLC. The structures of these vitamins are very close to that of 25-hydroxycholesterol. It is conceivable that the same technique may be applicable to the quantitation of 25-hydroxycholesterol without any significant modification. Recently, some studies have demonstrated the use of size exclusion together with subsequent ion-exchange column chromatography to isolate cholesterol oxides.[30-32] The procedure seems laborious, however, because of two consecutive chromatographies and moreover resulted in distributing cholesterol oxides into 3 different fractions as in the prefractionation procedure.[52]

C. PREPARATIVE THIN-LAYER CHROMATOGRAPHY

Thin-layer chromatography (TLC) is also frequently used to isolate sterols from lipid extracts. The basic principle of TLC is the same as that of column chromatography as long as the same adsorbent is used. Therefore, TLC may be viewed as open tubular-column chromatography. After developing lipids on TLC plates with suitable solvents, sterol bands are scraped, eluted into organic solvent, and concentrated for analytical chromatography. The applicability of TLC as a substitute for saponification was well demonstrated by the isolation of sterol fractions using preparative TLC to determine α-epoxide from human serum.[63] The lipid extract was first developed in hexane to remove hydrocarbons and then the plates were rerun in benzene/chloroform (4/1, v/v) to separate sterol esters, triglycerides, and free sterols from one another. Esterified sterols were collected, followed by hydrolysis at 40°C for 2 h and the derived sterols combined with the free sterol fraction prior to the final GC quantitation. Recently, Björkhem et at.[24] concentrated cholesterol epoxides and 5α-cholestane-3β,5,6β-triol from serum lipids by preparative TLC. Serum lipids containing added internal standards were developed in toluene/ethyl acetate (3/9, v/v) for cholesterol epoxides and in ethyl acetate for 5α-cholestane-3β,5,6β-triol. The appropriate chromatographic zones were scraped off and eluted with methanol for GC-mass spectrometry (MS) quantitation. Even though this procedure is much milder than saponification, the labor-intensive scraping and elution steps are a drawback.

TLC is also useful for preliminary purification of nonsaponifiable extracts. Nonsaponifiables recovered from saponification of serum lipids were further purified with TLC in 10% acetone in chloroform followed by scraping of the cholesterol oxides band and eluting with 80% dichloromethane in methanol.[42] The TLC purification after saponification was effective in eliminating most of the cholesterol from nonsaponifiables, which had to be removed because of its high concentration in the samples.[33]

IV. RESOLUTION AND QUANTITATION OF CHOLESTEROL OXIDES

Separation of cholesterol oxides into individual species requires that enriched sterols be subjected to a final analytical method capable of good resolution and sensitive detection. These requirements are even more demanding if quantitation is to be attempted. It was a TLC technique which established the spontaneous air oxidation of cholesterol, demonstrating formation of more than 30 oxides.[1] Color development of cholesterol oxides, such as the intense blue color of 7-hydroxycholesterols and the yellow color of 5α-cholestane-3β,5,6β-triol upon spraying with 50% aqueous sulfuric acid or other visualization reagents aids in the differentiation of cholesterol oxides.[1,64-66] Nevertheless, the resolving power of TLC is not comparable to that of other modern chromatographic methods, i.e., HPLC and GC and the detection sensitivity is also far below that of HPLC or GC detectors. Since the advent of more powerful techniques such as HPLC and GC, TLC is rather used for a preliminary purification of samples as already discussed.

Although some recent studies still relied on TLC for the quantitation of cholesterol oxides in butter oils[65] and used flame ionization detection (FID) after separation by TLC on a chromarod,[50] presently either HPLC or capillary GC is exclusively the method of choice because of the excellent resolving power with sensitive on-line detection. Using the mass spectrometer as a detector for GC or HPLC brings about further enhancement in analytical capability. The following discussion will review GC and HPLC methods for the simultaneous separation and quantitation of the more common cholesterol oxides, including 7α-hydroxycholesterol, 7β-hydroxycholesterol, 7-ketocholesterol, cholesterol α-epoxide, cholesterol β-epoxide, 5α-cholestane-3β,5,6β-triol, and 25-hydroxycholesterol. Later, the confirmation of chemical identity by mass spectrometry of the chromatographically separated peaks will follow.

A. GAS CHROMATOGRAPHY

Gas chromatography (GC) has been widely used for the resolution of cholesterol oxides in foods as well as in biological samples. FID is the choice for the detection of cholesterol oxides and is usually capable of detecting organic compounds in the nanogram range. Until about a decade ago, most GC analyses of cholesterol oxides had been performed with packed columns. The commonly used stationary phases were nonpolar to medium polar SE-30, QF-1, and OV-17, etc. With progress in column technology, high-resolution capillary columns have emerged. The same stationary phases used in the packed-column GC are coated or bonded on the wall of a fused silica capillary tube whose inside diameter is about 0.25 to 0.32 mm. This wall-coated (or cross-linked) open tubular column allows the carrier gas to go through the column quickly, thus curtailing the analysis time. Due to the limited loading capacity of a capillary column, only a small part of the sample injected is actually delivered into the column for the analysis (split injection). Nevertheless, this capillary column GC is superior to packed-column GC in detection sensitivity. The very slow flow rate of carrier gas, normally a few ml/min, requires supplemental gas flow in the detector cell (make-up gas). Thus, GC with capillary columns requires a little modification in the injection port and detector cell of the conventional GC instrument. During the past decade, the availability of capillary columns has facilitated significant progress in the analysis of cholesterol oxides.

As sterols have high boiling points, it is necessary to operate the GC at quite elevated temperatures. This high-temperature operation limits the choice for columns to only thermodurable ones. This has also hampered the use of GC in the quantitation of cholesterol oxides, because some oxides of biological importance have been reported to undergo thermal decomposition. To overcome the thermal lability of some cholesterol oxides, trimethylsilyl (TMS) derivatization has often been employed.

1. Resolution

Comprehensive documentation tabulation has been compiled on the relative retention times of many cholesterol oxides with reference to cholesterol from the GC analysis of aged cholesterol in which cholesterol oxides were gas chromatographed either as is or as the acetate or TMS ether derivatives.[67] Packed columns with 4 different stationary phases were evaluated; epimeric 7- hydroxycholesterols were found not separable from each other and moreover decomposed frequently. The decomposed product was later tentatively identified as cholesta-2,4,6-triene.[68] After TMS derivatization, 7-hydroxycholesterols were free from thermal decomposition and well resolved on SE-30, but 7α-hydroxycholesterol was completely fused with cholesterol.[67] This fusion of peaks remained on QF-1. 7-Ketocholesterol was also fused together with 5α-cholestane-3β,5,6β-triol on SE-30 but not on QF-1. These problems forced the running of the same sample twice (both before and after derivatization) or through two different columns to resolve the more common cholesterol oxides.[67]

Apart from the thermal lability of some oxides, structural similarity among a wide array of cholesterol oxide species requires that analytical columns possess a great deal of resolving power. This requirement seemed to be met quite successfully by a capillary column impregnated with OV101 when it separated 17 out of 22 cholesterol oxides as TMS ethers, including 7-hydroxycholesterols,[69] although two important oxides, 7-ketocholesterol and 5α-cholestane-3β,5,6β-triol, were completely fused to each other. Another GC study with a capillary column, SE-54, reported the successful baseline-to-baseline resolution between epimeric epoxides that were difficult to resolve by packed column GC.[70] Using a similar stationary phase, a DB-1 capillary column, epimeric epoxides were also completely separated not only from each other but also from other oxides without any difficulty.[13,71] Usually it seems α-epoxide elutes ahead of β-epoxide, but the reverse order has also been encountered.[50]

7-Hydroxycholesterols were found to still undergo thermal alteration during capillary-column GC as well.[72] Initially, these epimers were resolved, but soon after retention times became identical and far shorter than before (personal experience). This thermal alteration was commonly noticed during GC analysis of other oxides, i.e. 4-hydroxycholesterol and 25-hydroxycholesterol.[72] This observation agreed with the early documentation by Van Lier and Smith[67] that thermal alteration was quite common with dihydroxy cholesterol oxides. Based on the observation of no measurable difference in retention times among the above 4 dihydroxy cholesterol oxides (Table 1), it seems that structural alteration might have proceeded in the direction of offsetting the structural differences among them. Interestingly, 5α-cholestane-3β,5,6β-triol was still the last eluting species among the common cholesterol oxides as expected from its 3 hydroxyl groups. Whether it was also structurally altered was not known.

To avoid the thermal instability and incomplete separation of some cholesterol oxides,

TABLE 1
Relative Retention Times of Some Major Products of Cholesterol Oxidation

	DB-1		DB-5		DB-1701
	A[a]	B[b](TMS-ether)	C[c]	D[d](TMS-ether)	E[e]
5α-Cholestane[f]	—	0.647	0.625	0.585	0.582
Cholesta-3,5-dien-7-one	—	0.976	—	—	—
Cholest-5-en-3β-ol	1.000	1.000	1.000	1.000	1.000
Cholest-5-ene-3β,7α-diol	1.240	1.019	0.656[g]	0.980	0.641[g]
Cholest-5-ene-3β,25-diol	1.250	1.318	—[h]	1.522	—[h]
Cholest-5-ene-3β,7β-diol	1.257	1.155	0.656[g]	1.194	0.641[g]
5,6α-Epoxy-5α-cholestan-3β-ol	1.181	1.134	1.111	1.235	1.383
Cholest-5-ene-3β,4β-diol	—[h]	1.164	—[h]	1.227	—[h]
3β-Hydroxy-5α-cholestan-6-one	1.324	—	—	1.489	—
5α-Cholestane-3β,5,6β-triol	1.520	1.265	—	1.426	2.822
3β-Hydroxycholest-5-en-7-one	1.353	1.275	—	1.519	2.142

[a] A: DB-1, 15 m × 0.25 mm I.D. × 0.25 μm film thickness; temperature programming 200°C, 3°C/min, 260°C, 10 psi H_2.
[b] B: DB-1, 15 m × 0.25 mm I.D. × 0.10 μm film thickness; temperature programming 180°C, 3°C/min, 250°C, 8 psi H_2.
[c] C: DB-5, 30 m × 0.25 mm I.D. × 0.10 μm film thickness; temperature programming 220°C, 4°C/min, 260°C, 20 psi H_2.
[d] D: DB-5, 30 m × 0.25 mm I.D. × 0.10 μm film thickness; temperature programming 230°C, 2°C/min, 260°C, 13 psi H_2.
[e] E: DB-1701, 30 m × 0.25 mm I.D. × 0.15 μm film thickness; 260°C isothermal, 20 psi H_2.
[f] Not an oxidation product. Internal standard used in the study on the response linearity.
[g] Principal decomposition peak.
[h] No principal peak was identifiable.

From Park, S. W. and Addis, P. B., *Anal.Biochem.*, 149, 275, 1985. With permission.

TMS derivatization is often conducted. The reaction converts the hydroxyl groups of cholesterol oxides to the corresponding TMS ether derivatives as follows:

$$\text{Sterol–OH} + \text{TMS reagent} \rightarrow \text{Sterol–0–TMS}$$

For TMS etherification, various reagents are available which differ in reaction strength. The commercially available reagents include hexamethyldisilazane (HMDS), *N,O*-bis TMS acetamide (BSA), *N,O*-bis TMS trifluoroacetamide (BSTFA), *N*-TMS imidazole (TMSI), etc. Trimethylchlorosilane (TMCS) is a common catalyst. Cholesterol oxides or nonsaponifiable extracts are dissolved in pyridine and then reacted with one or a mixture of the above reagents. The routine procedure the authors employed was to dissolve nonsaponifiables (e.g., from 0.5 g tissue sample) in 150 μl pyridine, to which 100 μl derivatization reagent mixture (TMSI/BSA/TMCS, 3/3/2, v/v/v) was added. After leaving the reaction mixture at room temperature for about 30 min, 1 μl of reaction mixture was directly injected for GC. Previously, it was often necessary to evaporate excess derivatization reagents and then redissolve the dried derivatives in hexane prior to injection, because early eluting derivatives were frequently interfered with by the tailing of solvent and reagent peaks in packed-column GC. These preparation procedures are no longer needed in capillary-column GC due to its superior resolving power. It should be pointed out that the ether bond is susceptible to hydrolysis and also that water molecules compete with hydroxyl groups of cholesterol oxides for the TMS moiety from donor reagents. Therefore, both samples and pyridine should be reasonably free from water contamination to ensure complete derivatization.

Resolution of the TMS derivatives of the more common cholesterol oxides by capillary GC has been extensively studied in the past decade, and relative retention times have been

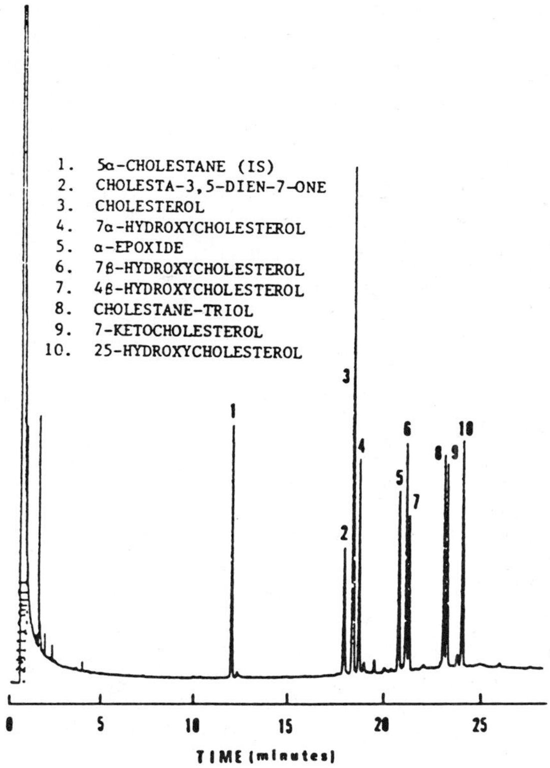

FIGURE 2. The capillary column gas chromatogram of TMS ether derivatives of major oxidized cholesterol derivatives. Gas chromatography was carried out on a fused-silica capillary DB-1 (15 m × 0.25 mm i.d., 0.10-μm film thickness). Temperature programming was from 180 to 250°C at 3°C/min. (From Park, S. W. and Addis, P. B., *Anal. Biochem.*, 149, 275, 1985. With permission.)

reported. As shown in Table 1, 7α-hydroxycholesterol was followed by cholesterol on nonpolar DB-5 (5% phenyl + 95% methyl silicone phase), whereas the order was reversed on DB-1 (100% methyl silicone phase). With DB-5, 7-ketocholesterol was almost fused with 25-hydroxycholesterol. Also, 7β-hydroxycholesterol was considerably overlapped by 4β-hydroxycholesterol. With DB-1, 7-ketocholesterol was well separated from 25-hydroxycholesterol and the interference with 7β-hydroxycholesterol by 4β-hydroxycholesterol became almost insignificant. Although 5α-cholestane-3β,5,6β-triol was closely followed by 7-ketocholesterol, the resolution value as the distance between maximas divided by the average width of the two peaks was still about 1.25, acceptable for reproducible quantitation (Figure 2). Later, these pairs were separated further apart by Pie et al.[33] who used a thicker film capillary column DB-5 (30 m, 1.0 μm film thickness) with prolonged analysis time (Figure 3). It is interesting to note that 25-hydroxycholesterol eluted ahead of 7-ketocholesterol on thick-film DB-5 unlike their relative order on DB-1[30,72] or 0.1-μm thick-film DB-5.[72] Although a number of capillary GC resolution studies have been reported, only a few seemed to have demonstrated satisfactory resolution among the more common cholesterol oxides for simultaneous analysis. In addition to Figures 2 and 3, another illustration is shown in Figure 4, where a cross-linked methyl silicone column ultra hp, the same as DB-1 in stationary phase, was used.

In most GC studies, cholesterol oxides are usually injected into the heated injector for vaporization before delivery to analytical columns for separation. The on-column injection

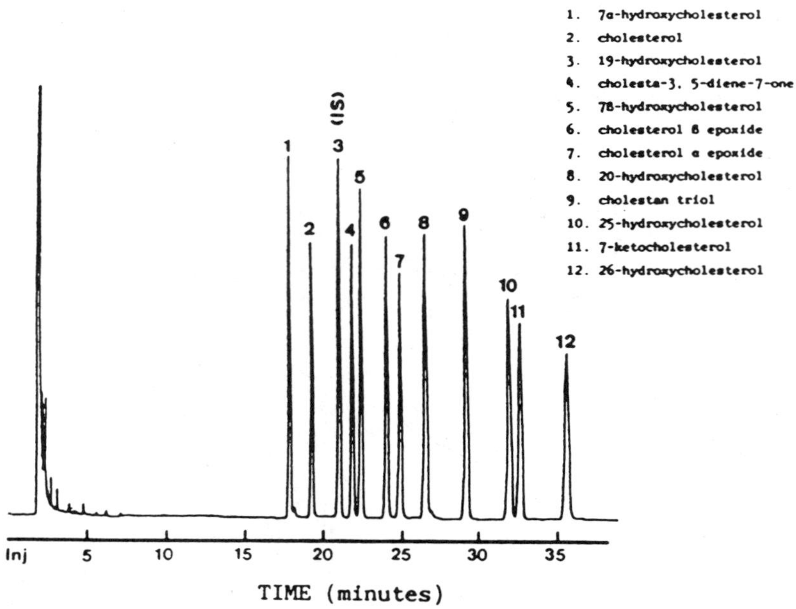

FIGURE 3. Gas-liquid chromatography profiles of standards as TMS ether derivatives on a 30-m DB-5 column (1.0-μm film thickness) at 280°C. (From Pie, J. E., Spahis, K., and Seillan, C., *J. Agric. Food Chem.*, 38, 973, 1990. With permission.)

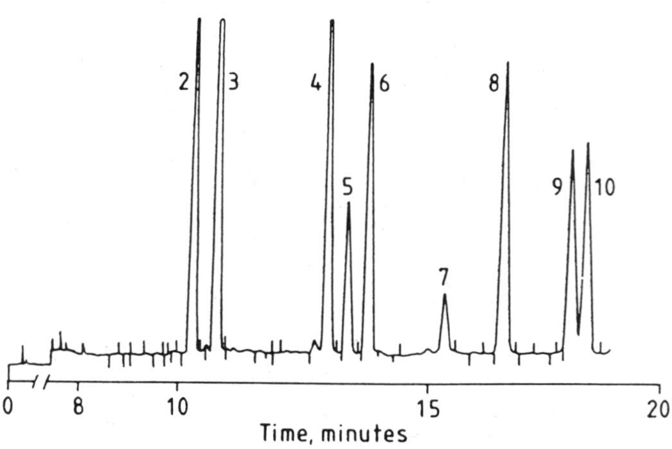

2- Cholest-5-ene-3β,7α-diol (7α-hydroxycholesterol)
3- Cholest-5-ene-3β-ol (cholesterol)
4- Cholest-5-ene-3β,7β-diol (7β-hydroxycholesterol)
5- 5,6β-Epoxy-5β-cholestan-3β-ol (5β-,6β-epoxycholestanol)
6- 5,6α-Epoxy-5α-cholestan-3β-ol (5α-,6α-epoxycholestanol)
7- Cholest-5-ene-3β,20α-diol (20α-hydroxycholesterol)
8- 5α-Cholestane-3β,5,6α-triol (cholestane-triol)
9- 3β-Hydroxycholest-5-en-7-one (7-ketocholesterol)
10- Cholest-5-ene-3β,25-diol (25-hydroxycholesterol)

FIGURE 4. Gas chromatogram of TMS derivatives of synthetic cholesterol oxides obtained with a cross-linked methyl silicone column, Ultra hp (25 m × 0.33 mm, 0.35 μm), at 280°C. (From Nourooz-Zadeh, J. and Appelqvist, L. A., *J. Food Sci.*, 52, 57, 1987. With permission.)

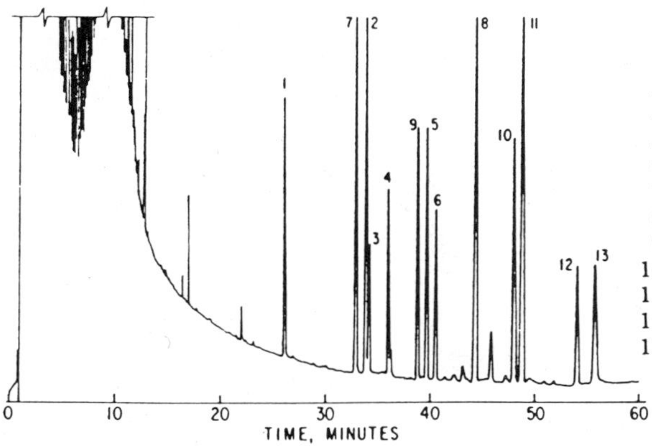

1 Tetradecyl tetradecanoate (ref.)
2 Cholesterol
3 3-Ketocholestanol
4 3,5-Cholestadien-7-one
5 5β,6β-Epoxycholestanol
6 5α,6α-Epoxycholestanol
7 7α-Hydroxycholesterol
8 25-Hydroxycholesterol
9 7β-Hydroxycholesterol
10 6-Ketocholestanol
11 7-Ketocholesterol
12 5-Hydroxy-6-ketocholestanol
13 5α,6β-Dihydroxycholestanol

FIGURE 5. GC of standard cholesterol oxides as TMS ethers on a bonded-phase 5% phenylsilicone column (25 m × 0.2 mm, 0.33 μm) with an uncoated fused silica (2 m × 0.32 mm i.d.) retention gap preceded for on-column injection. Column temperature was at 100°C for 5 min and then programmed to 260°C at 30°C/min and to 300°C at 0.5°C/min. (From Maeker, G. and Unruh, J., Jr., *J. Am. Oil Chem. Soc.*, 63, 767, 1986. With permission.)

is another option in which analytes are directly deposited on top of the analytical column without experiencing the exposure to split operation and a heated zone. Maeker and Unruh, Jr.[71] reported the separation of cholesterol oxides by using the on-column injection technique. A typical chromatogram obtained with on-column injection is illustrated in Figure 5. Although this on-column injection mode is free from the possible split-related problem like split discrimination, it suffered the longer tailing of the solvents peak and analysis time compared with the split injection mode.

2. Quantitative Analysis

Quantitative GC demands that all compounds of interest in the samples injected produce reproducible peaks for integration. Accordingly, one needs to control possible sources of error such as sampling error, column performance, detector performance, integration techniques, etc. When analytes undergo irreversible decomposition or adsorption in the system, the calibration curve (see below) will not pass through the origin. This situation can be corrected by the use of silanized glassware, a fused silica capillary column, TMS etherification of cholesterol oxides, etc. The measurement of peak area or height can be accomplished by electronic integrators. Variability in the injection amounts can be well compensated by the use of the internal standard (IS) method which relies on the response ratio of the analytes to the IS instead of the absolute analyte response. IS should not be an original constituent of the analytes and also should be separable from the analyte peaks. It is desirable that the chemical structure of the IS be similar to the analytes. The use of such an IS will make less significant even some sample loss (spillage) during the sample preparation, since the concentration ratio of the analytes to the IS can be reasonably assumed to remain constant.

The detector performance is usually the most critical factor for reproducible quantitative analysis. Each detector rsponds differently to different compounds. Accordingly, the response factor (RF) of each analyte should be evaluated prior to quantitative analysis. Moreover, as the GC operating conditions change, so does the detector response, and therefore this influence must be reflected in RF again. The relative response factor (RRF) can be defined as follows:

$$RRF_i = \frac{W_i/A_i}{W_{IS}/A_{IS}}$$

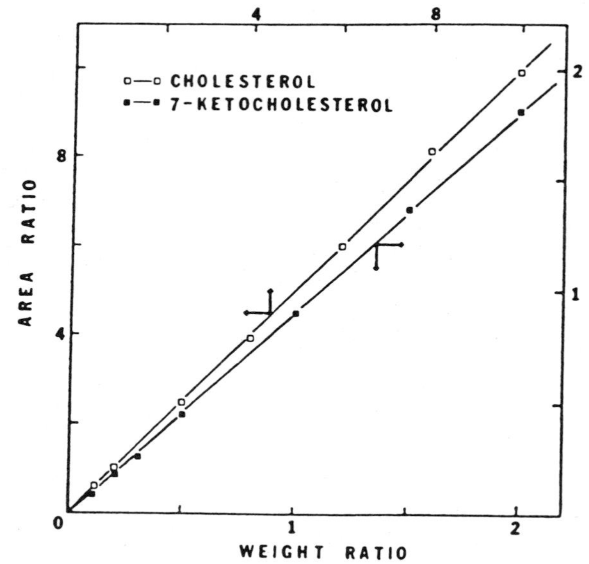

FIGURE 6. Response linearity of cholesterol and 7-ketocholesterol: area ratio (sterol/5α-cholestane) plotted against weight ratio (sterol/5α-cholestane) with various amounts of sterol and fixed amounts of 5α-cholestane. (From Park, S. W. and Addis, P. B., *Anal. Biochem.*, 149, 275, 1985. With permission.)

where W_i = weight of the *ith* cholesterol oxide; A_i = area response of the *ith* cholesterol oxide; W_{IS} = weight of IS; and A_{IS} = area response of IS. Peak height can also be used in lieu of peak area in the above equation. After the GC run of samples, the concentration of each oxide species can be calculated from the following equation:

$$\text{Concentration} = \frac{A_i}{A_{IS}} \, RRF_i \, \frac{IS}{SA}$$

where IS = amount of IS added to sample, and SA = sample amount.

It is prudent to make a few calibration mixtures which contain different amounts of analytes but a fixed amount of IS. Peak response ratios of each analyte to IS from these calibration mixtures are plotted against their known concentration ratios. A linear line with a high r^2 value should result, with the y-intercept passing through the origin or close to it as illustrated in Figure 6, otherwise one should review the instrument system to correct the source of error. With 5α-cholestane as the IS, RRFs of the more common cholesterol oxides were close to 1.0 except those of 7-ketocholesterol and α-epoxide (Table 2). Despite the critical role of the RF in quantitative analysis, it is rarely described in the literature as to how the calibration was performed. Nourooz-Zadeh and Appelqvist[30] applied all the same RRFs, i.e., 1.0, to all cholesterol oxides, citing RRFs reported by Park and Addis.[72] More recently, Van de Bovenkamp et al.[52] published their results on RRFs of the more common cholesterol oxides, which generally agreed with those of Park and Addis.[72] Pie et al.[33] used 19-hydroxycholesterol as the IS to obtain the RRFs.

B. HIGH-PERFORMANCE LIQUID CHROMATOGRAPHY

High-performance liquid chromatography (HPLC) has recently gained popularity in steroid chromatography, competing with the highly developed and generally accepted GC methods.[73] The use of ultrafine adsorbent particles in HPLC accounts for the high efficiency

TABLE 2
Response Linearity of Oxidized Cholesterol Derivatives[a]

Sterols as TMS ether	A_i/A_{IS} vs. W_i/W_{IS} Slope	Intercept	R_f^b
Cholest-5-en-3β-ol	0.99	−0.003	1.00 ± 0.015
Cholest-5-ene-3β,7α-diol	0.94	−0.020	1.09 ± 0.042
5,6α-Epoxy-5α-cholestan-3β-ol	0.92	−0.016	1.13 ± 0.039
Cholest-5-ene-3β,7β-diol	1.04	−0.017	0.98 ± 0.038
5α-Cholestane-3β,5,6β-triol	0.99	−0.014	1.03 ± 0.029
3β-Hydroxycholest-5-en-7-one	0.91	−0.009	1.13 ± 0.030
Cholest-5-ene-3β-25-diol	1.03	−0.003	0.98 ± 0.017

[a] Seven data points with two injections each (n = 7, r = 2).
[b] Response factor = $(W_i/A_i)(A_{IS}/W_{IS})$, presented as mean ± SD.

From Park, S. W. and Addis, P. B., *Anal. Biochem.*, 149, 275, 1985. With permission.

of this separation method. This also necessitates the application of high pressure to force solvents through packed columns. A normal-phase column with silica particles was the earlier development in HPLC column technology, which later has been replaced mostly with reversed-phase columns such as C18 bonded silica. Partition is the major separation mode in reversed-phase HPLC, in contrast to adsorption in normal-phase HPLC. This reversed-phase HPLC is useful especially for the resolution of polar compounds which take a long time for elution from normal-phase HPLC. For a mobile phase, normal-phase HPLC employs nonpolar solvents, while reversed-phase HPLC utilizes polar solvents. In the past few years, column technology has advanced considerably, offering many specialty columns tailored for separation of specific compounds. However, the normal-phase silica columns still seem to be most popular for the resolution of cholesterol oxides.

Despite the limited detectibility for most cholesterol oxides, HPLC has been utilized by many researchers. This is probably due to the general advantages of HPLC over GC, including better resolution of stereoisomers, no thermal alteration during the separation process, and nondestructive detection enabling collection of eluates if desired. Moreover, the poor detection sensitivity may be partly compensated for by the greater loading capacity of HPLC columns as compared to GC capillary columns.

In discussing HPLC separation, the following two values, k′ and α, are frequently used. The capacity factor, k′, is the retention volume of an analyte in terms of column void volume as unity. It is independent of flow rate and is characteristic of each analyte. The smaller the k′ value, the closer the proximity of the analyte elution to the solvent peak. The larger the k′ value, the longer the retention time and the larger the band broadening. The separation of two peaks as affected by different mobile phase is quantitatively expressed using the selectivity or separation factor, α value, which is the k′ ratio of two such peaks.

1. Resolution

Epimeric pairs of some cholesterol oxides such as 7-hydroxycholesterols and cholesterol epoxides were poorly resolved from each other by TLC and packed-column GC.[1,53,67] In such cases, the resolving power of HPLC as an alternative analytical tool has been well demonstrated by accomplishing complete separation between epimers of such pairs. Ansari and Smith[74] reported complete separation between epimeric cholesterol epoxides using both normal- and reversed-phase HPLC. With a μ-Porasil column and a hexane/2-propanol mobile phase, α-epoxide eluted ahead of the β-epimer and vice versa with a μ-Bondapak C18

TABLE 3
Relative Retention Volumes, K′, of Cholesterol and Related Compounds on a μPorasil Column

	Mobile phases		
Compound	Hexane/2-propanol (100:3)	Hexane/2-propanol (100:10)	Hexane/ethyl acetate (100:5)
5α-Cholestane	0.05		0.05
5-Cholestene	0.05		0.05
3,5-Cholestadiene	0.05		0.05
3,5-Cholestadien-7-one	0.14		0.70
5α-Cholestan-3-one	0.22		1.4
5-Cholesten-3-one	0.30		1.6
5-Cholesten-3α-ol	0.46		5.1
5β-Cholestan-3β-ol	0.70		6.5
5-Cholesten-3β-ol (cholesterol)	1.0	0.38	10.4
5α-Cholestan-3β-ol	1.1		
5,24-Cholestadien-3β-ol	1.2		
5,7-Cholestadien-3β-ol	1.2		
5-Cholesten-3β-ol-22-one	1.4		
5-Cholestene-3β,20-diol	1.5	0.38	
5-Cholestene-3β,25-diol	2.9	0.70	
5α-Cholestan-5,6α-epoxy-3β-ol	4.4	1.0	
5β-Cholestan-5,6β-epoxy-3β-ol	5.0		
5α-Cholestan-3β-ol-6-one	5.8		
5α-Cholestan-3β-ol-7-one	7.3		
5-Cholesten-3β-ol-7-one	7.6	1.4	
5-Cholestene-3β,7β-diol	18.5	2.2	
5-Cholestene-3β,7α-diol	21.7	2.7	
5α-Cholestane-3β,6β-diol	21.7	3.1	
5α-Cholestane-3β,5,6β-triol	>50.0	17.2	

Note: $K' = (V - V_o)/V_o$, where V is the retention volume of the compound and V_o the void volume of column.

From Tsai, L. S. and Hudson, C. A., *J. Am. Oil Chem. Soc.*, 58, 931, 1981. With permission.

column and a acetonitrile/water mobile phase. The separation of 7α-hydroxycholesterol from the 7β-epimer was completed by normal-phase HPLC but these epimers were almost fused to each other by reversed-phase HPLC. When reversed-phase HPLC was performed according to Ansari and Smith,[74] incomplete separation of epimeric 7-hydroxycholesterols in cheese products resulted.[65] In addition to the resolution between the above epimeric pairs, Ansari and Smith[74] also reported relative retention data for cholesterol hydroperoxides and other oxides.

Based on the complete separation between epimeric cholesterol epoxides by normal-phase HPLC, an HPLC method for epimeric epoxide quantitation was devised.[53] Hexane/2-propanol was concluded to be the choice for the mobile phase in comparison to hexane/tetrahydrofuran or hexane/ethyl acetate.[75] By examining k′ values of 24 steroid compounds (Table 3), Tsai and Hudson[75] generalized retention characteristics of the more common cholesterol oxides with reference to their functional groups. They found that k′ values of cholesterol oxides were affected not only by the nature but also by the position of functional groups. In the increasing order of k′ value were epoxy, ketone, and hydroxyl groups introduced to the cholesterol molecule. Introduction of such functional groups on the ring moiety rather than the side chain was more influential, as was manifested by the increasing order of k′ values of 25-hydroxycholesterol, cholesterol epoxides, 7-ketocholesterol, 7-hydroxycholesterols, etc. When the second hydroxyl group was introduced, the k′ value of the resulting dihydroxyl oxide was synergistically affected, as indicated by a k′ increase far greater when both hydroxyl groups were in close proximity than when they were apart. This synergism was ascribed to the increased accessibility of the second hydroxyl group once the molecule was adsorbed through hydrogen bonding of the other hydroxyl group.

Tsai and Hudson[75] observed that steric effects are another factor influencing the resolution. The unusually low k′ value of 20-hydroxycholesterol was thought to be due to the steric hindrance by the bulky ring structure and methyl group, preventing the tertiary hydroxyl group from hydrogen bonding. The elution of 7β-hydroxycholesterol ahead of 7α-hydroxycholesterol was also attributed to the steric hindrance effect caused by interference of hydrogens on C15, which was more severe with hydrogen bonding of the equatorial hydroxyl group of the former than the axial one of the latter. The greater k′ value of cholesterol β-epoxide than that of the α-epimer could be due to the higher synergistic effect between the epoxy and 3β-hydroxy group in the β-epimer.[75] The epoxy group of β-epoxide is on the same side as the 3β-hydroxy group adjacent to each other, while the epoxy group of the α-epimer is oriented downward from the multiring plane, opposite to the equatorial 3β-hydroxy group.

Although HPLC separation of cholesterol oxides was well studied in the experiments of Ansari and Smith,[74] and Tsai and Hudson,[75] chromatographic conditions in these studies may not be directly applicable to HPLC analysis of cholesterol oxides in food or biological specimens. The long analysis time would cause band broadening of late-eluting components,[74] whereas too fast a flow rate would be likely to cause incomplete separation of some cholesterol oxides.[75] Moreover, mobile phase strength had to be changed from 3% 2-propanol in hexane to 100% 2-propanol to elute the very polar 5α-cholestane-3β,5,6β-triol. Therefore, most later studies optimized mobile phase strength to specifically analyze a few target cholesterol oxides. Tsai and Hudson[27,28] used hexane/2-propanol (97/3, v/v) for the quantitation of epimeric cholesterol epoxides from spray-dried egg yolk. Park and Addis[22] found 7% 2-propanol in hexane well suited for quantitation of 7-ketocholesterol, 7β-hydroxycholesterol, and 7α-hydroxycholesterol in some food products. The latter mobile phase yielded k′ values of 2.2, 3.8, and 4.6 for above C7 cholesterol oxides, respectively and 5.9 for 7-ketopregnenolone used as an internal standard. A typical chromatogram is presented in Figure 7.

Another mobile phase, 4% 2-propanol in hexane, was evaluated in an attempt to include 25-hydroxycholesterol analysis.[22] Serious interference, however, was frequently encountered in lipid extracts purified by silica gel chromatography. When 2-propanol was reduced to 2.5%, k′ values for C7 cholesterol oxides became too large, which was unfavorable for quantitation. Accordingly, simultaneous measurement of 25-hydroxycholesterol with C7 cholesterol oxides was not pursued. Kou and Holmes[29] developed an HPLC method for the determination of 25-hydroxycholesterol in plasma and some foods. A normal-phase silica column was used with 2.5% 2-propanol in hexane mobile phase at 1.7 ml/min flow rate. Chromatograms of sterol fractions from plasma and egg yolk powder samples are shown in Figure 8. The interference with 25-hydroxycholesterol by the tailing of very large early-eluting peaks was noticeable even after rigorous purification of samples as previously discussed. The HPLC procedure of Park and Addis[22] was adopted by De Vore,[57] who used a 3-μm particle size column to quantify 7-ketocholesterol in ground beef products. Herian and Lee[76] relied on normal-phase HPLC with 9% 2-propanol in hexane for the measurement of epimeric 7-hydroxycholesterols in a dry egg nog mix.

As isocratic adsorption HPLC with 3% 2-propanol in hexane yielded poor resolution of some cholesterol oxides and failed to elute the most polar 5α-cholestane-5β,5,6β-triol,[75] Maeker et al.[77] studied solvent gradient elution. When the mobile phase was concavely programmed from 100% solvent A (3% 2-propanol in hexane) at injection to 98.5% A and 1.5% solvent B (2-propanol) over a 15-min period, then linearly over a 10-min period to

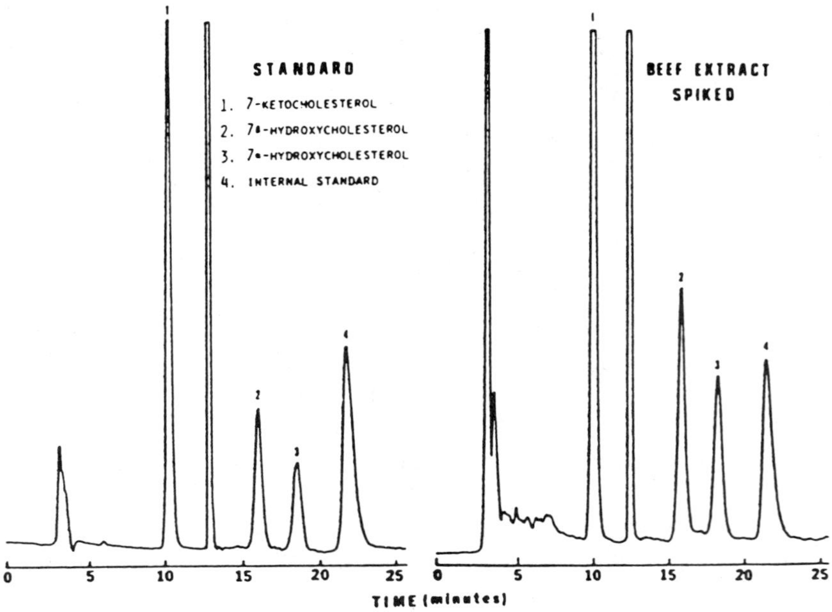

FIGURE 7. HPLC of standards and extract of beef spiked with standards at 50 ppm. Fifty μl of the sample in 2.5 ml mobile phase was injected onto a μPorasil column (3.9 mm × 30 cm) and eluted with 7.0% 2-propanol in hexane (v/v) at 1.0 ml/min. Elution was monitored at 233 nm until about 13 min and thereafter at 208 nm, at 0.032 AUFS. From Park, S. W. and Addis, P. B., *J. Food Sci.*, 50, 1437, 1985. With permission.)

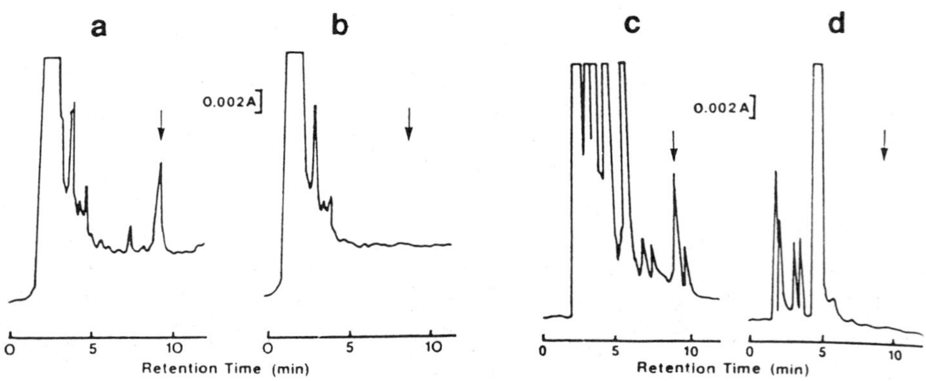

FIGURE 8. HPLC chromatograms of plasma from rats fed 25-hydroxycholesterol (a) and normal plasma (b); HPLC chromatograms of heated (c) and unheated (d) egg yolk powder. Arrows indicate where a 25-hydroxycholesterol standard elutes. (From Kou, I. L. and Holmes, R. P., *J. Chromatogr.*, 330, 339, 1985. With permission.)

97% A and 3% B, and finally to 50% A and B over the last 15-min period, excellent resolution was observed with good peak symmetry. A chromatogram obtained with solvent gradient operation is shown in Figure 9. The total flow rate was 1.5 ml/min during gradient elution. Maeker et al.[78] made a little adjustment in solvent programming to better suit their studies on the reaction of cholesterol epoxides with gastric juice. Solvent programming was also tried earlier by Ansari and Smith.[74]

Using a more efficient column, Sevanian and McLeod[5] accomplished satisfactory resolution of most of the more common cholesterol oxides without solvent programming. The

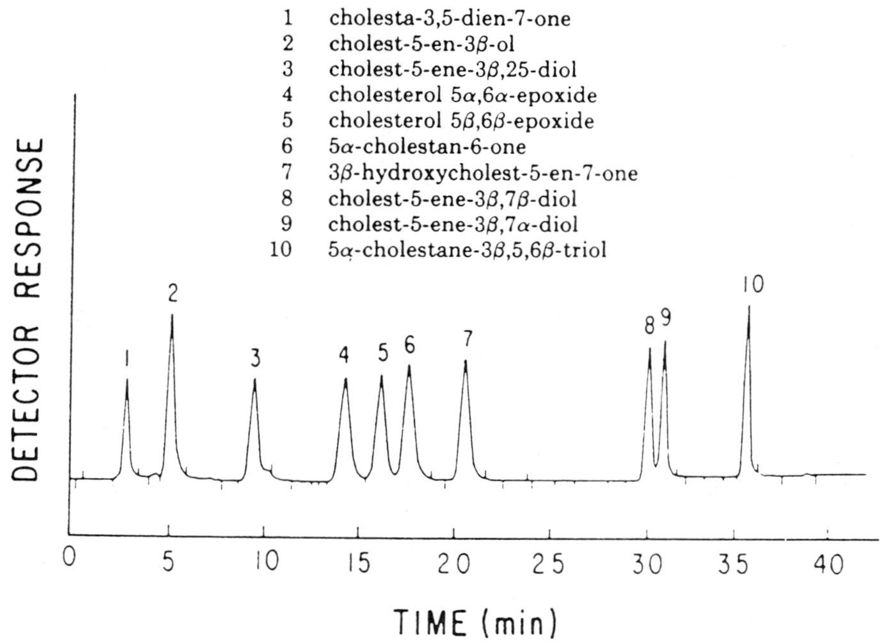

FIGURE 9. HPLC of cholesterol and nine oxidation products, each at 12.5 μg/100 μl. A normal-phase separation was performed on a 10-μm μPorasil column (3.9 mm × 30 cm) with HPLC flame ionization detector. Solvent gradient system at a total flow rate of 1.5 ml/min was: solvent A — hexane/2-propanol (100/3, v/v) at injection. The composition changed concavely over a 15-min period to 98.5% A + 1.5% solvent B (2-propanol), linearly over a 10-min period to 97% A + 3% B and then linearly over a 15-min period to 50% A + 50% B. (From Maeker, G., Nungesser, E. H., and Zulak, I. M., *J. Agric. Food Chem.*, 36, 61, 1988. With permission.)

chromatogram in Figure 10 was obtained with two 3-μm particle size columns (4.6 × 100 mm) in tandem with 4.2% 2-propanol in hexane at 1.5 ml/min. Due to the same normal-phase HPLC, the relative retention order of the cholesterol oxides tested was the same in both studies.[5,77] Cholesterol oxides that were not shown in Figure 10 are 7β-hydroxycholesterol and 5α-cholestane-3β,5,6β-triol. It is conceivable, however, that 7β-hydroxycholesterol would be eluted somewhere between 7-ketocholesterol and 7α-hydroxycholesterol, whereas 5α-cholestane-3β,5,6β-triol would be far behind 7α-hydroxycholesterol. The longer retention of polar compounds is a drawback of using isocratic adsorption HPLC. As far as simultaneous HPLC separation of the more common cholesterol oxides is concerned, HPLC procedures by Sevanian and McLeod[5] and Maeker et al.[77] appear to be the most successful. Their applicability to the analysis of cholesterol oxides in food or biological specimens, however, remains to be established because such samples frequently cause interference associated with complex sample constitutents even after rigorous sample purification.

2. Detection and Quantitative Analysis

Ultraviolet (UV) is most commonly used because of its high sensitivity and specificity. Most cholesterol oxides, however, are not strong UV absorbers, unlike other sterols of biological interest, and thus require detection at very low wavelengths, generally below 210 nm. This not only defeats the advantage of UV detection, i.e., specificity, but also limits the choice of solvents for the mobile phase due to the UV cut-off point. Among the more common cholesterol oxides, 7-ketocholesterol is the only exception to this limitation. The conjugate double bond of 7-ketocholesterol enables its UV detection at 233 nm down to the nanogram range.[22,57] Neither cholesterol epoxides nor 5α-cholestane-3β,5,6β-triol respond

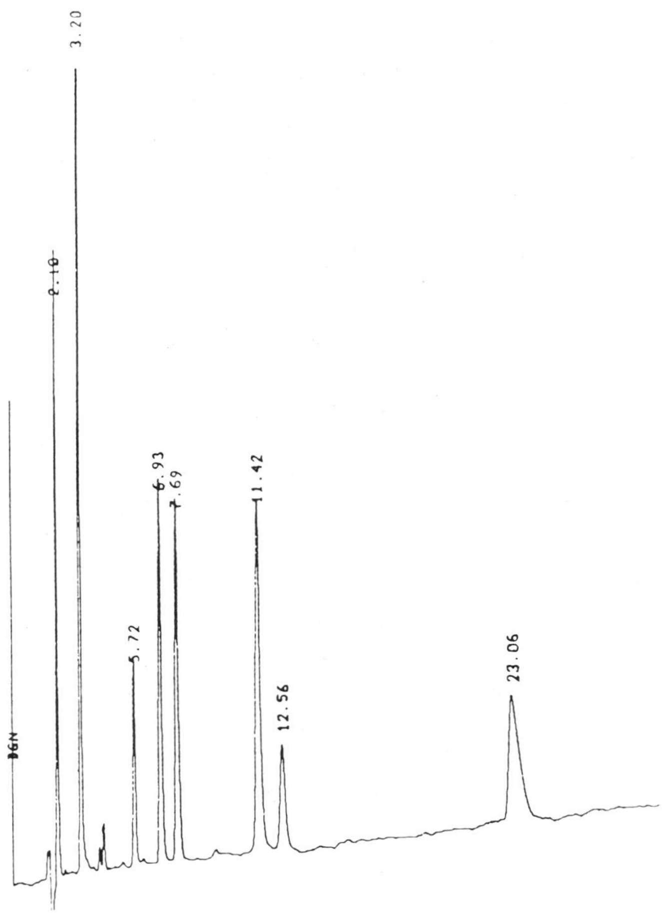

FIGURE 10. HPLC of cholesterol oxidation standards. Two 3-μm particle size columns (100 × 4.6 mm) in series were run with hexane/2-propanol (95.8/4.2, v/v) at a flow rate of 1.5 ml/min. The eluent was monitored by differential refractometry and the integrator response was set at A = 128 for the tracing shown. The oxidation products are indicated by their retention times and are as follows: 2.10, cholesta-3,5-dien-7-one; 3.20, cholesterol; 5.72, 25-hydroxycholesterol; 6.93, α-epoxide; 7.69, β-epoxide; 11.42, 7-ketocholesterol; 12.56, 7α-OOH; 23.06, 7α-hydroxycholesterol. (From Sevanian, A. and McLeod, L. L., *Lipids,* 22, 627, 1987. With permission.)

to UV detection even below 210 nm, and therefore have to be monitored by different types of detectors.

The mass spectrometer is currently a good choice for an HPLC detector, but its price still limits its feasibility. Although the refractive index (RI) detector is a universal one and can be used for non-UV absorbing cholesterol oxides, its poor sensitivity makes it less useful for the detection of cholesterol oxides at low levels. Nevertheless, RI detection seemed to be adequate for the quantitation of cholesterol epoxides in dry egg products[27,28] with a detection limit of about 5 μg. Sevanian and McLeod[5] reported a lower detection limit of about 0.5 μg for most cholesterol oxides, probably due to less band broadening of HPLC peaks. It should be noted that most UV detectors are capable of detecting 40 to 50 ng for cholesterol oxides,[22,29] which makes UV detection still the primary choice except for cholesterol epoxides and 5α-cholestane-3β,5,6β-triol. Recent developments in HPLC detection include FID. Column effluents are continuously fed onto a transport system moving through a furnace region where the solvents are evaporated and the nonvolatile cholesterol oxides are combusted for detection. Maeker et al.[77] reported cholesterol oxides could be detected by FID at 0.39 μg per component.

To compensate for the low sensitivity of cholesterol oxides by UV detection, Csiky[36] developed a one-step technique of enriching and detecting cholesterol oxides on the same analytical column and demonstrated its applicability by analyzing heated butter. The cholesterol oxides in a hexane extract of heated butter were enriched on an analytical nucleosil column by repeated sample injections with manipulation of the mobile phase strength, and then eluted for resolution. This single-step technique is by far the most simple and rapid method. The unsteady baseline, however, and the shift in retention indices of cholesterol oxides due to the displacement effect during on-column concentration may not make it readily applicable to the analysis of other foods of more complex composition than butter.

Derivatization is an entirely different approach toward circumventing the poor UV absorption of cholesterol oxides. Following the conversion of the hydroxyl groups of the sterols to benzoate or *p*-nitrobenzoate (PNB) esters, the derivatives acquired a remarkable enhancement in UV abosrption together with an absorption maxima shift from below 210 nm to 254 nm.[79] This derivatization reaction was used for the UV detection of epimeric epoxides from air-aged USP cholesterol[74] and for the detection of both epimeric epoxides as well as 5α-cholestane-3β,5,6β-triol from the plasma and liver of rabbits.[4,80] More recently, Sugino et al.[81] quantified cholesterol epoxides in spray-dried eggs after converting to PNB esters. The detection limit was about 10 ng per injection. Unfortunately, this derivatization procedure requires a few steps involving heating and extraction. These additional procedures seem to make its use less attractive for the routine quantitation of cholesterol oxides. It should be confirmed that the derivatization reaction goes to completion and that no hydrolysis of such derivatives occurs, since they undergo some harsh chemical conditions during the preparation. With the use of stigmasterol as IS, UV responses of epimeric epoxides were linear when they were run as PNB esters.[81]

Quantitative HPLC analysis requires that one ensure system response linearity as in GC quantitation. Especially when column effluents are monitored by UV detection, one cannot assume similar detector response for all cholesterol oxides because the UV detector response varies greatly depending upon the nature of the UV chromophore. The UV absorbance for 7-ketocholesterol was observed to be two times greater than those for dihydroxy cholesterol oxides near the absorption maxima (Figure 11). It is important, therefore, to know the RF for individual cholesterol oxides. The same approach stated for GC calibration is also equally true for HPLC calibration. 7-Ketopregnenolone was used as IS for quantitation of C7 cholesterol oxides because it reacted similarly to cholesterol oxides during preparation procedures and also separated well from C7 oxides.[22] In most cases of HPLC quantitation, however, the use of IS is relatively rare in spite of its many advantages. This may be partly accounted for by the limited probability of finding an appropriate IS.

Most quantitative studies utilize an external standard method in which one records the detector response output as a function of the injection amount. The plot of detector response vs. injection amount yields a standard curve as shown in Figure 12. Cholesterol oxides quantified using external standard method include 7-ketocholesterol[57] and 25-hydroxycholesterol,[29] both by UV detection, and epimeric epoxides[27,28,53] and most of the more common cholesterol oxides[5] by refractometry. Noteworthy is the performance of FID, a relatively new development for HPLC. When tested with 10 common cholesterol oxides, a good response linearity by FID was observed over the tested range of 100 to 0.39 μg per component.[77] In view of the simultaneous separation of the more common cholesterol oxides with improved and universal detection, it is unfortunate that the RRFs were given relative to cholesterol instead of a true IS.[77] Nevertheless, the value of these data does not diminish because they can be used for external standard quantitation.

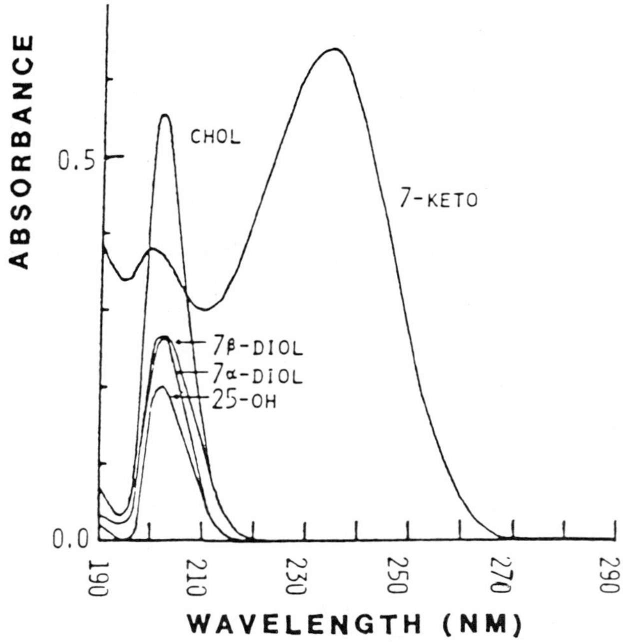

FIGURE 11. Ultraviolet absorption spectra of cholesterol and cholesterol oxidation product standards. 20 μg/ml solutions in 93:7 hexane:isopropanol (v/v). (From Csallany, A. S. et al., *Lipids,* 24, 645, 1989. With permission.)

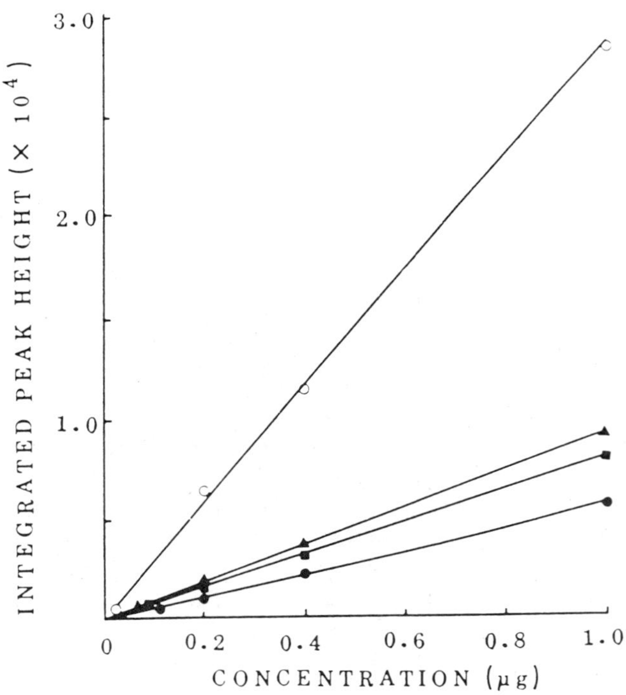

FIGURE 12. Standard curves of cholesterol oxidation products. ○, 7-ketocholesterol; ■, 7α-hydroxycholesterol; ▲, 7β-hydroxycholesterol; ●, 25-hydroxycholesterol. Each data point represents mean of three injections at each concentration. Correlation coefficients for all lines were at least 0.999. (From Csallany, A. S. et al., *Lipids,* 24, 645, 1989. With permission.)

C. MASS SPECTROMETRY

In chromatography, identification of cholesterol oxide species is based upon comparing the retention time of a separated peak with that of an authentic standard. Because a little variation in retention time is in practice not uncommon, especially in HPLC, rerunning the same sample after addition of a standard of the same presumed compound and observing the augmentation of the peak at issue (co-chromatography) can improve the confidence of identification. Subjecting the same sample to chromatography with columns of different stationary phases and comparing retention times of sample peaks with those of the standard again also improves confidence of identification. Because of the possibility, however, that more than one compound can elute with a retention time the same as or very close to that of the target compound, identity confirmation of a chromatographically separated peak needs to be performed in a more stringent manner, i.e., by means other than retention time comparison. To this end, mass spectrometry (MS) is commonly used and provides far more precise information on chemical identity.

In MS, analytes are introduced into the ionization chamber either via the chromatographic inlet or by a direct insertion probe. The so-called hyphenated techniques, GC-MS or LC-MS are on-line operations that enable very convenient and sensitive detection. A compound collected through chromatography can also be introduced into the ionization chamber using the direct inlet probe. The stream of molecules introduced in an ionization chamber is bombarded by high-energy electrons emitted from a heated filament, usually 70 eV. As a result, very small amounts of molecules are ionized and the ionized molecules may remain as such or undergo characteristic fragmentation producing a variety of ions. These ion beams are separated through a mass analyzer which employs either magnetic (single-focusing) or electromagnetic (double-focusing) sectors or quadrupole mass filters according to their mass/charge (m/z) ratio. Upon reaching the electron multiplier detector, an ion beam from the mass analyzer generates secondary electrons through a cascading effect whose intensity is proportional to the abundance of the ion. Molecular ion ($M^{+\cdot}$) is a radical cation formed from the ionization of the analyte molecule with the loss of one electron. The molecular ion peak, therefore, appears at an m/z value numerically equal to the nominal molecular weight of the compound. The most intense peak in the mass spectrum is called the base peak, against which the relative intensities of all other ions are normalized.

An insightful application of MS would deduce the structure of the unknown intact molecule by correlating the observed mass spectrum with molecular structure. For the analysis of cholesterol oxides, however, the usual application of MS would confirm the chemical identity of chromatographically isolated oxides by comparing their mass spectra with those of authentic standards. Since mass spectra of the more common cholesterol oxides are spread throughout recent literature, it is worthwhile to provide a compilation in the review. Listing of mass spectra either as TMS ether derivatives or underivatized may also be useful as a convenient reference for spectra comparison. Discussions on some ion peaks characteristic of each oxide structure and thus of diagnostic importance are also included to aid in the structure confirmation based on the mass spectrum. At the end, selected ion monitoring is described to illustrate the quantitative analysis of cholesterol oxides by GC-MS.

1. Trimethylsilyl Ether Derivatives

All the mass spectra listed below have been recorded as trimethylsilyl (TMS) ether derivatives, since cholesterol oxides were converted to these prior to GC-MS analysis. All cholesterol oxides maintain structural features of cholesterol, i.e., 3β-ol and Δ^5 (except epoxides and their hydrolysis products) and therefore it would be appropriate to begin with discussing cholesterol mass spectrum.

Cholesterol as TMS ether — The mass spectrum is dominated by fragment/ions of m/z 129 and 329, the former being the base peak (Figure 13). Diekman and Djerassi[82] confirmed

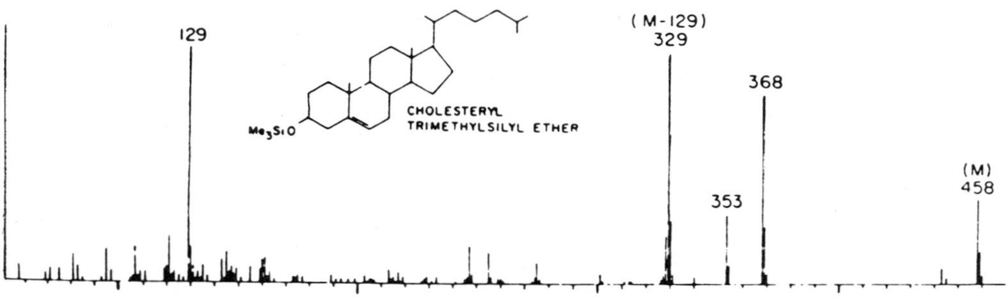

FIGURE 13. Mass spectrum of cholesterol as TMS ether. (From Brooks, C. J. W., Horning, E. C., and Young, J. S., *Lipids,* 3, 391, 1968. With permission.)

that m/z 129 originated from A-ring fragmentation between C1-10 and C3-4 by examining a deuterium-labeled derivative as shown below. Brooks et al.[83] reported that m/z 129 may arise from other structural types; however, the presence of the accompanying ion m/z 329 (M-129) is found as a major peak only from a 3β-TMS-Δ^5-sterol. The m/z 329 is not found in a Δ^4-3β-ol TMS sterol. 5-Cholesten-3α-ol, an epimer of cholesterol, displays an almost identical mass spectrum with only a small intensity difference.[83]

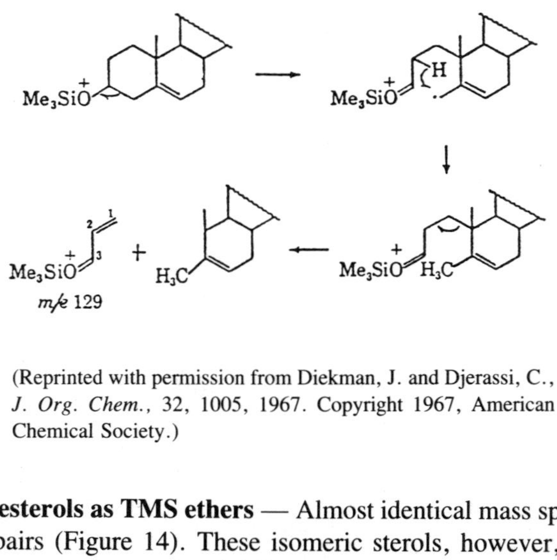

(Reprinted with permission from Diekman, J. and Djerassi, C., *J. Org. Chem.*, 32, 1005, 1967. Copyright 1967, American Chemical Society.)

7-Hydroxycholesterols as TMS ethers — Almost identical mass spectra were observed for these epimeric pairs (Figure 14). These isomeric sterols, however, can be clearly differentiated by chromatography as already discussed. Only a very small amount of M^+ is found. The base peak, m/z 456 (M-90), is due to facile elimination of trimethylsilanol — probably at C7, the allylic TMS moiety.[84] Although very weak, small peaks at m/z 366 (M-180) and 351 (M-180-15) are also observed. It is interesting to note that these epimers produced relatively very simple mass spectra at 70 eV. In contrast, another dihydroxy oxide species with the second hydroxyl group on the ring structure, 4β-hydroxycholesterol, produced a very different mass spectrum from those of 7-hydroxycholesterols. All three ions, m/z 546, 456, and 366, characteristic of bis-TMS ether sterols, were observed with high intensity over 40% in the mass spectrum of 4β-hydroxycholesterol.[72,84] The m/z 129 and 417 (M-129) were also strong. A little discrepancy in relative intensities was found in the literature. Park and Addis[72] observed m/z 456 (100%), 147 (98%), 129 (84%), and 366 (66%), whereas Brooks et al.[84] reported 366 (100%), 456 (90%), 147 (90%) and 129 (69%).

Side-chain hydroxycholesterols as TMS ethers — These species have a TMS moiety

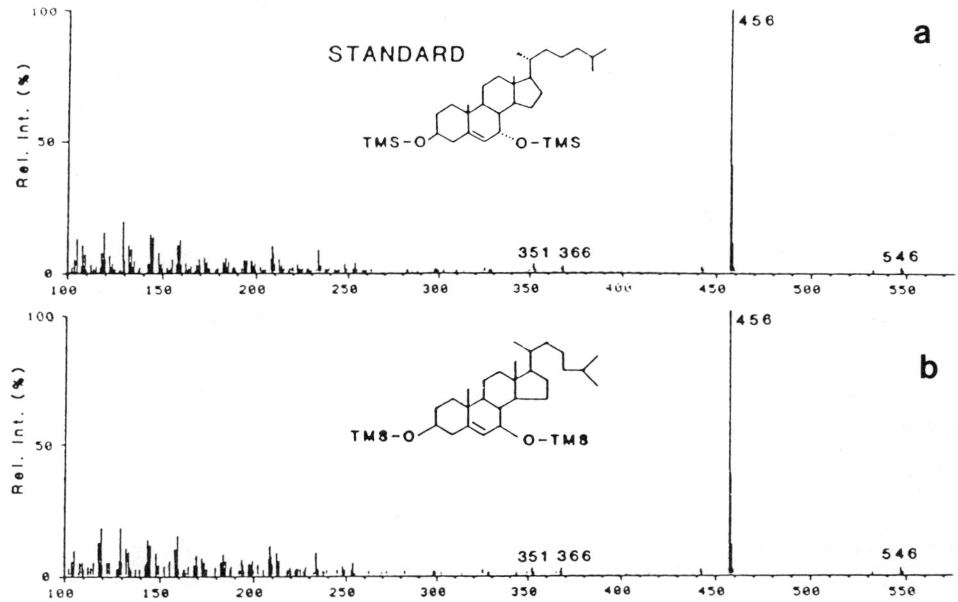

FIGURE 14. Mass spectrum of 7α-hydroxycholesterol (a) and 7β-hydroxycholesterol (b) as TMS ethers. ([a] from Park, S. W. and Addis, P. B., *J. Food Sci.*, 52, 1500, 1987; [b] from Park, S. W. and Addis, P. B., *J. Agric. Food Chem.*, 34, 653, 1986. With permission.)

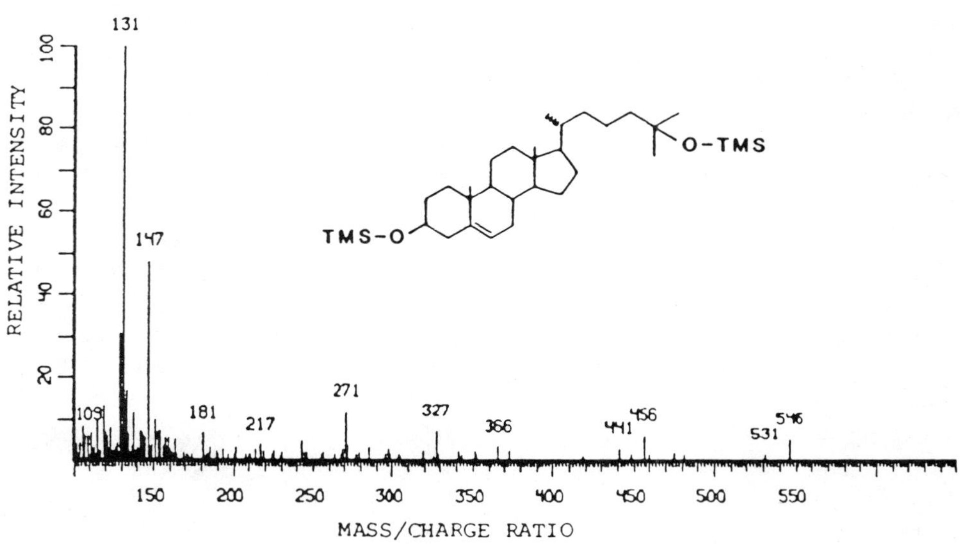

FIGURE 15. Mass spectrum of 25-hydroxycholesterol as TMS ether. (From Park, S. W. and Addis, P. B., *Anal. Biochem.*, 149, 275, 1985. With permission.)

at the side chain and therefore characteristic ions indicating the location of side-chain hydroxy constitutents are expected. Figure 15 shows the mass spectrum of 25-hydroxycholesterol, with the base peak at m/z 131 which originated from α-cleavage between C24 and 25 (cleavage A) as the dimethylcarbinol ether $(CH_3)_3Si{-}O{-}C(CH_3)_2$. There seemed to be no other major peaks of diagnostic consequence. Nevertheless, peaks at m/z 546, 456, and 366 due to the loss of the TMS moiety were confidently observed. The mass spectrum of 20-hydroxycholesterol as the TMS ether displayed m/z 201 as the base peak. The m/z 201 would be due to α-cleavage between C17 to 20 (cleavage C) which contains the complete side chain with the TMS group.[84] Also, α-cleavage between C20 and 21 (cleavage B) would account for the ion peak m/z 461 (M-85).

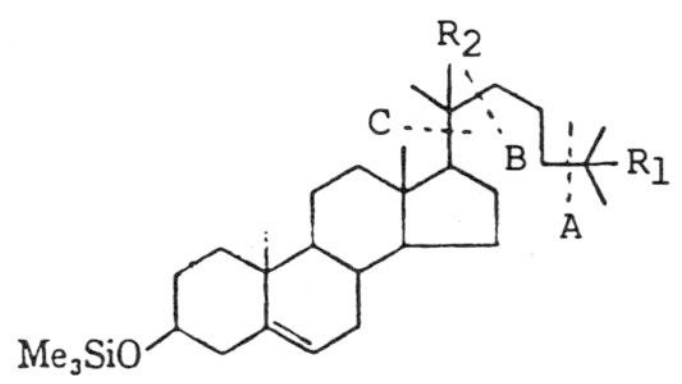

R_1 = OSiMe$_3$ and R_2 = OH

R_1 = OH and R_2 = OSiMe$_3$

Cholesterol epoxides as TMS ethers — Mass spectra of epimeric epoxides are more frequently found in the literature than other cholesterol oxides, probably due to their metabolic role as intermediates during the bioconversion of cholesterol to other sterols, as well as being autoxidation products. Minor discrepancies regarding the relative intensity order of some ions have occurred in literature mass spectra; these are presented in Figures 16 and 17 for comparison. Unlike the dihydroxy cholesterol oxides, both epoxides showed extensive fragmentation, leaving only a few ions in the high m/z region. Nevertheless, diagnostically characteristic ions, m/z 474 (M^+) and 384 (M-90) were confidently observed.

In α-epoxide, m/z 366 is another major ion. The m/z 459 (M-15) was reported to be more intense than m/z 456,[13,24,80] but Gray et al.[63] reported the opposite. The m/z 366 was followed by m/z 369,[13,63] but the reverse was also reported.[24] In β-epoxide, m/z 366 seemed to be absent and instead, m/z 369 was observed,[13] whereas Aringer and Eneroth[85] reported that m/z 366 was also more intense than m/z 369. This discrepancy needs to be further resolved.

7-Ketocholesterol as TMS ether — Molecular ion (m/z 472) was the base peak, probably due to the conjugated ketone group that is expected to confer stability on M^+.[72,83] Two other peaks, m/z 382 (M-90) and 369 (M-90-15) were also prominent (Figure 18). Although m/z 129, which is characteristic of Δ^5-3β-ol TMS ether sterols, was only noticeable in the mass spectrum by Brooks et al.,[83] it was the base peak in the report by Park and Addis.[13] Since Brooks et al.[83] applied lower electron energy, 20 eV, it seems plausible that M^+ is likely to be more intense, compared with 70 eV used by Park and Addis.[13] However, m/z 472 was also observed to be the base peak at 70 eV[72] and the m/z 467 (M-15) was also fairly intense (22%), which makes it hard to reconcile the relative intensity difference by different electron impact energy only. Another difference worthy of mention is that the magnetic sector was used when m/z 472 was observed for the base peak[72,83] whereas when the quadrupole was used, m/z 129 was the base peak.[13] Brooks et al.[83] listed two more major ions, m/z 174 and 142, whereas Park and Addis[13] found no useful ion peak in the low m/z region except m/z 129 due to extensive fragmentation.

5α-Cholestane-3β,5,6β-triol as TMS ether — This is the only species among the more common cholesterol oxides that has three hydroxyl groups. It is of interest to explore how many hydroxyl groups are involved in TMS etherification. A complete etherification of all 3 hydroxyl groups of 5α-cholestane-3β,5,6β-triol was reported[83] by heating at 60°C for 60 h after the addition of a mixture of TMSI/BSA/TMCS (3/3/2). When the derivatization was done at room temperature for 1 h, bis-TMS ether sterol resulted, suggesting that the 5α-hydroxyl group is sterically hindered. Most of major ions were common in both bis- and tris-TMS ether sterols mass spectra, but their relative intensity orders were quite different from each other. The base peak for bis-TMS ether was m/z 129 and that for tris-TMS ether, m/z 456.[84]

Park and Addis[72] initially reported that all three hydroxyl groups were derivatized by etherification reaction at room temperature for 30 min but later, however, indicated that it

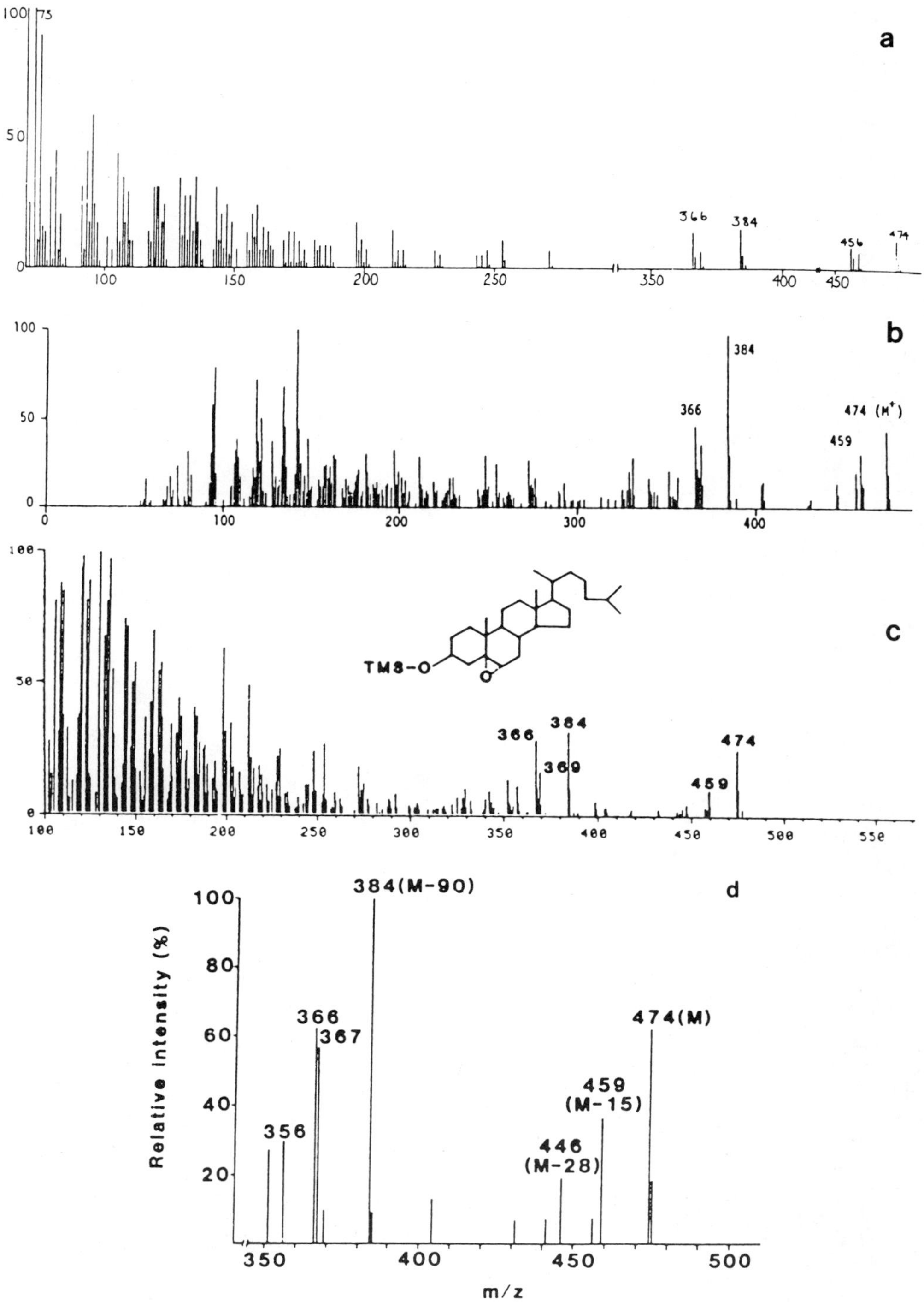

FIGURE 16. Mass spectrum of cholesterol α-epoxide as TMS ether. ([a] From Gray, M. F. et al., *Lipids,* 6, 836, 1971. [b] From Watabe, T. and Sawahata, T., *Biochem. Biophys. Res. Commun.*, 83, 1396, 1978. [c] From Park, S. W. and Addis, P. B., *J. Agric. Food Chem.,* 34, 653, 1986. [d] From Björkhem, I. et al., *J. Lipid Res.*, 29, 1031, 1988. With permission.)

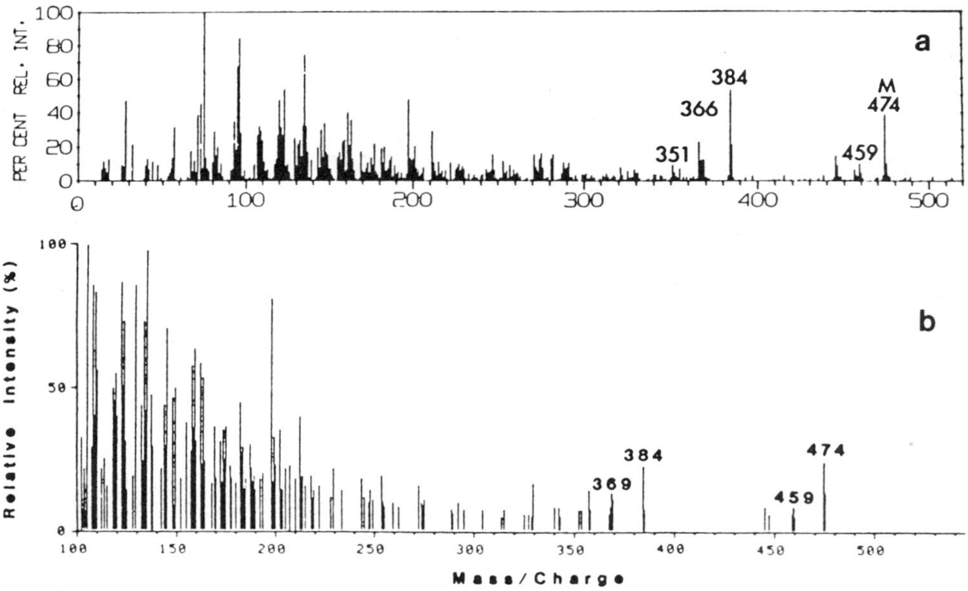

FIGURE 17. Mass spectrum of cholesterol β-epoxide as TMS ether. ([a] From Aringer, L. and Eneroth, P., *J. Lipid Res.*, 15, 389, 1974. [b] From Park, S. W. and Addis, P. B., *J. Agric. Food Chem.*, 34, 653, 1986. With permission.)

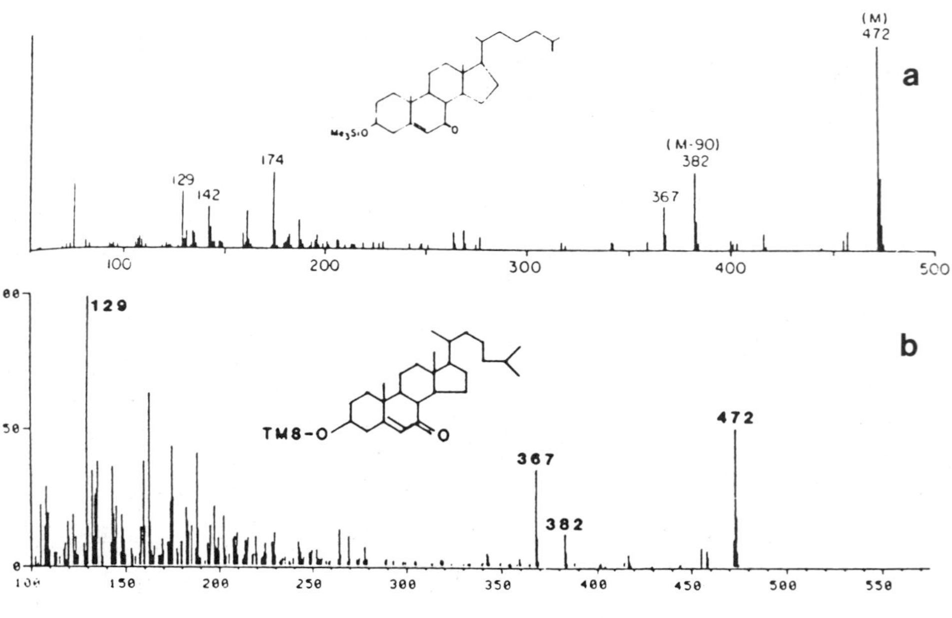

FIGURE 18. Mass spectrum of 7-ketocholesterol as TMS ether. ([a] From Brooks, C. J. W. et al., *Lipids*, 3, 391, 1968. [b] From Park, S. W. and Addis, P. B., *J. Agric. Food Chem.*, 34, 653, 1986. With permission.)

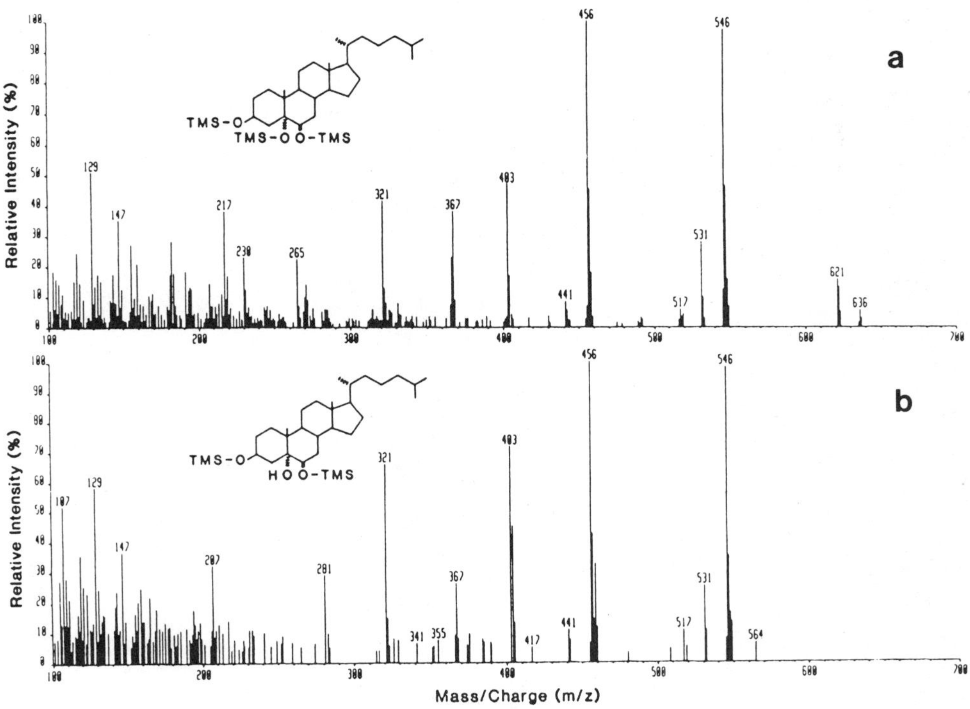

FIGURE 19. Mass spectra of 5α-cholestane-3β,5,6β-triol TMS ethers. Spectra are (a) tris- and (b) bis-trimethylsilyl ethers, respectively. (From Park, S. W. and Addis, P. B., *J. Am. Oil Chem. Soc.*, 66, 1632, 1989. With permission.)

was actually a bis-TMS not a tris-TMS ether derivative.[86] As the derivatization was done at 70°C for an extended period, the tris-TMS ether derivative peak emerged at the expense of the bis-TMS ether derivative peak as traced by GC. When both peaks were subjected to MS, almost identical mass spectra were displayed (Figure 19). However, m/z 636 (M^+) and 621 (M-15) from the tris-TMS ether derivative and m/z 564 (M^+) from the bis-TMS ether derivative were clearly observed, providing critical information for differentiation between these two TMS ethers. It was frustrating that the mass spectrum of the bis-TMS ether derivative from the later study[86] agreed little in terms of relative intensity order with that of bis-TMS ether derivative from the earlier study.[72] For the tris-TMS ether derivative, excellent agreement was found among literature mass spectra.[84-86] Björkhem et al.[24] implied that 5α-cholestane-3β,5,6β-triol was derivatized into tris-TMS ether sterol. The partially presented mass spectrum with m/z 403 (base peak), 456, and 546 in decreasing order of relative intensity, however, seems to agree more closely with the mass spectrum of bis-TMS than tris-TMS ether derivative. Brooks et al.[84] proposed that the ion m/z 321 arose from B- and C-ring cleavage at C6 and C7, C8 and C9, and C11 and C12 shown as below.

2. Underivatized Cholesterol Oxides

When HPLC is used for separation of cholesterol oxides, each species is usually collected as a purified fraction, concentrated if necessary, and then introduced into the ion source directly. This is called the direct inlet probe technique. When underivatized cholesterol oxides are subjected to MS, loss of water provides a major pathway for M^+ decomposition, characteristic of a cyclic aliphatic alcohol[87] leaving little or no M^+ as such.

7-Ketocholesterol — Due to the stability conferred by the conjugated double bond, M^+ would be expected to be quite stable, and indeed was the base peak (Figure 20). Other

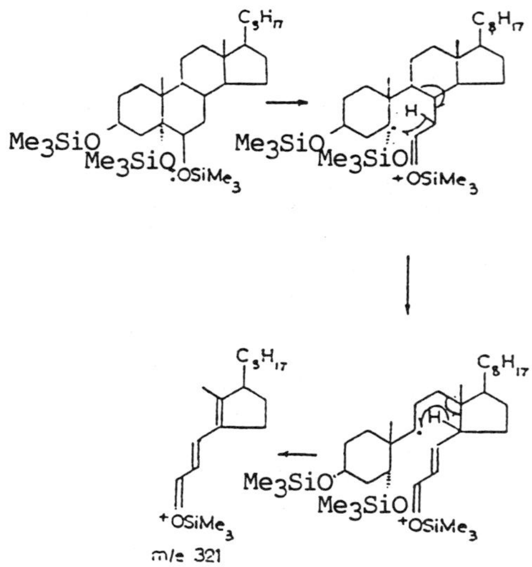

(From Brooks, C. W. J., Henderson, W., and Steel, G., *Biochim. Biophys. Acta,* 296, 431, 1973. With permission.)

peaks observed include m/z 382 (M-18), 367 (M-18-15), 287 (M-side chain), 205, 192, and 174, etc.[22] An almost identical mass spectrum was confirmed by De Vore.[57]

7-Hydroxycholesterols — Both epimers displayed only a slight fragmentation as in the corresponding TMS ethers (Figure 20). Due to the facile elimination of the allylic C7-hydroxyl group, only a small amount of M^+ was observed, about 5%.[22] The base peak was m/z 384 (M-18) followed by 366 (M-2H_2O), 351 (M-2H_2O-CH_3) and 247 (M-42-side chain). It seems that the α-epimer is a little stronger in relative intensities of m/z 366 and 247 than the β-epimer.[65]

Cholesterol epoxides — Both isomers show quite extensive fragmentation, and a good agreement has been found among literature mass spectra.[27,63,65] Some important ions include m/z 402 (M^+), 384 (M-18), 369 (M-18-15), 271 (M-18-side chain), 247 (M-42-side chain), 229 (M-18-42-side chain), etc. (Figure 21). Although both mass spectra looked almost identical to each other as usual for epimeric pairs, Tsai and Hudson[27] reported that m/z 331 occurs only in α-epoxide but not in the β-epimer and also that m/z 356 was prominent in β-epoxide but weak in α-epoxide. They assigned m/z 331 to M-C_4H_7O resulting from A-ring cleavage between C1 and C10 and C4 and C5, and suggested that the appearance of m/z 331 only in α-epoxide would be accounted for by the configurational difference of the A/B junction between α-epoxide *(trans)* and β-epoxide *(cis).*

Other oxides — From the direct insertion probe analysis of 25-hydroxycholesterol, Kou and Holmes[29] observed M^+ (m/z 402) and two other major ions, m/z 384 (M-18) and 369 (M-18-15). For 5α-cholestane-3β,5,6β-triol, Finochiaro et al.[65] listed m/z 420 (M^+, 3%) m/z 402 (M-18, 100%), 387 (M-H_2O-CH_3, 5%), 384 (M-2H_2O, 83%), 369 (M-2H_2O-CH_3, 24%), etc.

3. Selected Ion Monitoring

The course of separation in combined GC-MS is recorded in the total ion chromatogram by monitoring ion beams between the ion source and mass analyzer. For quantitative analysis, it is advantageous in terms of simplicity, selectivity, and sensitivity to dedicate the use of the mass spectrometer to acquiring and recording ion currents only at certain selected ions, which is called selected ion monitoring (SIM). Selected ions may be characteristic of each

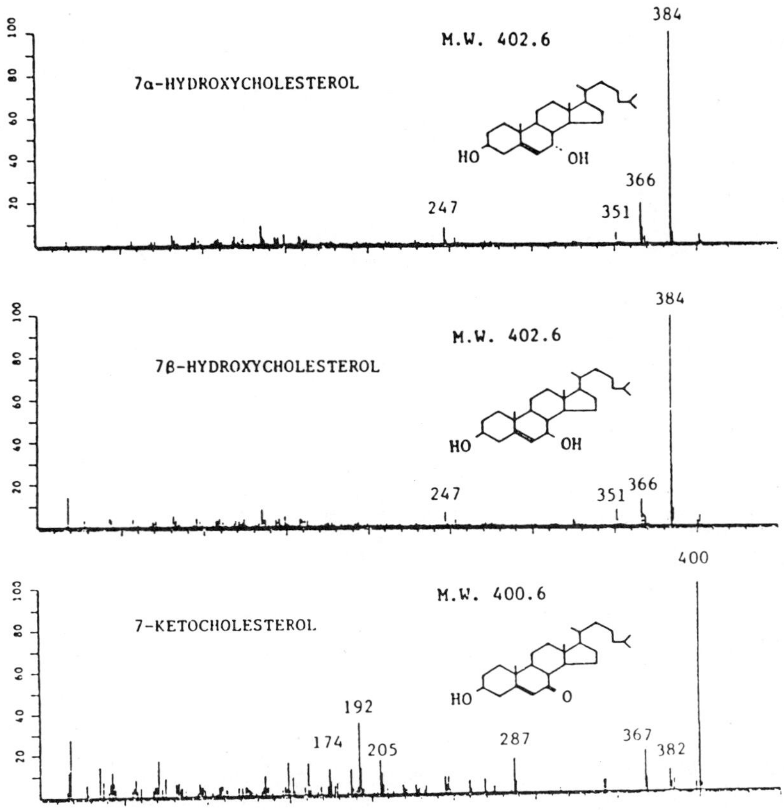

FIGURE 20. Mass spectra of C7 cholesterol oxides. (From Park, S. W. and Addis, P. B., *J. Food Sci.*, 50, 1437, 1985. With permission.)

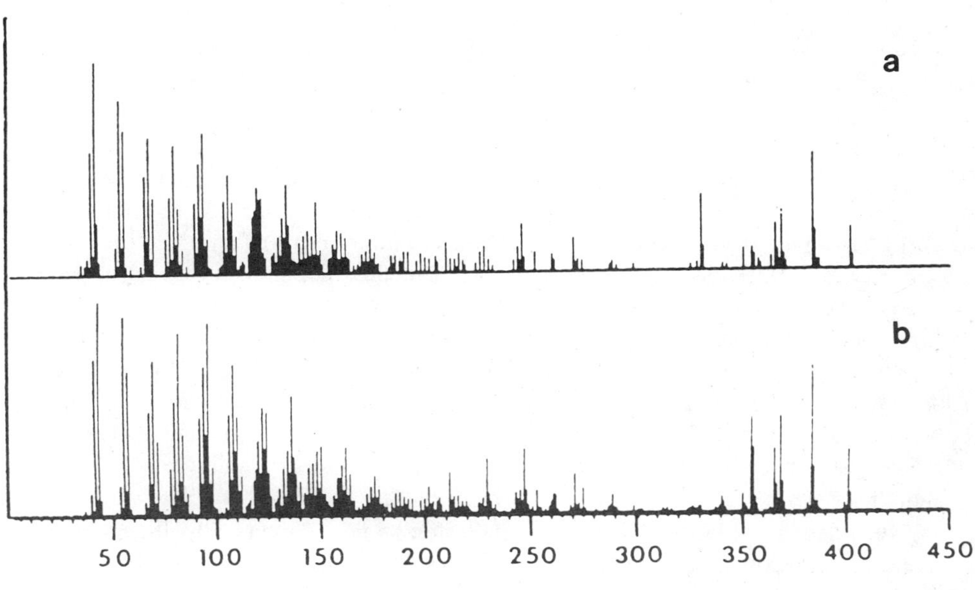

FIGURE 21. Mass spectrum of cholesterol α-epoxide (a) and β-epoxide (b). (From Tsai, L. S. and Hudson, C. A., *J. Food Sci.*, 49, 1245, 1984. With permission.)

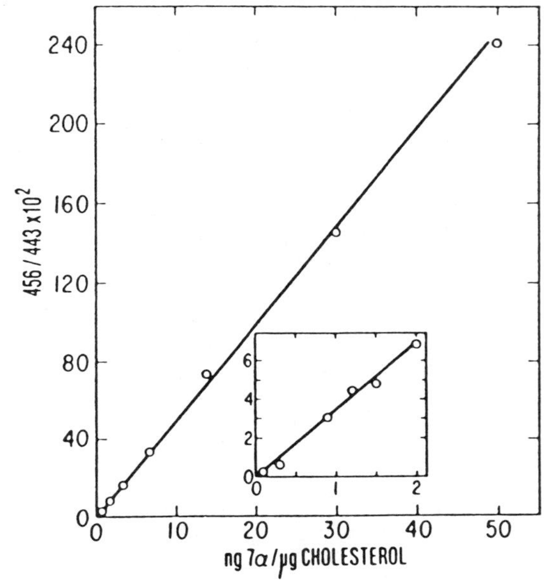

FIGURE 22. SIM standard curve for bis-TMS ether of 7α-hydroxycholesterol obtained by measuring m/z 456/443. (From Sanghvi, A. et al., *J. Lipid Res.*, 22, 720, 1981. With permission.)

analyte such as M^+ or base peak or other unique ions, all preferably with good intensity. In SIM quantitation, only a single ion from the mass spectrum of each analyte is used; software with modern mass spectrometers, however, allow one to enter more selected ions for qualitative purposes. Depending upon the complexity of the task, it would be prudent to enter a few more selected ions at the expense of a little sacrificed detection sensitivity to ensure identity of the oxide species under quantitation. Since combined GC-MS is the usual practice in SIM applications, the discussion here is restricted to those with a GC inlet.

Using IS, SIM quantitation of cholesterol oxides is performed in the same manner as GC or HPLC quantitation by comparing peak areas under selected ions from the analytes and IS, respectively. Accordingly, prior to SIM quantitation, calibrating the instruments by running mixtures of the analytes and IS is essential. An example of a standard curve for SIM quantitation of 7α-hydroxycholesterol is illustrated in Figure 22. The plot of the ratios of the peak areas in the ion current profiles, i.e., peak area at m/z 456 over m/z 443 (IS) against known concentration ratios of 7α-hydroxycholesterol over IS yielded the linear line with r^2 between 0.98 and 0.99.[88] For the simultaneous measurement of tissue levels of cholesterol oxides using SIM, monitoring of multiple selected ion current profiles is illustrated in Figure 23.[89] This study followed selected ions m/z 546 (M^+) for 7-hydroxycholesterols and 25-hydroxycholesterol, 472 (M^+) for 7-ketocholesterol and 443 (M-15) for cholesterol. Although m/z 546 is unique to dihydroxy cholesterol oxides as TMS ethers, it would not be a wise choice for SIM, because the intensity is very weak as previously stated. Later, m/z 546 was replaced with m/z 456, the base peak of 7-hydroxycholesterols for their SIM quantitation.[88]

When a deuterium-labeled analog of a cholesterol oxide is used as the IS, no separation in terms of retention time between oxide and IS is expected. They may be completely separated, however, in terms of m/z by SIM, because mass spectrum is three dimensional compared to the chromatogram. This can be better illustrated by Figure 24, which was recorded for SIM quantitation of α-epoxide as the TMS ether.[24] The molecular ion

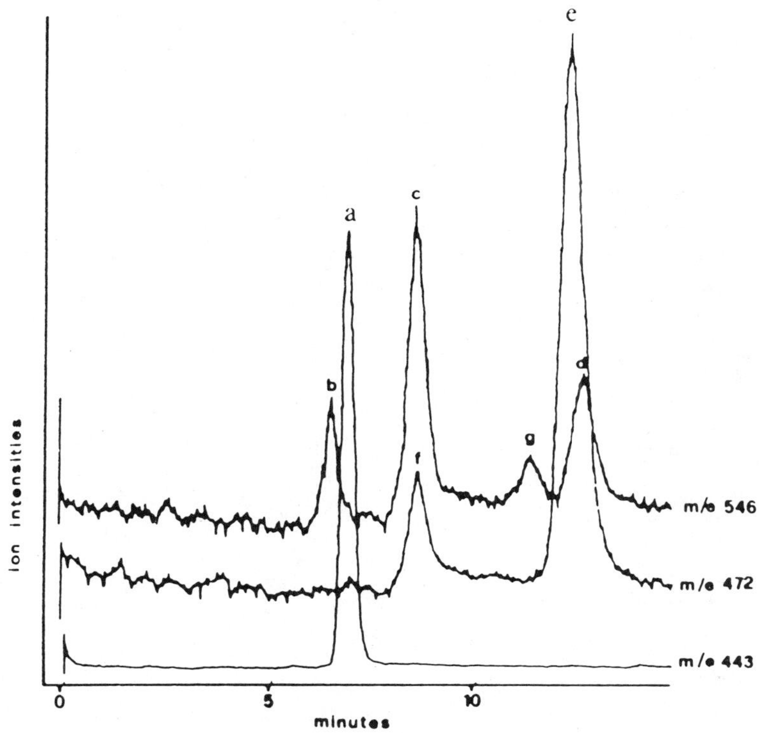

FIGURE 23. Multiple selected ion current profile of TMS ethers of cholesterol (a); 7α-hydroxycholesterol (b); 7β-hydroxycholesterol (c); 25-hydroxycholesterol (d); 7-ketocholesterol (e); and unidentified compounds (f,g). (From Sanghvi, A. et al., *Anal. Biochem.*, 85, 430, 1978. With permission.)

(m/z 474) was monitored for α-epoxide, whereas m/z 481 was monitored for d_7-α-epoxide (IS). Gruenke et al.[42] monitored m/z 474 for β-epoxide and 479 for d_5-labeled β-epoxide (IS). The calibration curve shown in Figure 25 was obtained when the authors plotted the calibration data collected for β-epoxide as TMS ether.[42] The calibration line displayed excellent response linearity, (area of m/z 474/area of m/z 479) = 1.61 (d_0-β-epoxide amount / d_5-β-epoxide amount) + 0.011, and r^2 = 0.99. It is interesting to note that the slope deviated considerably from unity, suggesting that d_0- and d_5-β-epoxides as TMS ethers have different ionization and fragmentation efficiencies. Also, the y-intercept close to the origin but not passing through it indicated d_5-β-epoxide (IS) contained unlabeled β-epoxide, probably about 1.1%.

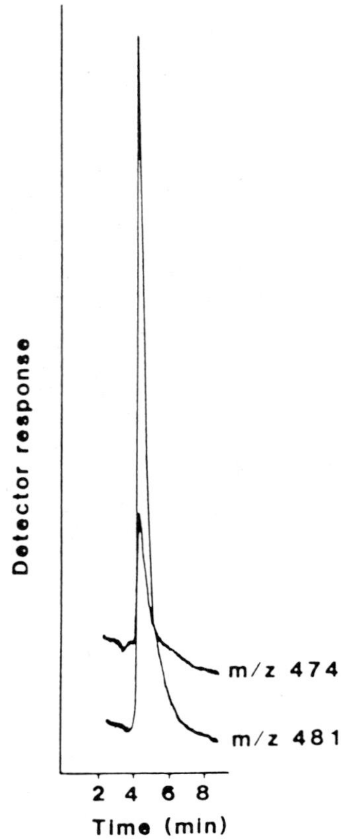

FIGURE 24. Selected ion monitoring of cholesterol α-epoxide as TMS ether derivative (m/z 474) of material isolated from serum of a healthy subject to which $^2H_7$-labeled cholesterol 5α,6α-epoxide (m/z 481) had been added as internal standard. (From Björkhem, I. et al., *J. Lipid Res.*, 29, 1031, 1988. With permission.)

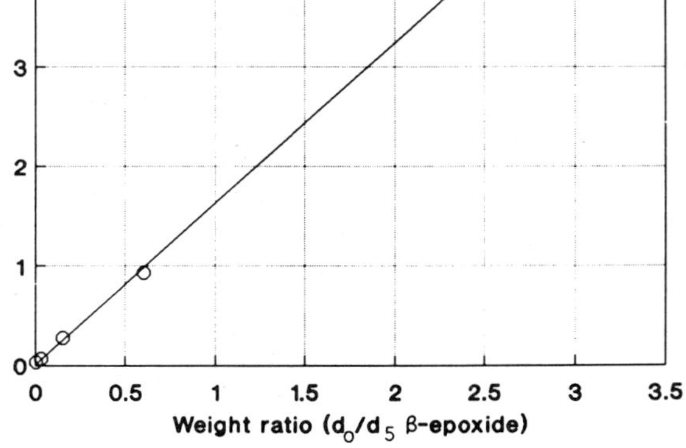

FIGURE 25. SIM standard curve for β-epoxide as TMS ether with d_5-β-epoxide TMS ether as the internal standard. (Data from Gruenke, L. D. et al., *Biomed. Environ. Mass Spectrom.*, 14, 335, 1987.)

REFERENCES

1. **Smith, L. L., Matthews, S., Price, J. C., Bachmann, R. C., and Reynolds, B.,** Thin-layer chromatographic examination of cholesterol autoxidation, *J. Chromatogr.*, 27, 187, 1967.
2. **Danielsson, H., Einarsson, E., and Johansson, G.,** Effect of biliary drainage on individual reactions in the conversion of cholesterol to taurocholic acid, *Eur. J. Biochem.*, 2, 44, 1967.
3. **Mosbach, E. H.,** Hepatic synthesis of bile acids, *Arch. Intern. Med.*, 130, 477, 1972.
4. **Watabe, T., Isobe, M., and Kanai, M.,** Cholesterol diet increases plasma and liver concentrations of cholesterol epoxides and cholestanetriol, *J. Pharm. Dyn.*, 3, 553, 1980.
5. **Sevanian, A. and McLeod, L. L.,** Cholesterol autoxidation in phospholipid membrane bilayers, *Lipids*, 22, 627, 1987.
6. **Kandutsch, A. A., Chen, H. W., and Heiniger, H.-J.,** Biological activity of some oxygenated sterols, *Science*, 201, 498, 1978.
7. **Jacobson, M. S., Price, M. G., Shamoo, A. E., and Heald, F. P.,** Atherogenesis in white carneau pigeons. Effects of low-level cholestane-triol feeding, *Atherosclerosis*, 57, 209, 1985.
8. **Sevanian, A. and Peterson, A. R.,** The cytotoxic and mutagenic properties of cholesterol oxidation products, *Food Chem. Toxicol.*, 24, 1103, 1986.
9. **Matthias, D., Becker, C.-H., Godicke, W., Schmidt, R., and Ponsold, K.,** Action of cholestane-3β,5α,6β-triol on rats with particular reference to the aorta, *Atherosclerosis*, 63, 115, 1987.
10. **Raaphorst, G. P., Azzam, E. I., Langlois, R., and Van Lier, J.,** Effects of cholesterol α and β epoxides on cell killing and transformation, *Biochem. Pharmacol.*, 36, 2369, 1987.
11. **Morin, R. and Peng, S.-K.,** The role of cholesterol oxidation products in the pathogenesis of atherosclerosis, *Ann. Clin. Lab. Sci.*, 19, 225, 1989.
12. **Ryan, T. C., Gray, J. I., and Morton, I. D.,** Oxidation of cholesterol in heated tallow, *J. Sci. Food Agric.*, 32, 305, 1981.
13. **Park, S. W. and Addis, P. B.,** Identification and quantitative estimation of oxidized cholesterol derivatives in heated tallow, *J. Agric. Food Chem.*, 34, 653, 1986.
14. **Park, S. W. and Addis, P. B.,** Further investigation of oxidized cholesterol derivatives in heated fats, *J. Food Sci.*, 51, 1380, 1986.
15. **Rhee, K. S., Dutson, T. R., Smith, G. C., Hostetler, R. L., and Reiser, R.,** Cholesterol content of raw and cooked beef longissimus muscles with different degrees of marbling, *J. Food Sci.*, 47, 216, 1982.
16. **Csallany, A. S., Kindom, S. E., Addis, P. B., and Lee, J. H.,** HPLC method for quantitation of cholesterol and four of its major oxidation products in muscle and liver tissues, *Lipids*, 24, 645, 1989.
17. **Van Delden, J. R., Cozijnsen, J. L., and Folstar, P.,** The determination of the cholesterol content in ready-to-serve foods by gas-liquid chromatography, *Food Chem.*, 7, 117, 1981.
18. **Kaneda, T., Nakajima, A., Fujimoto, K., Kobayashi, T., Kiriyama, S., Ebihara, K., Innami, T., Tsuji, K., Tsuji, E., Kinumaki, T., Shimma, H., and Yoneyama, S.,** Quantitative analysis of cholesterol in foods by gas-liquid chromatography, *J. Nutr. Sci. Vitaminol.*, 26, 497, 1980.
19. **Folch, J., Lees, M., and Stanley, G. H. S.,** A simple method for the isolation and purification of total lipids from animal tissues, *J. Biol. Chem.*, 726, 497, 1957.
20. **Bowden, J. P., Muschik, G. M., and Kawalek, J. C.,** The metabolic fate of cholesterol-5α,6α-epoxide in vivo, *Lipids*, 14, 623, 1979.
21. **Smith, L. L., Teng, J. I., Lin, Y. Y., and Seitz, P. K.,** Sterol metabolism. XLVII. Oxidized cholesterol esters in human tissues, *J. Steroid Biochem.*, 14, 889, 1981.
22. **Park, S. W. and Addis, P. B.,** HPLC determination of C-7 oxidized cholesterol derivatives in foods, *J. Food Sci.*, 50, 1437, 1985.
23. **Park, S. W. and Addis, P. B.,** Cholesterol oxidation products in some muscle foods, *J. Food Sci.*, 52, 1500, 1987.
24. **Björkhem, I., Breuer, O., Angelin, B., and Wikström, S.-H.,** Assay of unesterified cholesterol-5,6-epoxide in human serum by isotope dilution mass spectrometry. Levels in the healthy state and in hyperlipoproteinemia, *J. Lipid Res.*, 29, 1031, 1988.
25. **Bligh, E. G. and Dyer, W. J.,** A rapid method of total lipid extraction and purification, *Can. J. Biochem. Physiol.*, 37, 911, 1959.
26. **Vadji, M., Nawar, W. W., and Merritt, C., Jr.,** Identification of radiolytic compounds from beef, *J. Am. Oil Chem. Soc.*, 56, 611, 1979.
27. **Tsai, L. S. and Hudson, C. A.,** Cholesterol oxides in commercial dry egg products: isolation and identification, *J. Food Sci.*, 49, 1245, 1984.
28. **Tsai, L. S. and Hudson, C. A.,** Cholesterol oxides in commercial dry egg products: quantitation, *J. Food Sci.*, 50, 229, 1985.
29. **Kou, I. L. and Holmes, R. P.,** The analysis of 25-hydroxycholesterol in plasma and cholesterol containing foods by high performance liquid chromatography, *J. Chromatogr.*, 330, 339, 1985.

30. **Nourooz-Zadeh, J. and Appelqvist, L. A.,** Cholesterol oxides in Swedish foods and food ingredients: fresh eggs and dehydrated egg products, *J. Food Sci.,* 52, 57, 1987.
31. **Nourooz-Zadeh, J. and Appelqvist, L. A.,** Cholesterol oxides in Swedish foods and food ingredients: milk powder products, *J. Food Sci.,* 53, 74, 1988.
32. **Nourooz-Zadeh, J. and Appelqvist, L. A.,** Cholesterol oxides in Swedish food and food ingredients: lard and bacon, *J. Am. Oil Chem. Soc.,* 66, 586, 1989.
33. **Pie, J. E., Spahis, K., and Seillan, C.,** Evaluation of oxidative degradation of cholesterol in food and food ingredients: identification and quantification of cholesterol oxides, *J. Agric. Food Chem.,* 38, 973, 1990.
34. **Parks, O. W., Schwartz, D. P., Keeney, M., and Damico, J. N.,** Isolation of Δ^7-cholesten-3-one from butterfat, *Nature,* 210, 416, 1966.
35. **Flanagan, V. P., Ferretti, A., Schwartz, D. P., and Ruth, J. M.,** Characterization of two steroid ketones and two isoprenoid alcohols in dairy products, *J. Lipid Res.,* 16, 97, 1975.
36. **Csiky, I.,** Trace enrichment and separation of oxidation products by adsorption high performance liquid chromatography, *J. Chromatogr.,* 241, 381, 1982.
37. **Williams, L. D. and Pearson, A. H.,** Unsaponifiable fraction of pork fat as related to boar odor, *J. Agric. Food Chem.,* 13, 573, 1966.
38. **Luby, J. M., Gray, J. I., Harte, B. R., and Ryan, T. C.,** Photoxidation of cholesterol in butter, *J. Food Sci.,* 51, 904, 1986.
39. **Luby, J. M., Gray, J. I., and Harte, B. R.,** Effects of packaging and light source on the oxidative stability of cholesterol in butter, *J. Food Sci.,* 51, 908, 1986.
40. **Koopman, B. J., Van Der Molen, J. C., and Wolthers, B. G.,** Determination of some hydroxycholesterols in human serum samples, *J. Chromatogr.,* 416, 1, 1987.
41. **Doormaal, J. J., Smit, N., Koopman, B. J., Van Der Molen, J. C., Wolthers, B. G., and Doorenbos, H.,** Hydroxycholesterols in serum from hypercholesterolaemic patients with and without bile acid sequestrant therapy, *Clin. Chim. Acta,* 181, 273, 1989.
42. **Gruenke, L. D., Craig, J. C., Petrakis, N. L., and Lyon, M. B.,** Analysis of cholesterol, cholesterol-5,6-epoxides and cholestane-3β,5α,6β-triol in nipple aspirates of human breast fluid by gas chromatography/mass spectrometry, *Biomed. Environ. Mass Spectrom.,* 14, 335, 1987.
43. **Bergstrom, S. and Wintersteiner, O.,** Autoxidation of sterols in colloidal aqueous solution. The nature of the products formed from cholesterol, *J. Biol. Chem.,* 141, 597, 1941.
44. **Chicoye, E., Powrie, W. D., and Fennema, O.,** Isolation and characterization of cholesterol-5β,6β-oxide from an aerated aqueous dispersion of cholesterol, *Lipids,* 3, 335, 1968.
45. **Chicoye, E., Powrie, W. D., and Fennema, O.,** Synthesis, purification and characterization of 7-keto-cholesterol and epimeric 7-hydroxycholesterols, *Lipids,* 3, 551, 1968.
46. **Tsai, K. S.,** Quantification of cholesterol oxidation products by capillary gas chromatography, M.S. thesis, University of Minnesota, Minneapolis-St. Paul, MN, 1984.
47. **Sander, B. D., Addis, P. B., Park, S. W., and Smith, D. E.,** Quantification of cholesterol oxidation products in a variety of foods, *J. Food Protect.,* 52, 109, 1989.
48. **Sander, B. D., Smith, D. E., Addis, P. B., and Park, S. W.,** Effects of prolonged and adverse storage conditions on levels of cholesterol oxidation products in dairy products, *J. Food Sci.,* 54, 874, 1989.
49. **Pennock, H. F., Neiss, G., and Mahler, H. R.,** Biochemical studies on the developing avian embryo, *Biochem. J.,* 85, 530, 1962.
50. **Bascoul, J., Domergue, N., Olle, M., and de Paulet, A. C.,** Autoxidation of cholesterol in tallows heated under deep frying conditions: evaluation of oxysterols by GLC and TLC-FID, *Lipids,* 21, 383, 1986.
51. **Chicoye, E., Powrie, W. D., and Fennema, O.,** Photoxidation of cholesterol in spray-dried egg yolk upon irradiation, *J. Food Sci.,* 33, 581, 1968.
52. **Van de Bovenkamp, P., Kosmeijer-Schull, T. G., and Katan, M. B.,** Quantification of oxysterols in Dutch foods: egg products and mixed diets, *Lipids,* 23, 1079, 1988.
53. **Tsai, L. S., Ijichi, K., Hudson, C. A., and Meehan, J. J.,** A method for the quantitative estimation of cholesterol α-epoxide in eggs, *Lipids,* 15, 124, 1980.
54. **Miettinen, T. A., Ahrens, E. H., and Grundy, S. M.,** Quantitative isolation and gas-liquid chromatographic analysis of total dietary and fecal neutral steroids, *J. Lipids Res.,* 6, 411, 1965.
55. **Gerhardt, K. O. and Gehrke, C. W.,** Gas liquid chromatography of fecal neutral steroids, *J. Chromatogr.,* 135, 341, 1977.
56. **Hirsh, J. and Ahren, E. H.,** The separation of complex lipid mixtures by the use of silicic acid chromatography, *J. Biol. Chem.,* 233, 311, 1958.
57. **De Vore, V. R.,** TBA values and 7-ketocholesterol in refrigerated raw and cooked ground beef, *J. Food Sci.,* 53, 1058, 1988.
58. **Morgan, J. N. and Armstrong, D. J.,** Formation of cholesterol-5,6-epoxides during spray-drying of egg yolk, *J. Food Sci.,* 52, 1224, 1987.

59. **Morgan, J. N. and Armstrong, D. J.,** Wide-bore capillary gas chromatographic method for quantification of cholesterol oxidation products in egg yolk powder, *J. Food Sci.*, 54, 427, 1989.
60. **Missler, S. R., Wasilchuk, B. A., and Merritt, C., Jr.,** Separation and identification of cholesterol oxidation products in dried egg preparations, *J. Food Sci.*, 50, 595, 1985.
61. **Axelson, M., Schumacher, G., Sjövall, J., Gustafsson, B., and Lindell, J. O.,** Identification and quantitative determination of steroids in bovine corpus luteum during oestrous cycle and pregnancy, *Acta Endocrinol.*, 80, 149, 1975.
62. **Eisman, J. A., Shepard, R. M., and Deluca, H. F.,** Determination of 25-hydroxyvitamin D_2 and 25-hydroxyvitamin D_3 in human plasma using high-pressure liquid chromatography, *Anal. Biochem.*, 80, 298, 1977.
63. **Gray, M. F., Lawrie, T. D. V., and Brooks, C. J. W.,** Isolation and identification of cholesterol α-oxide and other minor sterols in human serum, *Lipids*, 6, 836, 1971.
64. **Taylor, C. B., Peng, S.-K., Werthessen, N. T., Tham, P., and Lee, K. T.,** Spontaneously occurring angiotoxic derivatives of cholesterol, *Am. J. Clin. Nutr.*, 32, 40, 1979.
65. **Finocchiaro, E. T., Lee, K., and Richardson, T.,** Identification and quantification of cholesterol oxides in grated cheese and bleached butter oil, *J. Am. Oil Chem. Soc.*, 61, 877, 1984.
66. **Girotti, A. W., Bachowski, G. J., and Jordan, J. E.,** Lipid peroxidation in erythrocyte membranes: cholesterol product analysis in photosensitized and xanthine oxidase-catalyzed reactions, *Lipids*, 22, 401, 1987.
67. **Van Lier, J. E. and Smith, L. L.,** Sterol metabolism. II. Gas chromatographic recognition of cholesterol metabolites and artifacts, *Anal. Biochem.*, 24, 419, 1968.
68. **Teng, J. I., Kulig, M. J., and Smith, L. L.,** Sterol metabolism. XXII. Gas chromatographic differentiation among cholesterol hydroperoxides, *J. Chromatogr.*, 75, 108, 1973.
69. **Korahani, V., Bascoul, J., and Crates de Paulet, A.,** Capillary column gas-liquid chromatographic analysis of cholesterol derivatives. Application to the autoxidation products of cholesterol, *J. Chromatogr.*, 211, 392, 1981.
70. **Gumulka, J., Pyrek, J. St., and Smith, L. L.,** Interception of discrete oxygen species in aqueous media by cholesterol: formation of cholesterol epoxides and secosterols, *Lipids*, 10, 197, 1982.
71. **Maeker, G. and Unruh, J., Jr.,** Cholesterol oxides. I. Isolation and determination of some cholesterol oxidation products, *J. Am. Oil Chem. Soc.*, 63, 767, 1986.
72. **Park, S. W. and Addis, P. B.,** Capillary column gas-liquid chromatographic resolution of oxidized cholesterol derivatives, *Anal. Biochem.*, 149, 275, 1985.
73. **Heftman, E. and Hunter, I. R.,** High-pressure liquid chromatography of steroids, *J. Chromatogr.*, 165, 283, 1979.
74. **Ansari, G. A. S. and Smith, L. L.,** High-performance liquid chromatography of cholesterol oxidation products, *J. Chromatogr.*, 175, 307, 1979.
75. **Tsai, L. S. and Hudson, C. A.,** High performance liquid chromatography of oxygenated cholesterols and related compounds, *J. Am. Oil Chem. Soc.*, 58, 931, 1981.
76. **Herian, A. M. and Lee, K.,** 7α- and 7β-hydroxycholesterol formed in a dry egg nog mix exposed to fluorescent light, *J. Food Sci.*, 50, 276, 1985.
77. **Maeker, G., Nungesser, E. H., and Zulak, I. M.,** HPLC separation and quantitation of cholesterol oxidation products with flame ionization detection, *J. Agric. Food Chem.*, 36, 61, 1988.
78. **Maeker, G., Nungesser, E. H., and Bunick, F. J.,** Reaction of cholesterol 5,6-epoxides with simulated gastric juice, *Lipids*, 23, 761, 1988.
79. **Fitzpatrick, F. A. and Siggia, S.,** High resolution liquid chromatography of derivatized non-ultraviolet absorbing hydroxy steroids, *Anal. Chem.*, 45, 2310, 1973.
80. **Watabe, T. and Sawahata, T.,** The recently proposed 20,22-epoxycholesterol as the intermediate in the conversion of cholesterol to pregnenolone by adrenal cortex mitochondria must be 5α,6α-epoxycholestan-3β-ol, *Biochem. Biophys. Res. Commun.*, 83, 1396, 1978.
81. **Sugino, K., Terao, J., Murakami, H., and Matsushita, S.,** High-performance liquid chromatographic method for the quantification of cholesterol epoxides in spray-dried egg, *J. Agric. Food Chem.*, 34, 36, 1986.
82. **Diekman, J. and Djerassi, C.,** Mass spectrometry in structural and stereochemical problems. CXXV. Mass spectrometry of some steroid trimethylsilyl ethers, *J. Org. Chem.*, 32, 1005, 1967.
83. **Brooks, C. J. W., Horning, E. C., and Young, J. S.,** Characterization of sterols by gas chromatography-mass spectrometry of the trimethylsilyl ethers, *Lipids*, 3, 391, 1968.
84. **Brooks, C. J. W., Henderson, W., and Steel, G.,** The use of trimethylsilyl ethers in the characterization of natural sterols and steroid diols by gas chromatography-mass spectrometry, *Biochim. Biophys. Acta*, 296, 431, 1973.
85. **Aringer, L. and Eneroth, P.,** Formation and metabolism in vitro of 5,6-epoxides of cholesterol and β-sitosterol, *J. Lipid Res.*, 15, 389, 1974.

86. **Park, S. W. and Addis, P. B.,** Derivatization of 5α-cholestane-3β,5,6β-triol into trimethylsilyl ether sterol for GC analysis, *J. Am. Oil Chem. Soc.,* 66, 1632, 1989.
87. **McLafferty, F. W.,** *Interpretation of Mass Spectra,* 3rd ed., University Science Books, Mill Valley, CA, 1980, 180.
88. **Sanghvi, A., Grassi, E., Bartman, C., Lester, R., Kienle, M. G., and Giovanni, G.,** Measurement of cholesterol 7α-hydroxylase activity with selected ion monitoring, *J. Lipid Res.,* 22, 720, 1981.
89. **Sanghvi, A., Kienle, M. G., and Galli, G.,** A rapid, sensitive method for simultaneous measurements of tissue levels of oxygenated derivatives of cholesterol, *Anal. Biochem.,* 85, 430, 1978.

Chapter 4

CHOLESTEROL OXIDE CONTENT OF FOODS

Paul B. Addis and Peter S. W. Park

TABLE OF CONTENTS

I. INTRODUCTION

Addis and Warner[1] concluded that "evidence suggests that cholesterol oxidation products (COPS) are far more damaging to vascular health than native cholesterol itself." If this conclusion is correct, the issue of the quantity of COPS in the human diet is exceedingly important. Unfortunately, data on COPS content of food are very limited in scope and are, in many cases, of questionable validity because of the analytical problems discussed in Chapter 3. Studies of food COPS may be conveniently divided into pre-1980 and post-1980. The literature prior to 1980 is spare and dubious, as noted in Chapter 3. Therefore, this chapter will focus primarily on published reports of food COPS content appearing since 1980.

The pace of publication of articles on COPS has accelerated over the past decade. However, the quantification of COPS content in foods, as eaten, is a monumental task. This is especially true if the experiments are performed as outlined by Park and Addis in Chapter 3. Inclusion of mass spectral data to accompany chromatographic separations adds greatly to the time and expense of the analysis but is a requisite for these types of studies. Although the size and scope of the task is great, COPS content of foods must be ascertained so that an average daily intake (ADI) may be calculated and foods with high COPS content may be limited in the diet until the underlying causative lipid oxidation problem can be controlled. Regulations (limitations) of COPS in foods will force food processors and restaurants to address the issue of how to deal with "run-away oil" and other problems which expose consumers to harmful COPS.

Although not addressed in this work, the issue of oxidation products derived from lipids other than cholesterol is an important one and has been reviewed elsewhere.[1] It is a mistake to assume that the recent change from animal/vegetable to purely vegetable shortening and oils by many fast-food firms lessens the need for concern. Indeed, the reverse is probably true: the change away from tallow increases the need for diligence because of the inherent oxidative instability of vegetable oils and the extensive evidence of deleterious biological properties possessed by oxidative breakdown products of polyunsaturated fatty acids.

This chapter will focus on the COPS content of foods but as stated earlier, relative to the enormous scope of the task, a small number of papers have been published. Also, many of the articles report data from pilot plant or laboratory-scale studies in which foodstuffs were artifically manipulated so as to convert some of the cholesterol present to COPS. Such studies will be reviewed briefly; important clues may be learned. We wish to focus to a greater extent on COPS content of foods as commercially processed so as to begin the tedious task of attempting to determine the degree to which consumers are exposed to COPS, the ADI. Several studies combined research on processing manipulation and storage with consumer "ADI". We will review the relevant literature on dairy products, meat and meat products, eggs, and fats, oils and French fried foods.

II. COPS IN VARIOUS TYPES OF FOODS

A. DAIRY PRODUCTS

Nourooz-Zadeh and Appelqvist[2] analyzed (1) cream, whole milk, and skim milk powders obtained from different Swedish manufacturing plants; (2) spray-dried skim milk powders stored in plastic bags (bulk, 25 kg); and (3) small consumer packages of spray-dried milk powder purchased from supermarkets. The method of Nourooz-Zadeh and Appelqvist[3] was used in the analysis and gas chromatography-mass spectrometry (GC-MS) was used to verify the findings.

The results indicated that in the freshly made product, no COPS were detected if "low or medium" heat treatments were used, both for spray- and roller-dried products (Table 1).

TABLE 1
Cholesterol Oxides in 12-Month-Old Whole Milk Powders

	Cholesterol oxides, ppm in lipids[a]							
Company[b]	αep	βep	7k	7α	7β	20α	25OH	Triol
A[c]	2.5	3.2	9.2	0.3	1.5	ND[d]	ND	TR[e]
C[c]	0.6	1.1	5.1	0.5	0.7	ND	ND	0.3
C[f]	0.5	1.3	5.6	1.0	1.3	ND	ND	TR

Note: The abbreviations used are αep, cholesterol-α-epoxide; βep, cholesterol-β-epoxide; 7k, 7-ketocholesterol; 7α, 7α-hydroxycholesterol; 7β, 7β-hydroxycholesterol; 20α, 20α-hydroxycholesterol; 25OH, 25-hydroxycholesterol; and triol, 5α-cholestane-3β,5,6β-triol.

[a] Means of duplicate analysis (extraction, enrichment, and GC).
[b] A, The Vanersborg plant ''low heat''; C, the Bjuv plant ''high heat'', all belonging to the Swedish Dairies' Association.
[c] Stored at the laboratory (about 20°C).
[d] Not detected, detection limit about 0.1 ppm in lipids.
[e] Traces (<0.1 ppm).
[f] Stored at the manufacturing plant (about 20°C).

Reprinted from Nourooz-Zadeh, J. and Appelqvist, L.-Å., *J. Food Sci.*, 53, 74, 1988. With permission.

Products produced by ''high heat'' processing displayed ''quantifiable amounts'' of 7α-hydroxycholesterol, 7β-hydroxycholesterol, cholesterol-α- and β-epoxides, and 7-ketocholesterol. Whole milk powders stored for 12 months showed variable quantities (0.3 to 9.2 ppm) of various COPS, including 0.3 ppm of the rarely seen 5α-cholestane-3β,5,6β-triol (triol), considered to be highly atherogenic (Table 1). Stored skim milk powders obtained from 14 samplings of several warehouses, so as to create a spread in storage time of 13 to 37 months, showed a range in total COPS of 19.6 to 77.9 ppm. The general trend was for higher COPS at longer storage times. It was also noted that the quantitative pattern of milk powders was fairly similar to that seen in powdered egg yolk in an earlier study by the authors.[3]

Sander et al.[4] examined a variety of foods for COPS content, including cheese, butter, and sour cream powders. All of the foregoing products were obtained from commercial ingredient suppliers and analyzed promptly by the methods of Park and Addis.[5-7] MS was utilized to confirm structures of COPS in selected products exhibiting at least moderately high levels of COPS. For comparison, 13 fresh dairy products were also analyzed.

Results obtained on powders indicated that, in general, the levels of COPS were modest. Of the 7 more commonly reported (Chapter 3) COPS, cholesterol α-epoxide (α-epoxide) was observed most frequently, appearing in 6 of 11 Cheddar powders (0 to 9 ppm). Sour cream powders appeared to be more oxidized, as evidenced by higher COPS levels, than the cheese powders. Fresh dairy products exhibited relatively low levels of COPS. Creamed cheese contained 9 ppm α-epoxide and 3 ppm 7-ketocholesterol, both confirmed by MS. Table 2 presents data from Cheddar and Parmesan powders. Data on Romano and blue cheese powders are not presented because these powders were almost free of COPS.[4]

The authors concluded that, with respect to the dairy products that they studied, it was difficult to discern any strong pattern of cholesterol oxidation.[4] α-Epoxide was seen most frequently. Some evidence existed that specific processing conditions or product-handling procedures affected COPS levels. Cheddar cheese #3 and Parmesan cheese #5 were the samples in the Cheddar and Parmesan categories that contained the highest COPS levels and both were obtained from the same manufacturer. Conversely, a different processor

TABLE 2
COPS in Dehydrated Cheddar and Parmesan Powders

	COPS, ppm						
Product (brand code)	**7α**	**7β**	**βep**	**αep**	**Triol**	**7k**	**25OH**
Cheddar powders							
(1)	—	—	4	7	—	—	—
(2)	—	7	—	3	—	—	—
(3)	6	9	3	5	17	14	3
(4)	—	—	—	—	—	—	3
(5)	—	—	—	3	4	3	—
(6)	—	—	3	9	—	—	—
(7)	—	—	—	6	—	—	—
(8)—(11)	—	—	—	—	—	—	—
Parmesan powders							
(1)	—	—	—	4	—	—	—
(2)	—	6	—	3	—	4	—
(3)	—	—	—	4	—	—	—
(4)	—	9	2	5	—	—	—
(5)	—	—	—	4	9	16	4
(6)	—	—	6		—	—	—

Note: Refer to Table 1 for definitions of symbols used.

Modified from Sander et al.[4]

manufactured both Cheddar #9 and Parmesan #6 and detectable levels of COPS were not seen in either sample.

Sander et al.[8] studied processing stage and storage conditions in relation to COPS content in butter and Cheddar cheese. Butter was studied at the following stages: cream, pasteurized cream, buttermilk, unsalted butter, and salted butter. Small quantities of α-epoxide were tentatively identified at 2, 1, 4, 2, and 1 ppm at the five processing stages, respectively. Little evidence was obtained that any of the other COPS were present. Similar results were seen for Cheddar cheese processing. At the stages of milk, whey, curd (after whey drained), curd after milling, curd with salt, and out of press, the α-epoxide levels noted were: 9, 1, 8, 6, 1, and 0 (or not detected). Even smaller levels of β-epoxide and 7-ketocholesterol were noted. Clearly no pattern existed or was discernable in these data other than the fact that fresh butter and Cheddar cheese are virtually free of COPS.

Butter was stored for 6 months at −26, 4, and 16°C. Only α-epoxide showed a moderately consistent pattern of increase with storage temperature, exhibiting 2, 10, and 13 ppm in −26, 4, and 16°C butter at 6 months. When Cheddar cheese was stored for 7 months, a fairly realistic storage period, minimal COPS were seen. Therefore, neither the storage of butter nor the ripening of Cheddar cheese results in significant elevation of COPS levels.[8]

In contrast, a clarified butter product used by Asians, ghee, has been suggested to contain as much as 12.3% sterols as oxides by Jacobson.[9] Although the procedure used to quantify COPS in ghee employed hot (2-hour) saponification (damaging to 7-ketocholesterol), and MS confirmation was not implemented, it does appear that long-term storage could result in significant oxidation. Some factors which may contribute uniquely to ghee oxidation include the very long storage times, warm temperatures, light, and oxygen. Jacobson speculated that COPS in ghee is related to the high prevalence of coronary disease in Indian immigrants to London and to Trinidad who lack the usual risk factors.[9]

The study by Jacobson,[9] among others, prompted Sander et al.[10] to investigate the effects of prolonged and adverse storage conditions on levels of COPS in salted and unsalted butter

TABLE 3
Mineral Content of Salted and Unsalted Butteroil

Sample	Mineral content, ppm[a]					
	P	K	Al	Fe	Na	Cu
Salted butter	228	320	0.74	3.39	6248	0.31
Unsalted butter	301	432	1.40	2.76	96	0.07
Salted butteroil	<1.37	98	0.45	3.39	1099	0.05
Unsalted butteroil	4.37	41	0.26	2.40	2	<0.04

[a] Mean of duplicate determination.

Modified from Sander et al.[10]

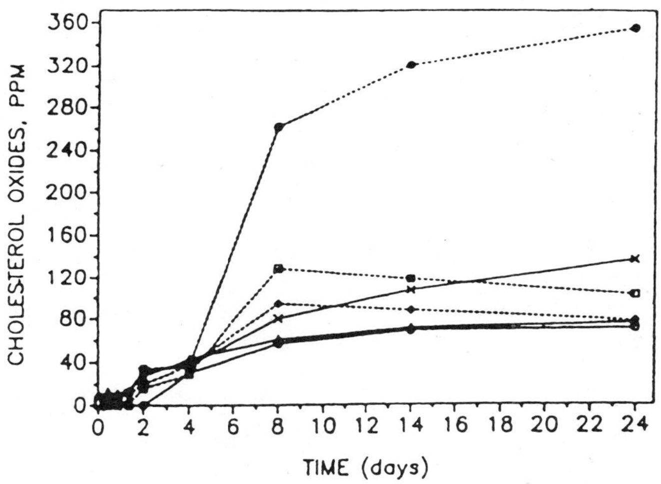

FIGURE 1. Formation of the C7 oxidation products of cholesterol during prolonged heating (110°C) of salted and unsalted butteroil. ○——○, 7α-hydroxycholesterol in salted butteroil; □----□, 7α-hydroxycholesterol in unsalted butteroil; △——△, 7β-hydroxycholesterol in salted butteroil; ◇----◇, 7β-hydroxycholesterol in unsalted butteroil; ×——×, 7-ketocholesterol in salted butteroil; ⊕----⊕, 7-ketocholesterol in unsalted butteroil. (From Sander, B. D. et al., *J. Food Sci.*, 54, 874, 1989. Reprinted with permission of the Institute of Food Technologists.)

oil and powdered cheeses. Butter oil was prepared from salted and unsalted butter, heated at 110°C for 24 d, and sampled at specific intervals. Fresh samples contained low levels of 7β-hydroxycholesterol and α-epoxide, with salted butter oil exceeding unsalted (10-fold) for α-epoxide.

After heating for several days, the three C_7 COPS and both α- and β-epoxide had reached quite high levels. Levels of 7-ketocholesterol reached ~260 ppm after 8 d at 110°C in unsalted butter oil. In the early stages of heating (0 to 4 d), salted butter oil exhibited higher levels of COPS than did unsalted butter oil, an expected finding. Salted butter oil was subjected to mineral analysis and, as expected, contained higher levels of prooxidant metals (Table 3). However, at four days the trend began to reverse itself and unsalted butter oil exceeded salted butter oil by a significant margin (Figures 1 and 2). These unexpected results prompted the authors to investigate further the effects of salt on COPS formation in butter oil.

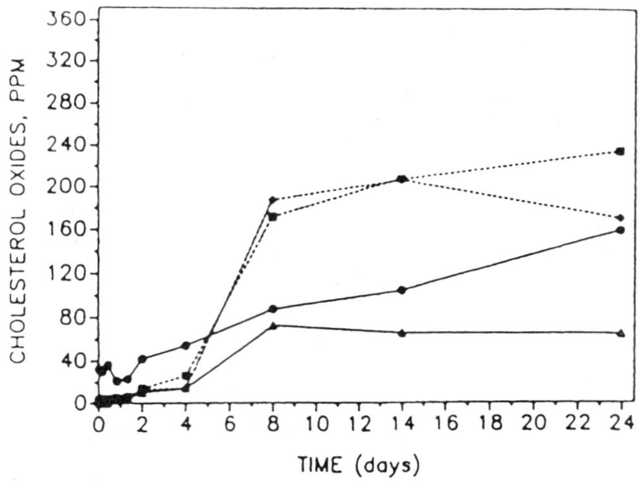

FIGURE 2. Formation of α- and β-epoxide during prolonged heating (110°C) of salted and unsalted butteroil. ⊕——⊕, α-epoxide in salted butteroil; □----□, α-epoxide in unsalted butteroil; △----△, β-epoxide in salted butteroil; ◇——◇, β-epoxide in unsalted butteroil. (From Sander, B. D. et al., *J. Food Sci.*, 54, 874, 1989. Reprinted with permission of the Institute of Food Technologists.)

TABLE 4
Effect of Various Salt Concentrations on the Formation of COPS in Butteroil

	COPS, ppm						
Treatment time	**7α**	**βep**	**αep**	**7β**	**Triol**	**7K**	**25OH**
Unsalted, initial	13	10	24	14	20	19	6
Unsalted, 4 d	34	31	25	44	9	64	2
0.1% NaCl, 4 d	44	38	31	47	7	75	4
0.2% NaCl, 4 d	21	21	13	31	14	58	3
0.3% NaCl, 4 d	—	9	18	17	6	14	—
Unsalted, 8 d	98	110	92	90	7	158	3
0.1% NaCl, 8 d	86	105	98	90	13	160	3
0.2% NaCl, 8 d	76	79	65	77	6	125	5
0.3% NaCl, 8 d	28	20	16	29	2	36	—
Unsalted, 16 d	109	193	191	111	17	308	6
0.1% NaCl, 16 d	100	164	164	99	22	276	5
0.2% NaCl, 16 d	72	104	94	82	10	156	3
0.3% NaCl, 16 d	66	102	91	81	8	146	3

Note: Chromatographic peaks confirmed by GC-MS. Refer to Table 1 for definition of abbreviations used.

From Sander, B. D., Smith, D. E., Addis, P. B., and Park, S. W., *J. Food Sci.*, 54, 874, 1989. With permission.

Unsalted butter oil was prepared as described earlier. NaCl (reagent-grade) was added to the unsalted butter oil at 0.1, 0.2, and 0.3%. Butter oil samples were heated 0, 4, 8, and 16 d at 105°C, sampled, and analyzed for COPS. The results (Table 4) agreed with the preceding experiments in that higher levels of salt appeared to produce an impressive antioxidant effect. Sander et al.[10] speculated that the apparent antioxidant effect of salt may be due to the ionic-dipolar interaction between Na^+ and water and Cl^- and water. Butter oil was analyzed for water content and was found to contain less than 1.0%. Salt, by binding to water, could reduce the mobility of prooxidant metals present and reduce their catalytic potential.

Sander et al.[10] also studied long-term storage of cheese powders at 4°C. An 18-month storage period resulted in increases of α-expoxide and 7-ketocholesterol which were, with few exceptions, very modest and in many cases, no increase at all was seen. Exposure of Cheddar cheese powder to fluorescent light for up to 12 weeks resulted in an increase in α-epoxide, β-epoxide, and 7-ketocholesterol but not 7β- and 7α-hydroxycholesterols, triol, and 25-hydroxycholesterol. α-Epoxide levels showed a clear trend to reach a maximum at three weeks and then decline thereafter; β-epoxide showed a similar trend.

The studies of Sander et al.[4,8,10] suggest that dairy products in general are fairly resistant to cholesterol autoxidation, even if storage times are prolonged and conditions adverse. The basis for this resistance probably stems from the following three characteristics: low transition metal content (Fe, Co, Cu, etc.), modest cholesterol level, and an environment of highly saturated fat.[8] Of course, no one can state the precise level of COPS in the diet which could exert deleterious health effects in humans. Therefore, it seems prudent to attempt to minimize COPS content of dairy products. To do so, curtailment of lipid autoxidation in general is necessary. The use of antioxidants would appear to be an appropriate method to reduce levels of toxic COPS as well as improve flavor, quality, and safety of the product. Wade et al.[11] studied ascorbyl palmitate (AP), *d,l*-α-tocopherol, butylated hydroxyanisole (BHA), and butylated hydroxytoluene (BHT) in anhydrous milk fat and noted variable results which suggested that BHA, BHT, and AP were fairly effective antioxidants. Combinations of the antioxidants were even more effective than separate antioxidants.

Over the past decade there have been several earlier articles published on COPS in dairy products. None really attempted to measure COPS in products as received by the consumer but are, nevertheless, useful in revealing some of the mechanisms responsible for the oxidative degradation of cholesterol. Some of these studies will be reviewed briefly.

Luby and co-workers[12,13] published two studies on the photooxidation of butter. Exposure of unpackaged butter to fluorescent light resulted in qualitative evidence of cholesterol oxidation, mostly at the surface and only after 8 days of exposure. These researchers also demonstrated that packaging materials delayed oxidation of butter compared to unpackaged controls but only aluminum foil performed satisfactorily.

Finocchiaro et al.[14] studied the occurrence of cholesterol oxides in grated cheese and bleached butter oil. Ungrated cheeses produced from bleached milk and freshly bleached butter oil (benzoyl peroxide treated, 500 μg/g in oil) contained no detectable oxidized sterols. Grating of cheese and storage of bleached butter oils (90 days and 1 year) resulted in elevated levels of COPS. Bleached cream produced by a simulated industrial process contained nondetectable levels of COPS.

In general, dairy products appear to be a very modest source of COPS in the human diet. Storage of dairy products in substantial packaging materials and favorable (cool or cold) temperatures, coupled with low iron and polyunsaturated fatty acid contents and a modest cholesterol level (except for butter and butter oil) resulted in low COPS levels, in some cases even after prolonged periods under adverse conditions. The area could use some further research, as will be discussed later, but it would rate the lowest in terms of urgency among the four areas of foods to be discussed.

B. MEAT PRODUCTS

Meat products, in general, contain high levels of iron, sufficient polyunsaturate content, and moderate cholesterol content so that during processing and storage the potential exists for extensive cholesterol oxidation. Nevertheless, reports published to date indicate that most meat products are modest sources of COPS. We will attempt to discuss some of the reasons for this phenomenon. However, we caution the reader that a tremendous amount of research will be required in the future before (1) an accurate assessment can be made of the overall contribution of meat products to dietary COPS; (2) the processes and storage conditions most likely to permit COPS formation are determined; and (3) effective measures are developed that will retard COPS formation. Furthermore, it must be emphasized that COPS are only part of the problem. Fatty acid hydroperoxides, malonaldehyde, and other degradation products will often occur even in the absence of cholesterol oxidation and the possibility exists that the toxicity of ''rancid'' meat may be significant in spite of nondetectable levels of COPS.

Park and Addis[5] applied high-performance liquid chromatography (HPLC) to a variety of foods, including several meats. Freshly obtained beef skeletal muscle, liver, and brain contained nondetectable levels of the C_7 COPS. However, ''health food'' brain and liver products contained 14 to 70 ppm COPS. Fast-food chicken, chicken crust, and hamburger were all free of COPS at the 1-ppm detection limit. Beef jerky and liver sausage contained nondetectable levels of COPS, which was thought to be due to the excellent antioxidant properties of nitrite.[15] However, the reader is cautioned that, because of the limited nature of the study in question, the results need further verification.

Higley et al.[16] applied dry-column extraction and HPLC of COPS from meat products and reported only trace quantities of COPS in most products. However, a few samples were reported to have exceedingly high levels of COPS. Cooked bratwurst was noted to contain 1640 ppm 7α-hydroxycholesterol and 1335 ppm triol. Duplicates of the same sample contained either trace or below-detection levels of these sterol oxides. The levels reported in this study of COPS exceeded the cholesterol content of meat (~750 ppm). The authors concluded that the chromatographic peaks for some COPS were not pure, a factor leading to spuriously high levels.[16]

An unusual form of lipid oxidation occurs in uncured meat products which is triggered by cooking. The resulting ''warmed-over flavor'' or WOF is an off-flavor that represents one of the most challenging technical problems faced by the meat and poultry industries because of the increased demand for precooked products. WOF results from heme-iron or inorganic-iron catalysis of lipid (mostly phospholipid) oxidation. Phospholipids of meat are a significant fraction of the total lipids and are highly polyunsaturated (see Allen and Foegeding[17] for review). Research of Sharma et al.[18,19] demonstrated that cooking and storage results in decreases in phospholipids and neutral lipids and increases in free fatty acids and thiobarbituric acid reactive substances (TBARS) values. Phospholipids decreased more rapidly at 4°C than at −10°C, consistent with the concept that, to some degree, freezing arrests WOF development during frozen storage.

Cholesterol is an important component of cell membranes and, as such, would be close to the subcellular location of lipid oxidation during WOF development. Therefore, Park and Addis[7] studied the possible accumulation of COPS during the development of WOF in beef and turkey and during the broiling of beef steak. Longissimus dorsi muscles were broiled to rare and well done (60° and 80°C internal, respectively). Charbroiling was also studied. Some typical chromatograms are shown in Figure 3. A comparison of raw, rare, well done, and charbroiled revealed only the slightest of COPS accumulation in cooked beef. Only 7β-hydroxycholesterol and 7-ketocholesterol were noted.

To study the possible relationship between COPS and WOF, Park and Addis[7] employed a simulated industrial process precooking procedure for beef semimembranosus muscle (vacuumization in cooking bag, overnight cook at 91°C to internal temperature of 79°C). After chilling to 4°C, muscle was sliced into 1.5 cm thick slices, wrapped in polyvinyl chloride (PVC) film, and stored up to 12 days at 4°C. The results were somewhat unexpected in that neither TBARS nor COPS showed any significant increase during storage for 12 days.

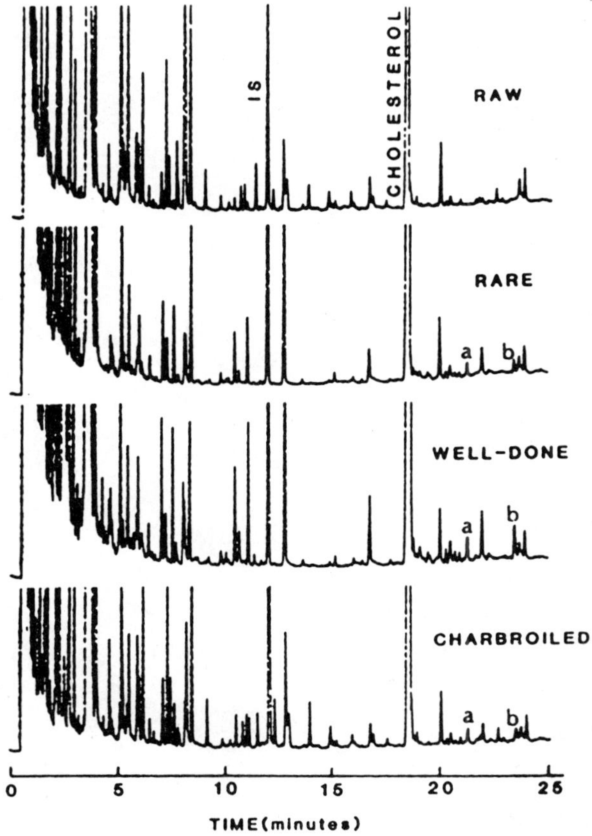

FIGURE 3. A comparison of gas chromatographic profiles of various beef steaks. Peaks as trimethylsilyl ether sterols: (a) 7β-hydroxycholesterol, (b) 7-ketocholesterol. The levels of COPS formed were too small to permit MS confirmation. (From Park, S. W. and Addis, P. B., *J. Food Sci.*, 52, 1500, 1987. Reprinted with permission of the Institute of Food Technologists.)

The authors resorted to comminution, precooking, and storage of turkey and beef as a method to stimulate elevated TBARS. The results demonstrated that turkey was more susceptible to WOF development than beef and that in both turkey and beef, increases in TBARS were associated with increased COPS (Table 5). The data indicate that a small but increasing percentage of the cholesterol present oxidized as TBARS increased, confirming the authors suggestion that WOF may be related to COPS. However, rather high TBARS were required to see >1% cholesterol oxidation. Beef and turkey probably represent minor sources of COPS in the human diet but more research needs to be conducted on precooked products, comminuted products, and precooked, comminuted products before a definitive statement can be made.

De Vore[20] used essentially the HPLC procedure developed by Park and Addis[5] to study in detail the question posed in the foregoing paragraph. TBARS and 7-ketocholesterol levels were determined on raw and cooked ground beef patties after 4°C storage for 0, 2, and 4 days. Mean values for 7-ketocholesterol at 0, 2, and 4 days storage were 0.10, 0.24, and 0.42 and 0.06, 2.8, and 4.8 ppm for raw and cooked patties, respectively. TBARS levels increased with storage time and were significantly correlated to 7-ketocholesterol levels in raw ($p \leq 0.0018$) and cooked ($p \leq 0.0001$) beef patties (Figures 4 and 5).

TABLE 5
TBARS and COPS in Comminuted, Cooked, and Stored Beef and Turkey

	Beef		Turkey	
Days at 4°C	**TBARS**	**Total COPS, % cholesterol**	**TBARS**	**Total COPS, % cholesterol**
0	0.52[a]	ND[b]	1.66[a]	ND[b]
3	1.88[c]	0.45[d]	3.53[c]	1.20[d]
8	2.22[e]	1.74[f]	4.68[e]	2.90[f]

Note: Mean values of three determinations. Means in a column with different superscript letter are significantly different at the 1% level.

[a] Not detected.

Modified from Park and Addis.[7]

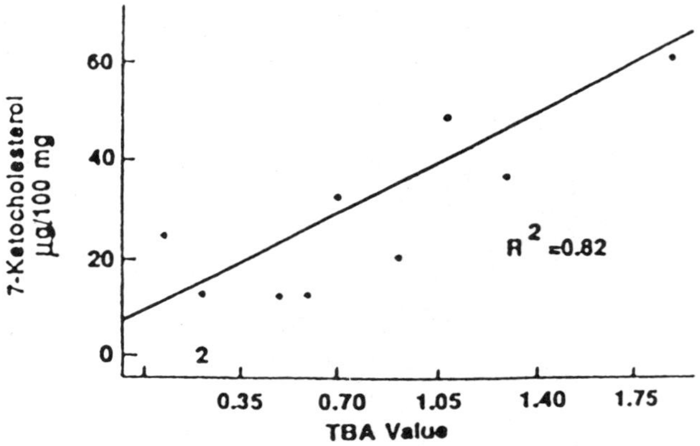

FIGURE 4. Correlations of TBA value with 7-ketocholesterol for raw ground beef stored 0, 2, and 4 days at 4°C. (From De Vore, R. A., *J. Food Sci.*, 53, 1508, 1988. Reprinted with permission of the Institute of Food Technologists.)

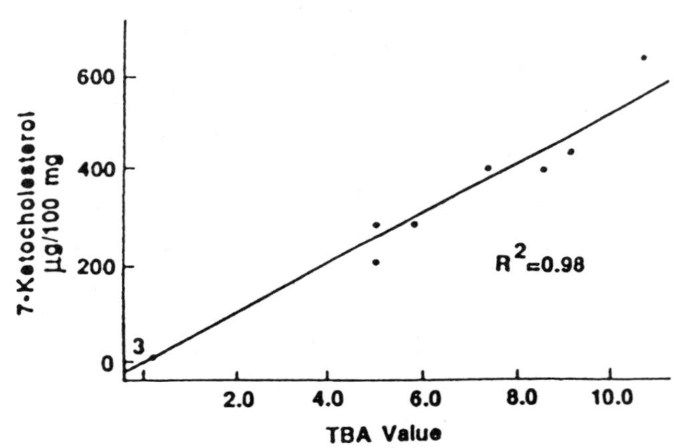

FIGURE 5. Correlation of TBA value with 7-ketocholesterol for cooked ground beef stored 0, 2, and 4 days at 4°C. (From De Vore, R. A., *J. Food Sci.*, 53, 1508, 1988. Reprinted with permission of the Institute of Food Technologists.)

Sander et al.[4] reported COPS values for different brands of freeze dried chicken, beef, and turkey. Although a great deal of variation existed, some impressive levels were seen. The "cleanest" sample of chicken contained only 8 ppm α-epoxide whereas the "worst" sample contained 10, 43, 3, and 22 ppm of β-epoxide, α-epoxide, 7β-hydroxycholesterol, and 7-ketocholesterol. Similar levels were seen in beef and turkey samples. For most people, freeze-dried foods represent a minor part of the diet. But for some, military personnel and others stationed in remote areas, ardent campers, and other persons not having access to fresh foods, freeze-dried (or drum-dried) foods could represent a fairly significant source of dietary COPS. Pie et al.[22] noted increases in COPS during frozen and household cooking storage of meat.

Torres et al.[21] noted trace quantities of COPS in Charqui, a popular salted and dried beef product of Brazil. Losses of unsaturated fatty acids occurred in both the triglyceride and phospholipid fractions. Increases in TBARS were seen which were indicative of oxidation-induced losses of triacylglycerols and phospholipids.

The results of the foregoing studies clearly indicate that the muscle phospholipid oxidation associated with WOF, freeze-drying, and storage results in cholesterol oxidation. The levels seen are modest in WOF but higher in freeze-dried products. Nevertheless, more data are needed on COPS as well as other lipid oxidation products (LOPS) in meat products; subsequently, appropriate toxicology data need to be obtained to estimate the risk, if any, of consumption of rancid meats.

It is perhaps premature to make an estimate of the potential seriousness of rancidity of dairy and meat products in terms of a potential health effect. Nevertheless, the authors feel that dairy products represent an extremely minor risk and meat products are, at most, a very minor risk but with less certainty surrounding the assessment. However, in our opinion, the next two areas to be discussed, powdered eggs and deep-fried foods, represent some potentially serious problems with regard to COPS content.

C. POWDERED EGGS

Egg yolks contain high levels of cholesterol and iron and are a moderately good source of polyunsaturated fatty acids, all factors which favor lipid peroxidation and COPS formation. Therefore, it is not surprising that powdered eggs are among the highest in terms of COPS content of any food. Tsai and Hudson[23,24] were among the very first researchers to employ modern analytical procedures for quantification of COPS in powdered eggs. Subsequently, several laboratories demonstrated the conversion of cholesterol to COPS in a number of egg systems. Van de Bovenkamp et al.[25] reported high levels of COPS in powdered eggs; low levels in fresh eggs were reported. Nabor and Biggert[26] showed that heating egg yolk for 100 hours at 100°C in air appeared to cause a fourfold increase in COPS over the small quantities seen in fresh egg. MS confirmation was not attempted. Missler et al.[27] quantified several COPS in scrambled egg mix produced by direct and indirect heat sources for spray drying. Table 6 summarizes the differences in COPS levels between the two spray-drying methods. The direct heat source resulted in much higher levels of COPS, in agreement with Tsai and Hudson.[23] In the direct method, air derived from gas-fired ovens is used to dry the egg mix. The authors[26] suggested that nitrogen dioxide, a byproduct of combustion, stimulated cholesterol oxidation. In the indirect method, air is heated by a steam injection process, eliminating both combustion and nitrogen dioxide. The use of MS confirmation in the study by Missler et al.[27] strengthens the conclusions made by the authors.

Herian and Lee[28] noted 7-oxy COPS in a dry egg nog mix exposed to fluorescent light.

Sugino et al.[29] reported that spray-dried egg contains α- and β-epoxides at levels of 0.1 and 0.2% of cholesterol, respectively. Reversed-phase HPLC of *p*-nitrobenzoyl derivatives of cholesterol and the epoxides was employed and MS confirmation was obtained.

A thorough study of the operating parameters of the indirect spray dryer, as they pertain to COPS (epoxides) production, was reported by Morgan and Armstrong.[30] Liquid egg yolk

TABLE 6
COPS in Egg Mixes Spray Dried by Direct and Indirect Heat Sources

COPS	Egg mix powder, ppm Direct	Indirect
7α-Hydroxycholesterol	7	2
7β-Hydroxycholesterol	19	2
Cholesterol-5α,6α-epoxide	50	22
Cholesterol-5β,6β-epoxide	37	2
25-Hydroxycholesterol	5	1
7-Ketocholesterol	37	2
5α-Cholestane-3β,5,6β-triol	13	12

Modified from Missler, S. R., Waselchuk, B. A., and Merritt, C., Jr., *J. Food Sci.*, 50, 595, 1985. With permission.

was spray dried with and without addition of H_2O_2. Epoxides were not detected in spray-dried yolk, despite large operating temperature ranges of 150 to 400°C inlet and 75 to 150°C outlet. Solids concentration ranged from 11.3 to 45%. H_2O_2 is sometimes used for purposes of enzymatic desurgarization of liquid egg yolk prior to spray drying for some specialty foods:

$$\text{Catalase} + 2H_2O_2 \rightarrow O_2 + 2H_2O \tag{1}$$

$$O_2 + \text{glucose} + \text{glucose oxidase} \rightarrow \text{gluconic acid} \tag{2}$$

According to the authors, H_2O_2 concentrations in egg yolk do not exceed 0.5% in commercial practice but 5% H_2O_2 in liquid yolk was used to accelerate cholesterol oxidation and produce epoxides at levels seen in commercial products. An outlet temperature of 90°C combined with inlet temperatures of 150 and 250°C produced little or no epoxides, even at 5% H_2O_2. Inlet temperatures had little effect on epoxide production at 5 or 10% H_2O_2. Outlet temperature was a factor affecting epoxide levels; higher temperatures produced more epoxides.[29]

Morgan and Armstrong[30] also attempted to demonstrate the role of oxidizing nitrogen oxide gases in catalyzing lipid oxidation. Nitrous oxide (N_2O) was introduced into the indirect spray dryer at the electric heater, producing (at >300°C) several nitrogen oxides which normally would occur in a gas-fired dryer. Antioxidants were also studied. The results (Table 7) indicated that N_2O treatment was marginally effective at increasing epoxides, that storage up to 7 months resulted in significant increases in epoxides and that BHT reduced epoxide formation during storage. However, as the data indicate, some extremely high levels of epoxides occurred in some samples, notably those without BHT but with added H_2O_2 and stored for 4 or 7 months.[30]

Many of the studies reviewed reported only cholesterol α- and β-epoxide; yet the opportunity to form other COPS during spray-drying is obvious. Using a method that quantifies several major COPS with a single GC injection,[6] Sander et al.[4] reported very high levels of COPS in addition to the epoxides in powdered eggs, including 7-keto- and 7β-hydroxycholesterol; levels of triol and 25-hydroxycholesterol were much lower (Table 8). The data shown in Table 8 were derived from samples shipped immediately after production to minimize storage effects. GC-MS confirmed all major peaks in three of the samples, the only three on which MS confirmation was attempted.[4]

TABLE 7
Effect of Storage[a] on Concentrations of Cholesterol 5,6-Epoxides in Spray-Dried Egg Yolk

Treatment	Cholesterol 5,6-epoxides[b] (ppm)		
	Initial	4 Months	7 Months
None	0	6	20
5% H_2O_2	16	123	123
10% H_2O_2	19	1030	1320
10% H_2O_2 + 67 ppm BHT	35	182	223
N_2O	7	8	—

[a] All test samples stored in glass jars with polyethylene caps at ambient temperature in the dark.

[b] Data are for individual test samples in each group and show most extreme storage effect.

Reprinted from Morgan, J. N. and Armstrong, D. J., *J. Food Sci.*, 52, 1224, 1987. With permission.

TABLE 8
Presence of COPS in Dehydrated Egg Products

Product	COPS, ppm						
	7α	βep	αep	7β	Triol	7K	25OH
Whole egg powder(1)	—	25[a]	70[a]	36[a]	--	13[a]	--
Whole egg powder(2)	—	24[a]	111[a]	41[a]	--	17[a]	--
Whole egg powder(3)	—	46	63	65	--	37	--
Whole egg powder(4)	—	18	32	38	--	12	10
Whole egg powder(5)	—	18	20	36	--	14	--
Whole egg powder(6)	—	16	12	22	2	20	8
Whole egg powder(7)	—	5	18	--	6	--	--
Egg yolk powder	—	28[a]	79[a]	51[a]		22[a]	2
Omelet mix	—	2	--	10	--	2	--

Note: See Table 1 for abbreviations of COPS. (7α Was not quantified because it was occluded by a very large cholesterol peak.) --, not detected; —, not quantified.

[a] Comfirmed by GC-MS.

From Sander, B. D., Addis, P. B., Park, S. W., and Smith, D. E., *J. Food Protection*, 52, 109, 1989. Reprinted with permission of the International Association of Milk, Food and Environmental Sanitarians.

Clearly, powdered egg yolk potentially represents a major source of COPS, relative to most other foods, in the human diet. It is not known whether the levels of intake in the current diet constitute a potential detrimental health effect. Few people consume pure powdered egg in large quantities. However, recent research by Emanuel et al.[31] has demonstrated clearly that COPS contained in powdered eggs, and presumably other foods as well, are absorbed by humans. Fresh and powdered eggs were analyzed by GC[6] and the results, consistent with earlier studies,[23-30] demonstrated very high levels of COPS in dried eggs. Fresh eggs displayed extremely low levels of COPS. Likewise, when fed to humans, COPS from powdered eggs were absorbed via chylomicrons, reached very high levels in plasma, and subsequently were cleared from the plasma. Fresh-egg feeding produced no such response.[31] The significance of the research of Emanuel et al.[31] is that a commonly held conception — that COPS are not absorbed — was clearly shown to be incorrect.

Data collected on COPS content of both fresh and powdered eggs, together with atherogenesis comparisons of cholesterol and COPS, permit some interesting observations on the putative role of egg consumption in the development of coronary heart disease. Addis and Warner[1] have provided detailed discussions and literature reviews on this subject and it is recommended that the reader consult those references. Briefly, for decades atherosclerosis has been induced in experimental animals by feeding powdered egg yolk (as a cholesterol source). In such studies no mention is made of the possibility of COPS being important atherogens; rather the blame is placed on cholesterol and, ultimately, the public is admonished to eat fewer eggs. Several excellent studies of the relative atherogenicity of cholesterol vs. COPS have been published (see Chapters 8 and 9) with a near consensus of researchers agreeing that COPS are far more deleterious than cholesterol. It is clearly time to use a fresh approach in studies designed to elucidate the role of lipids in atherosclerosis. The research reviewed here has demonstrated that it is possible to produce powdered eggs with high and low levels of COPS. Methods exist for retarding further oxidation of cholesterol during storage of feed (nitrogen-flushing, vacuum packaging, and frozen storage). The time has come to curtail the careless, injudicious use of dried eggs in feeding studies and, instead, analyze feed for content of COPS and intelligently use such information in the design of experiments and interpretation of experimental data.

D. HEATED OILS AND DEEP-FRIED FOODS

Addis and Warner[1] have stated that heated oils and foods deep fried in them are, by far, the single biggest source of LOPS in the human population and where animal fats are used, the biggest sources of COPS. Whereas this chapter primarily deals with COPS in foods, it must be recognized that other LOPS may be important in human pathologies as well. Therefore, Addis and Warner[1] caution that the recent, comprehensive change from tallow to vegetable shortening in the fast-food industry does not reduce the need for vigilance with respect to LOPS. On the contrary, vegetable oils and shortening are more labile than tallow and degradation of lipids will occur more rapidly. Also, if meat products are being fried, lipid exchange between oil and meat will occur; therefore, COPS could be formed in vegetable oils used to deep fry muscle foods.

One of the earliest studies of cholesterol oxidation in tallow was conducted by Ryan et al.[32] The authors noted that exposure of tallow to 180°C intermittently (8 hours daily) for 300 hours resulted in extensive degradation of cholesterol and formation of suspected oxysterols. Identity of oxysterols was tentative. Park and Addis[33,34] conducted extensive studies of cholesterol oxidation in heated tallow. COPS formation was noted during continuous heating of tallow at 155°C. 7-Ketocholesterol formation was linear with time up to 376 hours of heating. Interestingly, at >376 hours of heating, 7-ketocholesterol began to show a marked decline for reasons which were not apparent.[34] The authors also reported that ascorbyl palmitate reduced but did not prevent COPS formation in heated tallow.

Yan and White[35] used essentially the GC method of Park and Addis[6,33] to study cholesterol oxidation in lard that was spiked with either two or ten times the amount of cholesterol originally present in the lard and then heated at 180 ± 5°C internally (for 10 hours per day). The same trend for the formation of C_7 oxides and α-epoxide that was reported by Park and Addis[33,34] was confirmed by Yan and White.[35]

There is a serious scarcity of data on the levels of LOPS, including COPS, in deep-fried foods, both food service items and retail products such as potato chips, corn chips, and similar items. Lee et al.[36] studied sterol oxides in commercial potato chips and fast-food French fries. HPLC and thin-layer chromatography (TLC) were used but results were not confirmed by MS. Oxides of both cholesterol and β-sitosterol were quantified after the

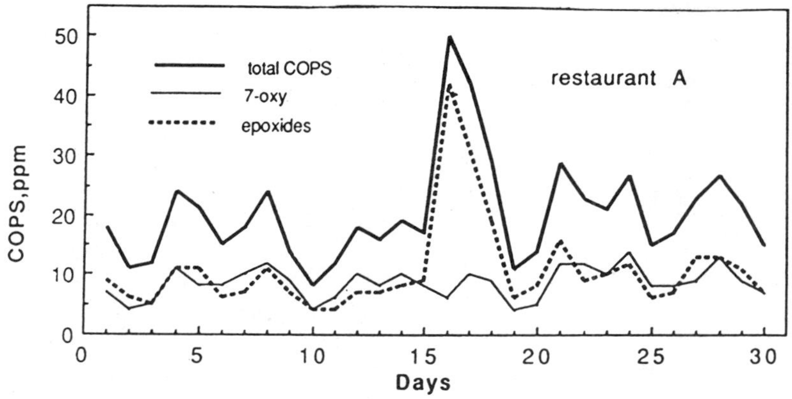

FIGURE 6. COPS content of food service French fries (Restaurant A). One serving of French fries was purchased daily for 30 consecutive days. Symbols: —— 7-oxy, 7αOH + 7βOH + 7K; ----- epoxides, αep + βep; —— COPS, total COPS. (From Zhang, W. B., Addis, P. B., and Krick, T. P., *J. Food Sci.*, 56, 716, 1991. With permission.)

tentative identifications were made. The authors concluded that ingestion of sterol oxides from French fries is possible whereas ingestion of sterol oxides from potato chips is unlikely. Further research is needed on phytosterol oxides, usually overlooked by most researchers. A review of phytosterols and phytosterol oxides has been published.[40]

A comprehensive study of COPS in fast-food French fries was conducted by Zhang et al.[38] French fries were purchased from two local restaurants for 30 consecutive days and analyzed for COPS by the method of Park and Addis.[33] Fresh tallow samples were analyzed as negative controls and consistently produced ‘‘zero’’ values. Restaurant A employed a 90% tallow:10% cottonseed oil mixture.

The results, displayed in Figure 6, showed highly variable levels of COPS with time but some impressively high levels were recorded (total COPS mean ± SD = 20 ± 9 ppm). Careful inspection of these curves reveal that the analytical method is not the source of variation. It is likely, but not proven, that operating factors such as filtration frequency and efficacy, quantity of fries processed, oil add back and time of oil discard, and similar factors are involved in the high degree of variation noted.[39] At day 15, the COPS levels displayed an exponential increase to 50 ppm followed by a steep 4-day decline. Epoxides and 7-oxy COPS derivatives account for approximately equal quantities of the total COPS except for the period from day 15 to day 19.

The quantities of COPS in restaurant A French fries is impressively high, especially around days 15 to 19. These findings prompted Zhang et al.[38] to monitor a second local restaurant, B (Figure 7), also using 90/10 tallow/cottonseed oil. Restaurants A and B belonged to different companies but both used tallow. As was the case in restaurant A, B displayed impressive COPS levels but with an important new phenomenon, the finding of unexpectedly high values for triol. (Only unconfirmed traces were noted in French fries from restaurant A.) Triol is rarely seen in food products, especially at levels high enough to confirm by MS. Triol levels recorded from restaurant B exceeded 10 ppm several times during the 30-day sampling period and the presence of triol was confirmed by MS as were 7-α-, 7-β-hydroxy-, and 7-ketocholesterol. Interestingly, levels of epoxides were lower in restaurant B than in A whereas triol levels were the reverse, lower in A than B.[38] This finding is consistent with the known pathways of triol formation:

$$\alpha ep + \beta ep + H^+, H_2O = \text{protonated epoxide} = \text{triol}$$

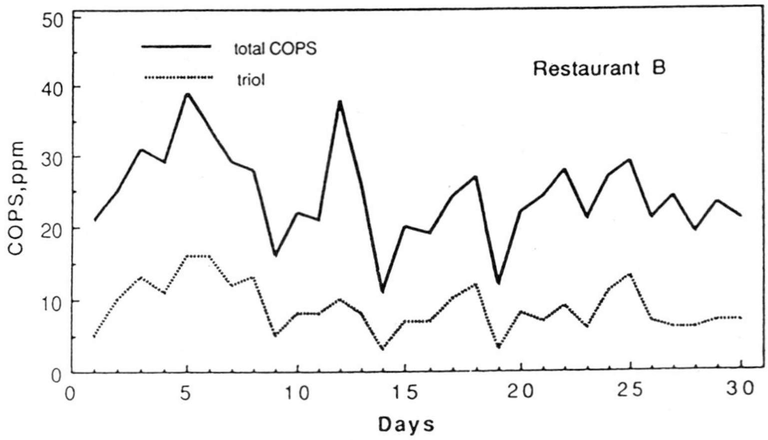

FIGURE 7. COPS content of food service French fries (Restaurant B). One serving of French fries was purchased daily for 30 consecutive days. Symbols: ——, total COPS; ·······, triol. (From Zhang, W. B., Addis, P. B., and Krick, T. P., *J. Food Sci.*, 56, 716, 1991. With permission.)

The study by Zhang et al.[38] is significant for three reasons: (1) it showed deep-fried foods are a major source of COPS (and other LOPS) in the diet; (2) for the first time, a common food has been shown to contain significant levels of triol, one of the most atherogenic of all the COPS (see Chapters 8 and 9 for review); and (3) this study and the study by Bascoul et al.,[39] to be reviewed next, plus our unpublished observations, indicate that heated oils are frequently in a poor state of quality to the extent that serious attention should be paid to this problem. (Instead, Alar is in the news!)

Bascoul et al.[39] measured COPS in tallow used in deep fryers during operation. About 25% of the original cholesterol was degraded to COPS, including triol. Results were not confirmed by MS but appear to be consistent with the results of Zhang et al.[38] Levels of COPS were 100 to 300 ppm in the oil.

Part of the problem of poor heated oil quality stems from lack of good quality control methodology. Zhang and Addis[40] studied the correlations among several methods of oil evaluation in heated 90/10 tallow/cottonseed oil. COPS, Food Oil Sensor (FOS), free fatty acids (FFA), and TBARS were determined. Of the quality control methods studied, FOS appeared to correlate best with COPS. FFA test strips, widely used in vegetable shortenings and tallow, for this experiment (in tallow) did not predict either COPS or FFA. The test strips are inexpensive and easy to use and are, therefore, widely employed.

III. CONCLUSIONS

An urgent need exists for the comprehensive study and eventual elimination of COPS in food products. This goal is attractive not only from the standpoint of consumer safety but also of food quality. The very existence of COPS in food, especially large quantities of COPS, is *prima facie* evidence of the failure of quality assurance programs designed to prevent rancidity. Cholesterol is slow to oxidize relative to polyunsaturated fatty acid. The existence of significant levels of COPS signals the severe degradation of fatty acids of two, three, four, five, and six double bonds if such fatty acids are present in the same food.

There also needs to be cooperative investigations among the laboratories involved in COPS analysis of foods. Many highly contrasting methods are being used with little or no consideration about agreement or lack of agreement among the methods (see Chapter 3). We suggest that a series of interlaboratory comparisons be conducted on common food samples employing both identical (where possible) and differing methodologies. It is critical that data collected in a controversial area be reliable and reproducible and the employment of widely divergent methodologies makes difficult the compilation of data that evokes confidence in toxicologists making risk assessments and the regulatory authorities.

The possible regulation of lipid oxidation product levels in foods is an extremely interesting and challenging notion. Little is currently being done in the U.S. but regulation of the quality of heated frying oils in restaurants is common in Europe. In some countries, frying oils are strictly regulated and the penalties for violations are severe. As more evidence is obtained on the deleterious effects of consumption of COPS, the possibility exists that pressure for such regulations will increase. Ironically, we note that as this manuscript was being prepared a controversy developed concerning the addition of BHA to the Proposition 65 list in California! Other synthetic antioxidants are being scrutinized for possible toxic and carcinogenic effects and an intense new area of research has developed concerning the search for and identification, isolation, and characterization of naturally occurring antioxidants. The foregoing controversies concerning synthetic antioxidant together with the recognized potential detrimental effects of both dietary and *in vivo*-produced lipid oxidation products indicate that some extremely interesting future discussions and debates will occur. The reader is referred to two recent reviews which include information on heated oil regulation and health benefits of antioxidants.[1]

REFERENCES

1. **Addis, P. B. and Warner, G. J.,** The potential health aspects of lipid oxidation products in foods, in *Free Radicals and Food Additives,* Aruoma, O. I. and Halliwell, B., Eds., Taylor and Francis, London, 1990, chap. 5.
2. **Nourooz-Zadeh, J. and Appelqvist, L.-Å.,** Cholesterol oxides in Swedish foods and food ingredients: milk powder products, *J. Food Sci.,* 53, 74, 1988.
3. **Nourooz-Zadeh, J. and Appelqvist, L.-Å.,** Cholesterol oxides in Swedish foods and food ingredients: fresh eggs and dehydrated egg products, *J. Food Sci.,* 52, 57, 1987.
4. **Sander, B. D., Addis, P. B., Park, S. W., and Smith, D. E.,** Quantification of cholesterol oxidation products in a variety of foods, *J. Food Protection,* 52, 109, 1989.
5. **Park, S. W. and Addis, P. B.,** HPLC determination of C-7 oxidized cholesterol derivatives in foods, *J. Food Sci.,* 50, 1437, 1985.
6. **Park, S. W. and Addis, P. B.,** Capillary column gas-liquid chromatographic resolution of oxidized cholesterol derivatives, *Anal. Biochem.,* 149, 275, 1985.
7. **Park, S. W. and Addis, P. B.,** Cholesterol oxidation products in some muscle foods, *J. Food Sci.,* 52, 1500, 1987.
8. **Sander, B. D., Smith, D. E., and Addis, P. B.,** Effects of processing stage and storage conditions on cholesterol oxidation products in butter and cheddar cheese, *J. Dairy Sci.,* 71, 3173, 1988.
9. **Jacobson, M. S.,** Cholesterol oxides in Indian ghee: possible cause of unexplained high risk of atherosclerosis in Indian immigrant populations, *Lancet,* 2 (8560), 656, 1987.
10. **Sander, B. D., Smith, D. E., Addis, P. B., and Park, S. W.,** Effects of prolonged and adverse storage conditions on levels of cholesterol oxidation products in dairy products, *J. Food Sci.,* 54, 874, 1989.
11. **Wade, V. N., Al-tahiri, R., and Crawford, R. J. M.,** The autoxidative stability of anhydrous milk fat with and without antioxidants, *Milchwissenschaft,* 41, 479, 1986.
12. **Luby, J. M., Gray, J. I., Harte, B. R., and Ryan, T. C.,** Photooxidation of cholesterol in butter, *J. Food Sci.,* 51, 904, 1986.
13. **Luby, J. M., Gray, J. I., and Harte, B. R.,** Effects of packaging and light source on the oxidative stability of cholesterol in butter, *J. Food Sci.,* 51, 908, 1986.
14. **Finocchiaro, E. T., Lee, K., and Richardson, T.,** Identification and quantification of cholesterol oxides in grated cheese and bleached butteroil, *J.A.O.C.S.,* 61, 877, 1984.
15. **Willemot, C., Fillion-Delorme, N., and Wood, D. F.,** Effect of nitrite on lipid degradation during storage of cooked pork meat, *Can. Inst. Food Sci. Technol. J.,* 20, 70, 1987.
16. **Higley, N. A., Taylor, S. L., Herian, A. M., and Lee, K.,** Cholesterol oxides in processed meats, *Meat Sci.,* 16, 175, 1986.

17. **Allen, C. E. and Foegeding, E. A.,** Some lipid characteristics and interactions in muscle foods — a review, *Food Technol.*, 35, 253, 1981.
18. **Sharma, N., Kowale, B. N., and Joshi, H. B.,** Effect of cooking and storage on phospholipids in broiler, *J. Food Sci. Technol.*, 19, 1, 1982.
19. **Sharma, N., Kowale, B. N., and Joshi, H. B.,** Effect of cooking and storage on neutral lipids in broiler and quail, *J. Food Sci. Technol.*, 19, 7, 1982.
20. **De Vore, V. R.,** TBA values and 7-ketocholesterol in refrigerated raw and cooked ground beef, *J. Food Sci.*, 53, 1508, 1988.
21. **Torres, E., Pearson, A. M., Gray, J. I., Ku, P. K., and Shimokomaki, M.,** Lipid oxidation in charqui (salted and dried beef), *Food Chem.*, 32, 257, 1989.
22. **Pie, J. E., Spahis, K., and Seillan, C.,** Cholesterol oxidation in meat products during cooking and frozen storage, *J. Agric. Food Chem.*, 39, 250, 1991.
23. **Tsai, L. S. and Hudson, C. A.,** Cholesterol oxides in commercial dry egg products: isolation and identification, *J. Food Sci.*, 49, 1245, 1984.
24. **Tsai, L. S. and Hudson, C. A.,** Cholesterol oxides in commercial dry egg products: quantification, *J. Food Sci.*, 50, 229, 1985.
25. **Van de Bovenkamp, P., Kosmeijer-Schull, T. C., and Katan, M. B.,** Quantification of oxysterols in Dutchfoods: egg products and mixed diets, *Lipids*, 23, 1079, 1988.
26. **Naber, E. C. and Biggert, M. D.,** Analysis for and generation of cholesterol oxidation products in egg yolk by heat treatment, *Poultry Sci.*, 64, 341, 1985.
27. **Missler, S. R., Waselchuk, B. A., and Merritt, C., Jr.,** Separation and identification of cholesterol oxidation products in dried egg preparation, *J. Food Sci.*, 50, 595, 1985.
28. **Herian, A. M. and Lee, K.,** 7α- and 7β-Hydroxycholesterols formed in a dry egg nog mix exposed to fluorescent light, *J. Food Sci.*, 50, 276, 1985.
29. **Sugino, K., Terao, J., Murakami, H., and Matshushita, S.,** High-performance liquid chromatographic method for the quantification of cholesterol epoxides in spray-dried egg, *J. Agric. Food Chem.*, 34, 36, 1986.
30. **Morgan, J. N. and Armstrong, D. J.,** Formation of cholesterol-5,6-epoxides during spray-drying of egg yolk, *J. Food Sci.*, 52, 1224, 1987.
31. **Emanuel, H. A., Hassel, C. A., Addis, P. B., Bergmann, S. D., and Zavoral, J. H.,** Plasma cholesterol oxidation products (oxysterols) in human subjects fed a meal rich in oxysterols, *J. Food Sci.*, 56, 843, 1991.
32. **Ryan, T. C., Gray, J. I., and Morton, I. D.,** Oxidation of cholesterol in heated tallow, *J. Sci. Food Agric.*, 32, 305, 1981.
33. **Park, S. W. and Addis, P. B.,** Identification and quantitative estimation of oxidized cholesterol derivatives in heated fats, *J. Agric. Food Chem.*, 34, 653, 1986.
34. **Park, S. W. and Addis, P. B.,** Further investigation of oxidized cholesterol derivatives in heated tallow, *J. Food Sci.*, 51, 1380, 1986.
35. **Yan, P. S. and White, P. S.,** Cholesterol oxidation products in heated lard enriched with two levels of cholesterol, *J. Am. Oil Chem. Soc.*, 67, 927, 1990.
36. **Lee, K., Herian, A. M., and Higley, N. A.,** Sterol oxidation products in French fries and in stored potato chips, *J. Food Prot.*, 48, 158, 1985.
37. **Finocchiaro, E. T. and Richardson, T.,** Sterol oxides in foodstuffs: a review, *J. Food Prot.*, 46, 917, 1983.
38. **Zhang, W. B., Addis, P. B., and Krick, T. P.,** Quantification of 5α-cholestane-3β,5,6β-triol and other cholesterol oxidation products in fast food French fries, *J. Food Sci.*, 56, 716, 1991.
39. **Bascoul, J., Domergue, N., Olle, M., and Crastes de Paulet, A.,** Autoxidation of cholesterol in tallows heated under deep frying conditions: evaluation of oxysterols by GLC and TLC-FID, *Lipids*, 21, 383,
40. **Zhang, W. B. and Addis, P. B.,** Prediction of levels of cholesterol oxides in heated tallow by dielectric measurement, *J. Food Sci.*, 55, 1673, 1990.
41. **Addis, P. B. and Hassel, C. A.,** Safety issues with antioxidants in foods, in *Symposium on Food Safety Evaluation*, Finley, J. W. and Armstrong, D., Eds., American Chemical Society, Washington, D.C.,

Chapter 5

CHOLESTEROL OXIDES IN PLASMA AND TISSUES

Robert J. Morin and Shi-Kaung Peng

TABLE OF CONTENTS

I. CHOLESTEROL OXIDES IN PLASMA

A. EXPERIMENTAL ANIMALS

Several cholesterol oxides were isolated from rabbit plasma by Watabe et al.[1] and then identified by gas chromatography-mass spectrometry (GC-MS). In control rabbits, cholesterol 5α,6α-epoxide concentration was 0.09 g/ml, cholesterol 5β,6β-epoxide was 0.22 g/ml and cholestane-3β,5α,6β-triol was 0.07 g/ml. After feeding of purified cholesterol (from which oxidation products had been removed), plasma levels had increased to 0.24 g/ml for the α-epoxide, 0.50 for the β-epoxide, and 0.32 for cholestanetriol. These results were interpreted as indicating endogenous oxidation of cholesterol. The similarity of the ratio of the α/β isomers of 2:1 in both the control and the cholesterol-fed rabbits suggested that these epoxides originated from lipid peroxidation-mediated epoxidation, since this ratio is observed after *in vitro* peroxidation of cholesterol.

Using high-performance liquid chromatography (HPLC) Kou and Holmes[2] were unable to detect 25-hydroxycholesterol at levels above the detection limit of 10 ng/ml. After feeding of 10 mg of 25-hydroxycholesterol to the rats, however, plasma levels were in the range of 630 ng/ml. Of this amount, 60% was esterified to fatty acids.

Another cholesterol oxide, 7-ketocholesterol, has been identified by mass spectrometry in plasma of domestic chickens.[3] This compound has also been identified in the adrenal venous blood of the baboon *(Papio hemadryas)* and in the adrenal and peripheral blood of the rhesus monkey *(Macacus rhesus)*.[4] 7β-Hydroxycholesterol has been isolated from pregnant mare serum.[5]

B. HUMAN PLASMA

Cholesterol oxides have been isolated from human serum by thin-layer chromatography (TLC) and identified and quantitated by GC-MS by Gray et al.[6] Normal serum was found to contain 25 to 50 μg/dl of cholesterol α-epoxide, whereas sera from patients with type II hypercholesterolemia contained 250 to 4,550 μg/dl of this oxide. Cholesta-4-en-3-one and cholesta-4,6-diene-3-one were present in the range of 10 to 30 μg/dl, but 7-hydroxycholesterol and 7-ketocholesterol were below detectable limits. In a subsequent study, Kuksis et al. reported a concentration of 100 μg/dl of cholesterol α-epoxide in normal human serum.[7]

More recently, using the highly accurate method of isotope dilution-mass spectrometry, Bjorkhem et al.[8] found that the total α- and β-cholesterol epoxide concentration in normal human sera was in the range of 6.7 to 29.3 μg/dl (Table 1). Contrary to the results reported by Gray et al.,[6] the level of cholesterol epoxides in hyperlipemic sera was even lower than normal. Since Bjorkhem et al.[8] added an antioxidant during the extraction procedure, it is possible that the higher levels reported by Gray et al. were attributable to artifactual *in vitro* autoxidation, which might occur more extensively in hypercholesterolemia due to the increased level of substrate. Although there was no correlation with serum cholesterol levels or sex in the Bjorkhem et al. study, they did find increasing epoxide levels with increasing age. Cholestanetriol was found to be below the detection limit of 1.0 μg/dl. Kudo et al.,[9] using special precautions to suppress autoxidation during sample processing, found no detectable levels of either epoxide or cholestanetriol in normal human plasma after correction for autoxidation.

The levels of cholesterol derivatives oxygenated at the 7 position have also been analyzed in human sera. Early studies by Sobel et al. identified 7-hydroxycholesterol in sera.[10] After removal of cholesterol by HPLC, Koopman et al.[11] analyzed serum levels of 7α- and 7β-hydroxycholesterol by gas chromatography. The range of 7α-hydroxycholesterol in normal subjects was 1.98 to 16.6 μg/dl, and for 7β-hydroxycholesterol was 1.17 to 26.6 μg/dl. Using the more accurate method of isotope dilution-mass spectrometry, Bjorkhem et al.[12]

TABLE 1
Concentration of Cholesterol-5,6-Epoxide in Plasmas of Different Subjects[a]

Subjects	Age	Number	Concentration of cholesterol-5,6-epoxide μg/dl
			μg/dl
Healthy medical students	22—35	8	13.1 ± 2.9
Healthy subjects and patients with minor nonmedical diseases	20—40	5	10.8 ± 1.6
	41—60	4	17.2 ± 3.1
	61—80	10	19.3 ± 1.8
Patients with familial hypercholesterolemia (phenotype IIA)	20—66	12	6.9 ± 1.4
Patients with familial combined hyperlipidemia (phenotype IIB)	41—56	4	10.5 ± 5.8
Patients with familial combined hyperlipidemia (phenotype IV)	46—48	3	9.3 ± 6.4

[a] Mean + SEM.

Reprinted from Bjorkheim, I., Breuer, O., Angelin, B., and Wikstrom, S.-A., *J. Lipid Res.*, 29, 1031, 1988. With permission.

reported a mean ± S.E.M. of 3.0 ± 0.4 μg/dl concentration of 7α-hydroxycholesterol in 6 patients with gallstones. The ratio of 7α- to 7β-hydroxycholesterol ranged from 2 to 4. No significant amounts of either were present in the esterified form. 7β-Hydroxycholesterol is believed by Bjorkhem to originate from autoxidation of cholesterol, whereas 7α-hydroxycholesterol is derived from enzymatic oxidation of cholesterol during the biosynthesis of bile acids in the liver. After correction for autoxidation during sample processing, Kudo et al.[9] found 0.4 to 1.7 μg/dl of 7α-hydroxycholesterol in normal human plasma, but undetectable levels of 7β-hydroxycholesterol. Serum levels of 7α-hydroxycholesterol correlate well with cholesterol 7α-hydroxylase activity in liver biopsy samples.[12] In patients with terminal ileal resection of 20 to 100 cm, both liver cholesterol 7α-hydroxylase activity and serum levels of 7α-hydroxycholesterol are increased. Cholestyramine therapy, which increased bile acid synthesis, was found by Bjorkhem et al.[12] to increase serum 7α-hydroxycholesterol levels to 12.8 ± 2.0 g/dl and by Van Doormaal et al.[13] to increase to 49.5 μg/dl (25.0 to 117.6). In the latter study, 7β-hydroxycholesterol was also increased to 23.6 g/dl (0 to 76.4). These authors indicate that although the source of 7β-hydroxycholesterol is unknown, the negative correlation between its serum concentration and body mass index suggests a mode of regulated formation. Serum levels of 7α-hydroxycholesterol are also increased in patients with cerebrotendinous xanthomatosis and the cerebrohepatorenal syndrome, who have a defect in the enzymatic degradation of the cholesterol side chain.

Another 7-oxygenated derivative, 7-ketocholesterol, was found by isotope dilution-mass spectrometry to be less than 7.0 μg/dl in normal human sera.[14] Sera from some subjects contained less than the detectable limit of 1.5 μg/dl. Patients with cerebrotendinous xanthomatosis were also found to have higher-than-normal levels of 7-ketocholesterol. A similar sterol, cholesta-3,5-dien-7-one has been identified in hyperlipemic sera.[15]

If autoxidation of cholesterol occurs, more 25-hydroxycholesterol than 26-hydroxycholesterol is generated. 26-Hydroxycholesterol is believed to be produced primarily by mitochondrial enzymatic oxidation of the cholesterol side chain. Using isotope dilution-mass spectrometry, Javitt et al.[16] found 9.2 to 25.6 μg/dl of 26-hydroxycholesterol in normal human sera. Of this amount, 31 to 35% was present as free sterol and the remainder as fatty acid esters (similar to the ratio of free/esterified cholesterol). After correction for autoxi-

TABLE 2
Percentage Plasma Distribution of ^{14}C-Labeled Cholesterol Oxides and Cholesterol 24 h After Oral Administration to Rabbits

	VLDL	LDL	HDL
Cholestane-3β,5α,6β-triol	53.7	20.7	25.5
7α,β-Hydroxycholesterol and 7-ketocholesterol	59.0	22.9	18.1
25-Hydroxycholesterol	33.6	53.4	13.0
Cholesterol	32.1	33.1	34.8

Note: Means of experiments with 6 rabbits/group. Abbreviations: VLDL, very low density lipoproteins; LDL, low density lipoproteins; HDL, high density lipoproteins.

Reprinted from Peng, S.-K., Phillips, G. A., Xia, G.-Z., and Morin, R. J., *Atherosclerosis,* 64, 1, 1987. With permission.

dation, Kudo et al.[9] found a range of 6.4 to 10.1 μg/dl in normal human plasma. Patients with cerebrotendinous xanthomatosis were found by Javitt et al.[16] to have serum levels in the range of 0 to 0.6 μg/dl, due to their reduced C27 steroid 26-hydroxylase activity.[17] In a study by Kou and Holmes,[18] older patients undergoing coronary bypass surgery had plasma levels of 26-hydroxycholesterol in the range of 33.5 to 95.0 μg/dl, significantly higher than the 14.0 to 26.0 μg/dl in their normal young subjects. This suggests either an age- or atherosclerosis-related increase in 26-hydroxycholesterol.

The human plasma levels of a number of fatty acyl esters of dihydroxy cholesterol derivatives have been analyzed by Smith et al.[19] The esters of the following sterols were identified and quantitated: 7α-hydroxycholesterol, 1.9 μg/dl; 7β-hydroxycholesterol, 1.1 μg/dl; 7-ketocholesterol, 0.6 μg/dl; 24-hydroxycholesterol, 0.7 μg/dl; 25-hydroxycholesterol, 0.5 μg/dl; and 26-hydroxycholesterol, 10.6 μg/dl. Since esterification does not occur during the extraction process, the parent oxysterols were unlikely to arise by artifactual autoxidation, and most likely originated *in vivo* from enzymatic oxidation or lipid peroxidation. In a study by Claude[20] no qualitative differences were noted in cholesterol oxide esters between normal serum and serum from xanthomatous subjects.

C. DISTRIBUTION OF CHOLESTEROL OXIDATION PRODUCTS IN PLASMA LIPOPROTEINS

Peng et al.[21] orally administered a single dose of [^{3}H] 25-hydroxycholesterol to squirrel monkeys and studied its distribution in plasma lipoproteins 24 hours later. The very-low-density lipoproteins (VLDL) were found to contain 34.7% of the labeled 25-hydroxycholesterol, low-density lipoproteins (LDL) 55.1%, and high-density lipoproteins (HDL) 10.2%. This distribution differed significantly from labeled cholesterol percentages of 3.4% in VLDL, 44.9% in LDL, and 51.7% in HDL. In a subsequent similar study, rabbits were fed a mixture of ^{14}C-labeled cholesterol oxidation products produced by autoxidizing ^{14}C-cholesterol in air at 60°C for 5 weeks.[22] As seen in Table 2, after 24 hours cholestane-3β,5α,6β-triol, 7α- and 7β-hydroxycholesterol, and 7-ketocholesterol were more selectively transported in VLDL, whereas most of the 25-hydroxycholesterol was present in LDL. The HDL contained relatively smaller amounts of the cholesterol oxides. The lipoprotein distribution of all the labeled cholesterol oxides was different from that of labeled cholesterol.

Brooks et al.[23,24] analyzed the content of cholesterol oxides in human type II hypercholesterolemic serum. As shown in Table 3, cholesterol α- and β-epoxides were found to be transported mostly in VLDL, 7-ketocholesterol equally in VLDL and LDL, and 26-hydrox-

TABLE 3
Percentage Plasma Distribution of Cholesterol Oxides in Human Type II Hypercholesterolemia

	VLDL	LDL	HDL
Cholesterol-β-epoxide	61.9	21.6	16.5
Cholesterol-α-epoxide	56.1	22.4	21.4
7-Ketocholesterol	42.9	42.9	31.3
26-Hydroxycholesterol	23.0	27.7	49.2

Reprinted from Brooks, C. J. W., McKenna, R. M., Colle, W. J., MacJacklan, J., and Vertch Lawrie, T. D., *Biochem. Soc. Trans.*, 11, 700, 1983. With permission.

TABLE 4
Cholesterol Oxides in Plasma Lipoprotein Fractions From Normal Fasting Humans[a]

	VLDL	LDL	HDL
Cholesterol-β-epoxide	17 (0—49)	257 (0—640)	28 (0—40)
Cholesterol-α-epoxide	71 (0—300)	70 (0—185)	113 (0—530)
7β-Hydroxycholesterol	0	16 (0—50)	4 (0—15)
Cholestanetriol	0	7 (0—20)	0
7-Ketocholesterol	9 (0—20)	63 (0—215)	19 (0—55)
25-Hydroxycholesterol	0	3 (0—20)	0

[a] Means of 6 subjects (ranges), μg/dl plasma.

Reprinted with permission from Addis, P. B., Emanuel, H. A., Bergmann, S. D., and Zavoral, J. H., *Free Radical Biol. Med.*, 7, 179, 1989. Copyright 1989, Pergamon Press PLC.

ycholesterol mostly in HDL. VLDL was found to contain 23% of the 26-hydroxycholesterol, which is in contrast to analyses by Javitt,[16] who could find none of this compound in normal human serum VLDL, and reported 61% in LDL and 39% in HDL. More recently, Addis et al.[25] have analyzed the cholesterol oxide distribution in lipoprotein fractions from normal fasted human plasmas. These results are reproduced in Table 4. In contrast to the study by Brooks et al.[23,24] The lower temperature saponification employed by the Addis group may have prevented degradation of these sterols.

II. CHOLESTEROL OXIDES IN TISSUES

A. ARTERIES

Early histochemical and chromatographic studies demonstrated the presence of cholesta-3,5-diene-7-one in human atherosclerotic aorta.[26,27] Although this was considered to be a possible artifact, its absence in brain after identical extraction conditions suggested otherwise. Other cholesterol oxides identified in human atherosclerotic aortas are listed in Table 4. The aortic content of 26-hydroxycholesterol has been studied in somewhat more detail. Greatly increased amounts were found in fatty streaks and atheromas as compared to normal aortic tissue.[24] Smith and Van Lier[34] noted that the intimal content of 26-hydroxycholesterol increased in parallel with increasing cholesterol and lipid content (Figure 1). Since 26-hydroxycholesterol is a product of enzymatic oxidation, it was unlikely that the levels observed were due to artifactual autoxidation occurring during extraction.

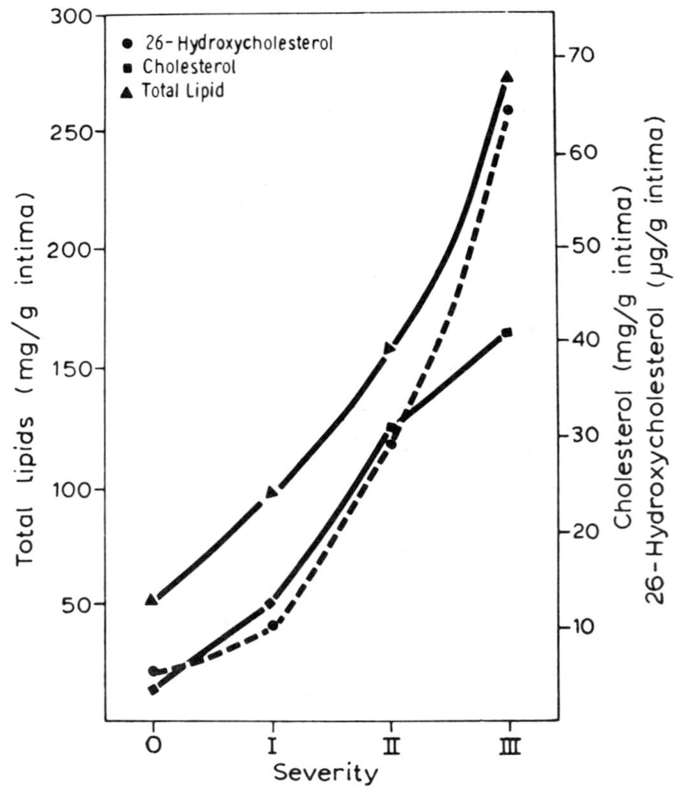

FIGURE 1. Relationship between absolute levels of total lipid, cholesterol, and 26-hydroxycholesterol and increasing severity of atherosclerosis in 36 intimal samples. (From Smith, L. L. and Van Lier, J. E., *Atherosclerosis*, 12, 1, 1970. With permission.)

It seems that 26-hydroxycholesterol may be unique to human aortic tissue, since it has not been found in other human tissues. It was also below detectable limits in aortas from baboons or White Carneau pigeons and was also not present in aortas from rabbits with atherosclerosis induced by cholesterol feeding or other means.[35] Biosynthesis of 26-hydroxycholesterol in human aortic intima has been demonstrated by Chobanian.[36]

More recently Smith et al.[39] performed detailed analysis of the aortic fatty acid ester content of several hydroxylated cholesterol derivatives. Total diesters of 25-hydroxycholesterol and 26-hydroxycholesterol were reported to be 28 µg/g, mostly in the form of palmitate and oleate esters. Total aortic diol esters had previously been reported by Aringer et al.[40] to be in the range of 120 $\pm$ 6 µg/g. Monoesters of the 7-hydroxycholesterol epimers were also identified. These sterol esters of cholesterol oxides are most likely formed *in vivo* and are unlikely to represent artifactual autoxidation products.

B. LIVER

Several cholesterol oxides are produced in the liver in the course of enzymatic oxidation of cholesterol for the production of bile acids. The enzyme cholesterol 7α-hydroxylase catalyzes production of 7α-hydroxycholesterol, which has been found in rat liver microsomes,[41-44] as well as human,[45-46] porcine,[47] and bovine livers[48-50] (Table 5). The epimer 7β-hydroxycholesterol and 7-ketocholesterol are also present in similar concentrations in liver. In human subjects with cerebrotendinous xanthomatosis,[45] who have a defect in bile and synthesis, 7α-hydroxycholesterol concentration as measured by isotope dilution-mass

TABLE 5
Cholesterol Oxides in Tissues

Tissues	Compound	Ref.
Aorta		
Rat	7-Ketocholesterol	35
Cat	7-Ketocholesterol	35
Bovine	7-Ketocholesterol	35
Horse	7-Ketocholesterol	35
Baboon	7-Ketocholesterol	35
Human	Cholest-3,5-diene-7-one	26,32
	Cholestane-3β,5α,6β-triol	27-32
	7α-Hydroxycholesterol	28,29
	7β-Hydroxycholesterol	28,29
	7-Ketocholesterol	28,40
	24-Hydroxycholesterol	28,29
	25-Hydroxycholesterol	28
	Dipalmitate ester (32 μg/g)	39
	Dioleate ester (24 μg/g)	39
	26 Hydroxycholesterol (13—494 μg/g)	37,38
	Dipalmitate ester (100 μg/g)	39
	Distearate ester (3 μg/g)	39
	Dioleate ester (94 μg/g)	39
Liver		
Rat	7α-Hydroxycholesterol (0.7—1.2 μg/g)	41-44,56
	7β-Hydroxycholesterol (0.15—0.6 μg/g)	41-44,56
	7-Ketocholesterol (1.2 μg/g)	44
	Cholest-3,5-diene-7-one	58
	Cholesterol-3β,5α,6β-triol	52,53
	Cholesterol-5α,6α-epoxide	51
	Cholesterol-5β,6β-epoxide	51
Porcine	7α-Hydroxycholesterol	47
Bovine	7β-Hydroxycholesterol	48-50
	Cholestane-3β,5α,6β-triol	49,50
Horse	Cholest-3,5-diene-7-one	50
Human	7α-Hydroxycholesterol (117—198 ng/mg protein)	45,55
	7-Ketocholesterol	57
	Cholesterol-5α,6α-epoxide	55
	Cholesterol-5β,6β-epoxide	55
	Cholesterol-24S,25-epoxide	59
	25-Hydroxycholesterol	57
	Cholestane-3β,5α,6β epoxide	55
	Cholest-4-ene-4-3-one	46
	7α,12α-Dihydroxy-4-cholestane-3 one	46
Adrenal		
Rat	7-Ketocholesterol	61
	Cholest-4-ene-3-one (32—165 μg/g)	60
Porcine	7-Ketocholesterol	67
	Cholest-3,5-diene-7-one	67
	7α-Hydroxycholesterol	62,63
	7β-Hydroxycholesterol	62,63
	20-Hydroxycholesterol	62
	20,22-Dihydroxycholesterol	62
	25-Hydroxycholesterol	66
	Cholesterol-5α,6α-epoxide	63,64
	Cholesterol-5β,6β-epoxide	63
	Cholestane-3β,5α,6β-triol	65
Brain		
Rat	Cholesterol-5α,6α-epoxide	68
	Cholestane-3β,5α,6β-triol	68

TABLE 5 (continued)
Cholesterol Oxides in Tissues

Tissues	Compound	Ref.
Human	7α-Hydroxycholesterol	71
	7β-Hydroxycholesterol	71
	7-Ketocholesterol	72
	24-Hydroxycholesterol	71
	Cholestane-3β,5α,6β-triol	69
Skin		
Mouse	Cholesterol-5α,6α-epoxide (0.3—1.0 μg/g)	73,74
	7α-Hydroxycholesterol	75
	7β-Hydroxycholesterol	75
	7-Ketocholesterol	75
	Cholesterol-5α,6α-epoxide	64
	Cholestane-3β,5α,6β-epoxide	75
Breast fluid		
Human	Cholesterol-5α,6α-epoxide	77-79
	Cholesterol-5β,6β-epoxide	77-79
Prostate		
Human	Cholesterol-5α,6α-epoxide (120 μg/g)	80
	Cholesterol-5β,6β-epoxide (10 μg/g)	80
Lung		
Chinese hamster	25-Hydroxycholesterol (1.5/g/cell)	81
Fibroblasts	24(S),25-Epoxycholesterol (7.2 fg/cell)	81
Erythrocytes		
Human	7-Hydroxycholesterols	82
	7-Ketocholesterol	82
	Cholesta-3,5-diene-7-one	82

spectrometry was 255 ng/mg protein, considerably higher than the normal range of 117 to 198. Liver microsomes from subjects with cerebrotendinous xanthomatosis produce 3 times more 4-cholestene-3-one than normal,[46] and the liver level of 7α,12α-dihydroxy-4-cholestene-3-one is 170 ng/mg protein as compared to the normal range of 3 to 27.

Rat liver microsomes have been shown to produce cholesterol 5,6-epoxide and cholesterol 5β,6β-epoxide.[51] The β-epoxide is formed 3 to 4 times faster than the α epimer. Rat liver also contains a cholesterol epoxide hydrolase which converts the α- and β-epoxides to cholestane-3β,5α,6β-triol.[52,53] Epoxidation of cholesterol is mediated by microsomal lipid peroxidation in the presence of a ferrous ion-ADP complex and NADPH.[54] Fatty acyl peroxy radicals generated from phospholipid hydroperoxides may act as direct oxidants in this system. Both the α- and β-epoxides and cholestanetriol have been identified in human liver, and the epoxide content is increased in subjects with Wolman's disease.[55]

Esters of the 7-diol cholesterol derivatives have been identified in rat liver,[56] and in normal human livers palmitate esters of 7-ketocholesterol (70 ng/g) and 25-hydroxycholesterol (100 ng/g) were present.[57] Livers of patients with Wolman's disease contain esters of 7α-hydroxycholesterol (70 to 80 μg/g), 7β-hydroxycholesterol (30 to 50 μg/g), 7-ketocholesterol (20 to 30 μg/g) cholesterol α-epoxide (10 μg/g), and cholesterol β-epoxide (20 μg/g).[55] These were deemed unlikely to be post-mortem autoxidation artifacts because esters of cholesterol oxides were not detected in comparable control post-mortem specimens of liver, adrenals, or brain or from tissues of patients with Tangier disease. There was a close similarity in the fatty acyl composition of the oxide steryl esters and cholesterol esters from the Wolman's livers, and unesterified cholesterol oxides were not detected. The authors suggest a possible enzymatic origin of these steryl esters via the NADPH-dependent lipid peroxidation mechanism.

C. ADRENAL

The adrenal cortex from several animal species has been demonstrated to synthesize and/or contain a number of cholesterol oxides (Table 5). Cholest-4-ene-3-one is produced in rat adrenals by oxidation of cholesterol catalyzed by 3-hydroxysteroid dehydrogenase.[60] The rat adrenal contains 32 to 165 μg/g of cholestenone. 7-Ketocholesterol has been identified in rat adrenal and was proven not to originate from artifactual autoxidation. Synthesis of 25-hydroxycholesterol has also been demonstrated in rat adrenal. Earlier studies by Cargill and Cook[62] demonstrated the presence of 7α- and 7β-hydroxycholesterol, 20-hydroxycholesterol, and 20,22-dihydroxycholesterol in bovine adrenal cortex. The 7α- and β-hydroxycholesterol epimers were confirmed by Bosisio et al.,[63] who also found that 7-ketocholesterol and cholesterol 5α,6α-epoxide and cholesterol 5β,6β-epoxide were produced in bovine adrenal cortex after incubation with a purified ^{14}C-labeled 26-hydroxycholesterol substrate. The β-epoxide was produced in greater amounts than the α isomer. Incubations were performed in the dark under nitrogen to minimize production of autoxidation products during the procedure. The absence of the common autoxidation products 20-hydroxycholesterol and 25-hydroxycholesterol suggested an enzymatic origin for the others. The enzymatic conversion of cholesterol to cholesterol 5α,6α-epoxide was demonstrated in bovine adrenal microsomes by Watabe and Sawahata.[64]

The reaction required NADPH, and could be inhibited by P-450 inhibitors, suggesting that the α-epoxide was formed by a monoxygenase-mediated reaction. A 5-min incubation period was used to minimize microsomal lipid peroxidation in the absence of ethylenediaminetetraacetate (EDTA). Under these conditions, no detectable amounts of the α-epoxide were produced. Using a 30-min incubation without EDTA yielded significant amounts of the β-epoxide. This could be blocked by the addition of EDTA, suggesting that β-epoxidation of the cholesterol double bond is a reaction linked to metal-catalyzed peroxidation of microsomal lipids. Watabe and Sawahata[65] also demonstrated the production of cholestane-3β,5α,6β-triol from cholesterol α-epoxide in bovine adrenal cortex catalyzed by a microsomal cholesterol epoxide hydrolase.

D. BRAIN

Rat brain microsomes have been demonstrated to produce cholesterol α-epoxide from a ^{14}C-labeled cholesterol palmitate substrate.[68] The epoxide was subsequently transformed to cholestanetriol. Although the oxidation of cholesterol required prior hydrolysis of the cholesterol ester, no cholesterol oxides were produced when free ^{14}C-labeled cholesterol was added, suggesting that only the cholesterol esters were transported to the sites of cholesterol oxidation. The oxidation could be inhibited by addition of EDTA, suggesting that the process involved an NADPH-dependent enzymatic lipid peroxidation mechanism. Cholestanetriol has also been identified in human brain tissue.[69] Several cholesterol oxides, such as 7α-hydroxycholesterol, 7β-hydroxycholesterol, 7-ketocholesterol, and 24-hydroxycholesterol have previously been identified in rabbit,[70] bovine,[70] and human brain,[71,72] although their origin from possible procedural autoxidation was not ruled out.

E. SKIN

Cholesterol α-epoxide has been identified in the skin of hairless mice in the range of 0.3 to 1.0 mg/g skin.[73,74] After irradiation with UV light, the skin levels of cholesterol α-epoxide increased to up to 8 times the control level. When human skin specimens were labeled with ^{14}C-cholesterol and irradiated with a UV lamp, a number of cholesterol oxides were produced, including cholesterol α-oxide, 7α- and 7β-hydroxycholesterol, 7-ketocholesterol, and cholestanetriol.[75] In human skin fibroblasts in culture, cholesterol α-oxide has been noted to induce chromosome damage.[76]

F. BREAST

Analysis of aspirated breast fluid from normal nonpregnant women revealed that 45% had detectable amounts of cholesterol 5α,6α-epoxide and cholesterol 5β,6β-epoxide, with a 3:1 ratio of the β- to the α-epimer.[77,78] Total cholesterol epoxides increased with age, from a mean of 187 mg/dl in the 20 to 29 age category, 1957 mg/dl in ages 30 to 39, and 3554 mg/dl in ages 40 to 49. High levels of cholesterol in breast fluid were correlated with higher levels of cholesterol epoxides. Levels of both epoxides were higher in white than in nonwhite women, and higher in current or past smokers. It was also found that cholesterol β-epoxide was also lower in women who were lactating, parous, or who had breast fed in the past. The lower levels in parous women persisted at least two years postpartum or postlactation. The presence of detectable amounts of cholesterol β-epoxide in breast fluid of women with proliferative benign breast disease was 8.5 times more frequent than in those without this disorder.[79]

G. PROSTATE

Cholesterol α- and β-epoxides were detected in one third of prostatic fluid specimens from 25 control men aged 23 to 75.[80] Mean total epoxide content of the fluids was 22.6 $\pm$ 3.3 μg/ml. Prior to surgery for benign prostatic hypertrophy or prostate cancer, 40 patients had prostatic fluid total cholesterol epoxide concentrations of 33.7 $\pm$ 7.9 μg/ml. There were no differences between those with benign prostatic hypertrophy only and those with cancer. Analysis of tissue extracts from prostatic lobes of these surgical patients showed cholesterol α-epoxide to be 120 $\pm$ 23.4 μg/g dry tissue and cholesterol β-epoxide to be 9.9 $\pm$ 5.2 μg/g.

Cholesterol α- and β-epoxides were detected in prostatic fluid from one third of 25 control men aged 23 to 75.[80] Mean total epoxide content of the fluids was 22.6 $\pm$ 3.3 μg/ml. Prior to surgery for benign prostatic hypertrophy or prostate cancer, 40 patients had prostatic fluid total cholesterol epoxide concentrations of 33.7 $\pm$ 7.9 μg/ml. There were no differences between those with benign prostatic hypertrophy only and those with cancer.

III. SUMMARY

A number of different cholesterol oxides have been identified in significant concentrations in plasma, cell secretions, and solid tissues. Although cholesterol α- and β-epoxides, cholestanetriol, and 7β-hydroxycholesterol have been identified in human serum using precise analytical techniques, the actual endogenous amounts of these compounds vs. possible artifactual increases during sample processing is still controversial. The identification of fatty acyl esters of the dihydroxy cholesterol oxides is evidence against their origin by artifactual autoxidation. There is somewhat more unanimity on the plasma levels of 7α-hydroxycholesterol and 26-hydroxycholesterol. The distribution of cholesterol oxides between the various lipoprotein fractions is also unresolved, with some reports indicating transport of the cholesterol epoxides and 26-hydroxycholesterol in human plasma primarily in VLDL, and others showing most of these oxides in LDL. A number of cholesterol oxides, as well as fatty acid acyl esters of 25- and 26-hydroxycholesterol and 7-hydroxycholesterols have been identified in human atherosclerotic aortas. Liver produces 7α-cholesterol and possibly 7β- and 7-ketocholesterol, as well as both cholesterol epoxide epimers. Esters of the diol derivatives have been found in normal human liver and in Wolman's disease livers. Adrenal cortex and brain have also been demonstrated to produce and contain most of the common cholesterol oxides. The presence of the cholesterol epoxide epimers in skin, breast fluid, and prostatic fluid may have carcinogenic significance.

REFERENCES

1. **Watabe, T., Isobe, M., and Kanai, M.,** Cholesterol diet increases plasma and liver concentrations of cholesterol epoxides and cholestanetriol, *J. Pharm. Dyn.*, 3, 553, 1980.
2. **Kou, I.-L. and Holmes, R. P.,** The analysis of 25-hydroxycholesterol in plasma and cholesterol-containing foods by high-performance liquid chromatography, *J. Chromatogr.*, 330, 339, 1985.
3. **Furr, B. J. A.,** Identification of cholesterol, 7-oxocholesterol, pregnenolone, progesterone, 20-hydroxy-pregn-4-en-3-one epimers and 5-androstane-3,17-dione in plasma and ovarian tissue of the domestic fowl, *Steroids*, 16, 471, 1978.
4. **Gontscharow, N. P., Wehrberger, K., and Schubert, K.,** Steroidstoffwechsel bei primaten X. Isolierung von 3 -hydroxy-5-cholesten-7-one aus nebennierenvenenblut und peripherem blut von macacus rhesus und papio hamadryas, *J. Steroid Biochem.*, 2, 389, 1971.
5. **Wintersteiner, O. and Ritzmann, J.,** The isolation of 7(β)-hydroxycholesterol from the serum of pregnant mares, *J. Biol. Chem.*, 136, 697, 1940.
6. **Gray, M. F., Lawrie, T. D. V., and Brooks, C. J. W.,** Isolation and identification of cholesterol α-oxide and other minor sterols in human serum, *Lipids*, 6, 836, 1971.
7. **Kuksis, A., Marai, L., Myhrer, J. J., and Geher, K.,** Identification of plant sterols in plasma and red blood cells of man and experimental animals, *Lipids*, 11, 581, 1976.
8. **Bjorkhem, I., Breuer, O., Angelin, B., and Wikstrom, S.-A.,** Assay of unesterified cholesterol-5,6-epoxide in human serum by isotope dilution mass spectrometry. Levels in the health state and in hyperlipoproteinemia, *J. Lipid Res.*, 29, 1031, 1988.
9. **Kudo, K., Emmons, G. T., Casserly, E. W., Via, D. P., Smith, L. C., St. Pyrek, J., and Schroepfer, G. J.,** Inhibitors of sterol synthesis. Chromatography of acetate derivatives of oxygenated sterols, *J. Lipid Res.*, 30, 1097, 1989.
10. **Sobel, A. E., Goldberg, M., and Slater, S. R.,** Estimation of 7-dehydrocholesterol, 7-hydroxycholesterol, and bile acids in serum, *Anal. Chem.*, 25, 629, 1953.
11. **Koopman, B. J., Van Der Molen, J. C., and Wolthers, B. G.,** Determination of some hydroxycholesterols in human serum samples, *J. Chromatogr.*, 416, 1, 1987.
12. **Bjorkhem, I., Reihern, E., Angelin, B., Ewerth, S., Akerlund, J.-E., and Einarsson, K.,** On the possible use of the serum level of 7α-hydroxycholesterol as a marker for increased activity of the cholesterol 7α-hydroxylase in humans, *J. Lipid Res.*, 28, 889, 1987.
13. **van Doormaal, J. J., Smit, N., Koopman, B. J., van der Molen, J. C., Wolthers, B. G., and Doorenbos, H.,** Hydroxycholesterols in serum from hypercholesterolaemic patients with and without bile acid sequestrant therapy, *Clin. Chim. Acta*, 181, 273, 1989.
14. **Bjorkhem, I.,** Assay of unesterified 7-oxocholesterol in human serum by isotope dilution-mass spectrometry, *Anal. Biochem.*, 154, 497, 1986.
15. **Gray, M. F., Morrison, A., Farish, E., Laurie, T. D. V., and Brooks, C. J. W.,** Characterization of 5α-lanost-8(9)-en-3β-ol in hyperlipaemic serum, *Biochim. Biophys. Acta*, 187, 163, 1969.
16. **Javitt, N. B., Kok, E., Burstein, S., Cohen, B., and Kutcher, J.,** 26-Hydroxycholesterol, *J. Biol. Chem.*, 256, 12644, 1981.
17. **Javitt, N. B., Kok, E., Cohen, B., and Burstein, S.,** Cerebrotendinous xanthomatosis: reduced serum 26-hydroxycholesterol, *J. Lipid Res.*, 23, 627, 1982.
18. **Kou, I.-L. and Holmes, R. P.,** The analysis of 26-hydroxycholesterol in plasma by high performance liquid chromatography, *Fed. Proc. Fed. Am. Soc. Exp. Biol.*, 45, 311, 1986.
19. **Smith, L. L., Teng, J. I., Lin, Y. Y., Seitz, P. K., and McGehee, M. F.,** Lipid peroxidations of cholesterol, in *Lipid Peroxides in Biology and Medicine*, Yagi, K., Ed., Academic Press, New York, 1982, 89.
20. **Claude, J. R.,** L-analyse des sterols des millieux biologiques par chromatographie en couche mince et apr chromatographie en phase gazeuse, *Clin. Chim. Acta*, 17, 371, 1967.
21. **Peng, S.-K., Taylor, B., Mosbach, E. H., Huang, W. Y., Hill, J., and Mikkelson, B.,** Distribution of 25-hydroxycholesterol in plasma lipoproteins and its role in atherogenesis, *Atherosclerosis*, 41, 395, 1982.
22. **Peng, S.-K., Phillips, G. A., Xia, G.-Z., and Morin, R. J.,** Transport of cholesterol autoxidation products in rabbit lipoproteins, *Atherosclerosis*, 64, 1, 1987.
23. **Brooks, C. J. W., McKenna, R. M., Colle, W. J., MacJacklan, J., and Vertch Lawrie, T. D.,** Profile analysis of oxygenated sterols in plasma and serum, *Biochem. Soc. Trans.*, 11, 700, 1983.
24. **Brooks, C. J. W. and Cole, J. W.,** Sterols in serum and in serum lipoprotein fractions, *J. Am. Oil Chem. Soc.*, 62, 622, 1985.
25. **Addis, P. B., Emanuel, H. A., Bergmann, S. D., and Zavoral, J. H.,** Capillary GC quantification of cholesterol oxidation products in plasma lipoproteins of fasted humans, *Free Radical Biol. Med.*, 7, 179, 1989.

26. **Kantiengar, N. L. and Morton, R. A.,** Cholesta-3:5-dien-7-one in human atherosclerotic aortas, *Biochem. J.,* 60, 25, 1955.
27. **Hardegger, v.E., Ruzicka, L., and Tagmann, E.,** Untersuchungen uber organextrakte, *Helv. Chim. Acta,* 26, 2205, 1943.
28. **Henderson, A. E.,** A histochemical and chromatographic study of normal and atheromatous human arteries, *J. Histochem. Cytochem.,* 4, 153, 1956.
29. **Henderson, A. E. and MacDougall, J. D. B.,** A histochemical and chromatographic study of the lipid distribution in human arteries, *Proc. Biochem. Soc.,* 57, xxi, 1954.
30. **McArthur, C. S.,** The acetone-soluble lipid of the atheromatous aorta, *Biochem. J.,* 36, 559, 1942.
31. **Robertson, D. M. J.,** The use of polarography in the examination of ketosteroids related to cholesterol, *Biochem. J.,* 61, 681, 1955.
32. **van Lier, J. E. and Smith, L. L.,** Sterol metabolism. I. 26-Hydroxycholesterol in human aorta, *Biochemistry,* 6, 3269, 1967.
33. **Fumagalli, R., Galli, G., and Urna, G.,** Cholestanol and 26-hydroxycholesterol in normal and atherosclerotic human aorta, *Life Sci.,* 10, 25, 1971.
34. **Smith, L. and Van Lier, J. E.,** 26-Hydroxycholesterol levels in the human aorta, *Atherosclerosis,* 12, 1, 1970.
35. **Smith, L. L. and Pandya, M. L.,** Cholesterol metabolism, *Atherosclerosis,* 17, 21, 1973.
36. **Chobanian, A. V.,** Sterol synthesis in the human arterial intima, *J. Clin. Invest.,* 47, 595, 1968.
37. **Brooks, C. J. W., Harlan, W. A., and Steel, G.,** Squalene, 26-hydroxycholesterol and 7-ketocholesterol in human atheromatous plaques, *Biochim. Biophys. Acta,* 125, 620, 1966.
38. **van Lier, J. E. and Smith, L. L.,** Sterol metabolism. I. 26-Hydroxycholesterol in the human aorta, *Biochemistry,* 10, 3269, 1967.
39. **Smith, L. L., Teng, J. I., Lin, Y., Seitz, P. K., and McGhee, M. F.,** Sterol metabolism. XLVII. Oxidized cholesterol esters in human tissues, *J. Steroid Biochem.,* 14, 889, 1981.
40. **Aringer, L., Eneroth, P., and Nordstrom, L.,** Side chain hydroxylation of cholesterol, campesterol, and β-sitosterol in rat liver mitochondria, *J. Lipid Res.,* 17, 263, 1976.
41. **Yokohama, H., Ohtsuka, I., Shiojiri, H., Katayama, K., and Ishikawa, S.,** A high-performance liquid chromatographic method for the assay of cholesterol 7α-hydroxylase activity, *Anal. Biochem.,* 157, 186, 1986.
42. **Sanghvi, A., Grassi, E., Bartman, C., Lester, R., Kienle, M. G., and Galli, G.,** Measurement of cholesterol 7α-hydroxylase activity with selected ion monitoring, *J. Lipid Res.,* 22, 720, 1981.
43. **Tsaconas, C., Padieu, P., Maume, G., Chessebeuf, M., Hussein, N., and Pitoizet, N.,** Gas chromatography-mass spectrometry of isobutyl ester trimethylsilyl ether derivatives of bile acids and application to the study of bile sterol and bile acid biosynthesis in rat liver epithelial cell lines, *Anal. Biochem.,* 157, 300, 1986.
44. **Sanghvi, A., Kienle, M. G., and Galli, G.,** A rapid, sensitive method for simultaneous measurements of tissue levels of oxygenated derivatives of cholesterol, *Anal. Biochem.,* 85, 430, 1978.
45. **Bjorkhem, I., Oftebro, H., Skrede, S., and Pedersen, J. I.,** Assay of intermediates in bile acid biosynthesis using isotope dilution-mass spectrometry: hepatic levels in the normal state and in cerebrotendinous xanthomatosis, *J. Lipid Res.,* 22, 191, 1981.
46. **Salen, G., Shefer, S., and Tint, G. S.,** Transformation of 4-cholesten-3-one and 7-hydroxy-4-cholesten-3-one into cholestanol and bile acids in cerebrotendinous xanthomatosis, *Gastroenterology,* 87, 276, 1984.
47. **MacPhillamy, G. B.,** On the nature of Haslewood's hepatols, *J. Am. Chem. Soc.,* 62, 3518, 1940.
48. **Haslewood, G. A. D.,** LXXXVIII. Metabolism of steroids. I. The isolation of 7-hydroxycholesterol and of the "hepatols" from ox liver, *Biochem. J.,* 33, 709, 1939.
49. **Haslewood, G. A. D.,** Metabolism of steroids. II. The isolation of cholestane-3:5:6-triol and other substances from ox liver extracts, *Biochem. J.,* 5, 708, 1941.
50. **Hall, A. N., Kulka, D., and Walker, T. K.,** Formation of arabinose, ribulose and tartronic acid from 2-keto-D-gluconic acid, *Biochem. J.,* 60, 274, 1955.
51. **Aringer, L. and Eneroth, P.,** Formation and metabolism in vitro of 5,6-epoxides of cholesterol and β-sitosterol, *J. Lipid Res.,* 15, 389, 1974.
52. **Levin, W., Michaud, D. P., Thomas, P. E., and Jerina, D. M.,** Distinct rapt hepatic microsomal epoxide hydrolases catalyze the hydration of cholesterol 5,6-oxide and certain xenobiotic alkene and arene oxides, *Arch. Biochem. Biophys.,* 220, 485, 1983.
53. **Sevanian, A. and McLeod, L. L.,** Catalytic properties and inhibition of hepatic cholesterol-epoxide hydrolase, *J. Biol. Chem.,* 261, 54, 1986.
54. **Watabe, T., Tsubaki, A., Isobe, M., Ozawa, N., and Hiratsuka, A.,** A mechanism for epoxidation of cholesterol by hepatic microsomal lipid hydroperoxides, *Biochim. Biophys. Acta,* 795, 60, 1984.
55. **Assmann, G., Fredrickson, D., Sloan, H. R., Fales, H. M., and Highet, R. J.,** Accumulation of oxygenated steryl esters in Wolman's disease, *J. Lipid Res.,* 16, 28, 1975.
56. **Boyd, G. S.,** Effect of linoleate and estrogen on cholesterol metabolism, *Fed. Proc. Fed. Am. Soc. Exp. Biol.,* 21, 86, 1962.

57. **Shefer, S., Cheng, F. W., Dayal, B., Hauser, G. S., Tint, G., Salen, G., and Mosbach, E. H.,** A 25-hydroxylation pathway of cholic acid biosynthesis in man and rat, *J. Clin. Invest.*, 57, 897, 1976.
58. **Kantiengar, N. L. and Morton, R. A.,** Formation of Cholesta-3:5-dien-7-one in rats, *Biochem. J.*, 60, 28, 1955.
59. **Spencer, T. A., Gayen, K., Phurwa, S., Nelson, J. A., Taylor, F. R., Kandutsch, A. A., and Erickson, S.,** 24(S),25-epoxycholesterol: evidence consistent with a role in the regulation of hepatic cholesterogenesis, *J. Biol. Chem.*, 260, 13391, 1985.
60. **Tallova, J. and Hakl, J.,** Determination of natural cholestenone in rat adrenal estracts by high performance liquid chromatography, *J. Chromatogr.*, 202, 487, 1980.
61. **Prost, M., Maune, B. F., and Padieu, P.,** Etude des sterols de la gland surrenaliene de rat, *Biochim. Biophys. Acta,* 360, 230, 1974.
62. **Cargill, D. I. and Cook, R. P.,** The free and esterified sterol present in bovine adrenal cortex and medulla, *Biochem. J.*, 93, 504, 1964.
63. **Bosisio, E., Galli, G., and Nicosia, S.,** Catabolism of bovine adrenal-cortex enzymes: in vitro formation of oxygenated sterols and side-chain cleavage products, *Eur. J. Biochem.*, 63, 491, 1976.
64. **Watabe, T. and Sawahata, T.,** The recently proposed 20,22-epoxycholesterol as the intermediate in the conversion of cholesterol to pregnenolone by adrenal cortex mitochronia must be 5,6-epoxycholestan-3β-ol, *Biochem. Biophys. Res. Commun.*, 83, 1396, 1978.
65. **Watabe, T. and Sawahata, T.,** Biotransformation of cholesterol to cholestane-3β,5α,6β-triol via cholesterol α-epoxide (5α,6α-epoxycholestan-3β-ol) in bovine adrenal cortex, *J. Biol. Chem.*, 254, 3854, 1979.
66. **Mason, J. I., Murry, B. A., and Aberhart, D. J.,** The synthesis and metabolism of [6-^{3}H]-25-hydroxycholesterol in rat adrenal tumor cells, *J. Steroid Biochem.*, 18, 765, 1983.
67. **Homberg, E.,** Fette Seif, *Anstrich,* 81, 148, 1979.
68. **Martin, C. M. and Nicholas, H. J.,** Metabolism of cholesteryl palmitate by rat brain in vitro; formation of cholesterol epoxides and cholestane-3β,5α,6β-triol, *J. Lipid Res.*, 14, 618, 1973.
69. **Fieser, L. F. and Bhatacharyva, B. K.,** Cholesterol and Companions. VI. Lathosterol, cholestanetriol and ketone 104 from a variety of sources, *J. Am. Chem. Soc.*, 75, 4418, 1953.
70. **Richter, V. R. and Dannenberg, H.,** Fraktionierung der, neutralen spurenlipoide aus Rinder- und Kanninchengehirm, Hoppe-Seyler's Z. Physiol. Chem., 350, 1213, 1969.
71. **Schubert, K., Rose, G., and Burger, M.,** Uber das Vorkommen von Dihydroxysterinen im menschichen gehirn, *Hoppe-Seylers Z. Physiol. Chem.*, 326, 225, 1961.
72. **Galli, G., Paoletti, E. G., and Weiss, J. F.,** Sterol precursors of cholesterol in adult human brain, *Science,* 162, 1495, 1968.
73. **Black, H. S. and Douglas, D. R.,** A model system for the evaluation of the role of cholesterol α-oxide in ultraviolet carcinogenesis, *Cancer Res.*, 32, 2630, 1972.
74. **Chan, J. T. and Black, H. S.,** Skin carcinogenesis: cholesterol-5α,6α-epoxide hydrase activity in mouse skin irradiated with ultraviolet light, *Science,* 186, 1216. 1974.
75. **Lo, W. B. and Black, H. S.,** Formation of cholesterol-derived photoproducts in human skin, *J. Invest. Dermatol.*, 58, 278, 1972.
76. **Parsons, P. G. and Goss, P.,** Chromosome damage and DNA repair induced in human fibroblasts by UV and cholesterol oxide, *Aust. J. Exp. Biol. Med. Sci.*, 56, 287, 1978.
77. **Petrakis, N., Gruenke, L. D., and Craig, J. C.,** Cholesterol and cholesterol epoxides in nipple aspirates of human breast fluid, *Cancer Res.*, 41, 2563, 1981.
78. **Gruenke, L. D., Wrensch, M. R., Petrakis, N. L., Mikke, R., Ernster, V. L., and Craig, J. C.,** Breast fluid cholesterol and cholesterol epoxides: relationship to breast cancer risk factors and other characteristics, *Cancer Res.*, 47, 5483, 1987.
79. **Wrensch, M. R., Petrakis, N. L., Gruenke, L. D., Mikk, R., Ernster, V. L., King, E. B., Hauck, W. W., Craig, J. C., and Goodson, W. H.,** III. Breast fluid cholesterol and cholesterol β-epoxide concentrations in women with benign breast disease, *Cancer Res.*, 49, 2168, 1989.
80. **Sporer, A., Brill, D. R., and Schaffner, C. P.,** Epoxycholesterols in secretions and tissues of normal benign, and cancerous human prostate glands, *Urology,* 20, 3, 1982.
81. **Saucier, S. E., Kandutsch, A. A., Taylor, F. R., Spencer, T. A., Phirwa, S., and Gayer, A. K.,** Identification of regulatory oxysterols, 24(S),25-epoxycholesterol and 25-hydroxycholesterol, in cultured fibroblasts, *J. Biol. Chem.*, 260, 14671, 1985.
82. **Irie, R., Iwanaga, M., and Yamakawa, T.,** The chemistry of lipid of posthemolytic residue or stroma of erythrocytes, *J. Biochem.*, 50, 122, 1961.

Chapter 6

Effects of Cholesterol Oxides on Cholesterol Metabolism

Robert J. Morin, Bing Hu, and Shi-Kaung Peng

TABLE OF CONTENTS

I. INHIBITION OF CHOLESTEROL BIOSYNTHESIS

Inhibitory effects of a number of cholesterol oxides on cholesterol biosynthesis have been demonstrated both *in vitro* and *in vivo*. Initial studies focused primarily upon the effects of 25-hydroxycholesterol (25-OH) and 7-ketocholesterol (7-keto) in liver and fibroblasts, but subsequent studies have identified a number of other inhibitory oxides, with effects upon most mammalian cells. The inhibitory effects of the cholesterol oxides on cholesterol biosynthesis generally are much greater than that of pure cholesterol. The regulatory enzyme hydroxymethylglutaryl coenzyme A (HMG-CoA) reductase (EC 1.1.1.34) appears to be the primary site of action, although there is also evidence for inhibitory effects at other stages of the biosynthetic pathway.

A. *IN VIVO* EFFECTS ON STEROL SYNTHESIS AND HMG-CoA REDUCTASE

Early studies by Tomkins, et al.[1] showed that feeding of cholesten-3-one, 0.45 g, to rats for one day resulted in greater than 90% inhibition of incorporation of acetate into cholesterol by liver slices. In a subsequent study, feeding this oxide to rats at 1% of the diet for 8 weeks resulted in 73% inhibition of sterol synthesis from acetate.[2] In mice, it appeared that low dietary doses of cholestenone inhibited both sterol synthesis and HMG-CoA reductase activity but higher doses were found to increase both.[3] When 25-OH was added to rat diets at 0.25% for 18 h, conversion of acetate to sterols by liver was decreased by 70 to 80%.[4] Erickson et al.[5] also found that 25-OH fed at 0.1% of diet reduced liver HMG-CoA reductase at 18 h, with no further inibition at 66 h. No inhibitory effects of 7-keto on hepatic sterol synthesis and HMG-CoA activity were detected in studies by Chen and Kandutsch[4] and Langdon et al.[6] Erickson et al.,[7] however, found that 7-keto fed to rats at 0.1% or 0.05% for 18 h did inhibit HMG-CoA reductase by 60 to 70%, but that no inhibition was evident at 66 h, indicating that tolerance had developed, perhaps due to induction of enzyme systems that catabolize the increased 7-keto. In mice, dietary 7-keto and 25-OH appear to inhibit sterol synthesis in the intestines after 1 d but not after 8 d.[8] Liver sterol synthesis was less markedly decreased after 1 d, and after 8 d had risen to above control levels. In chicks, dietary 7-keto produced a moderate decrease in HMG-CoA activity, but less than that of pure cholesterol.[9]

B. *IN VITRO* STUDIES ON STEROL SYNTHESIS AND HMG-CoA REDUCTASE

Inhibitory effects of the cholesterol oxides on cholesterol biosynthesis *in vitro* are more marked and prolonged than *in vivo*. Kandutsch et al.[10,13] earlier demonstrated inhibition of acetate incorporation into sterols and inhibition of HMG-CoA reductase by 7α- and 7β-hydroxycholesterol and 7-keto in mouse liver and mouse L fibroblast cultures. The effect was specific for cholesterol biosynthesis, since acetate incorporation into fatty acids or CO_2 was unchanged, and there was no effect on rates of protein or RNA synthesis. Purified cholesterol induced no inhibition in the cell cultures. Inhibition only occurred when the cholesterol oxides were added directly to the cell cultures, and not to the cell fractions or homogenates during the assays. It was essential to grow the cells in a serum-free medium. Subsequent experiments using liver cell cultures indicated that 25-OH and 20-OH were the most potent inhibitors of HMG-CoA reductase and sterol synthesis in mammalian cells[14] and that established cell lines were more sensitive to inhibition than primary cultures.[4] Cholesterol 5,6-epoxide at 50 *M* resulted in 55% inhibition of HMG-CoA reductase in rat liver microsomes.[15]

Mouse L cell fibroblasts appear to be more sensitive to inhibition by all the oxides than liver cells.[16] HMG-CoA reductase in embryonic rat fibroblasts was greater than 90% inhibited by 2.5 μg/ml of 25-OH.[17] The inhibitory effects were rapid, with 25-OH or 20α-OH decreasing enzyme activity to half within 1 to 1.3 hours. In cultured human fibroblasts,

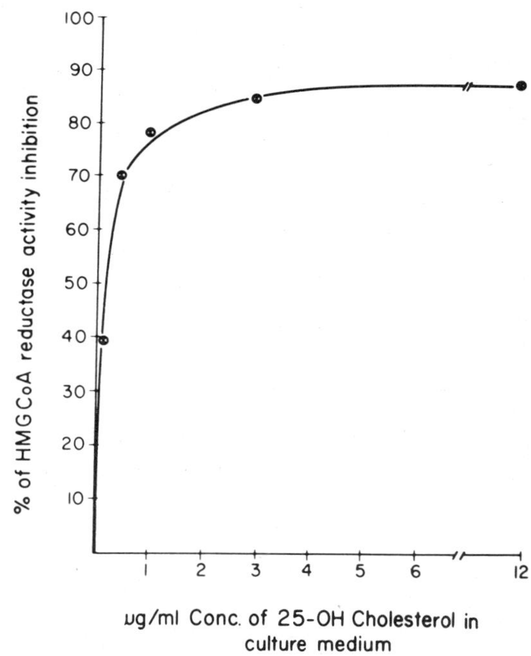

FIGURE 1. Inhibition of cultured rabbit aortic smooth muscle cell HMG-CoA reductase activity by varying concentrations of 25-hydroxycholesterol. (From Peng, S.-K., Tham, P., Taylor, C. B., and Mikkelson, B., *Am. J. Clin. Nutr.*, 32, 1033, 1979. With permission.)

7-keto was 100 times more potent than cholesterol, and the inhibition was much faster.[18] At 0.1 μg/ml 7-keto and 0.8 μg/ml of 25-OH, HMG-CoA reductase was inhibited by 50%.[17] 7α- and 7β-hydroxycholesterol were effective in the same range as 7-keto.[20] HMG-CoA reductase in cultured fibroblasts from familial hypercholesterolemic patients was also inhibited by 7-keto, indicating that the effect was not mediated by LDL (low-density lipoprotein) receptors.[18] Normal human fibroblasts incubated with LDL containing 25-hydroxycholesterol oleate also showed HMG-CoA reductase inhibition, but this required prior hydrolysis of the ester.[21]

In cultured rabbit and canine intestinal mucosa cells, both 7-keto and 25-OH induced inhibition of HMG-CoA reductase activity.[22,23] In the rabbit cells this resulted in depletion of cholesterol content, whereas in the canine cells no depletion occurred. The inhibitor responses were similar to those in liver cells, but less marked than in other cells, perhaps due to a more rapid metabolism and excretion of the cholesterol oxides. In human small intestinal epithelial cells, conversion of acetate to sterols was also inhibited by 7-keto at 10 μ*M*.[23]

Effects of cholesterol oxides on cholesterol biosynthesis in arterial smooth muscle cells may be of particular relevance to atherosclerosis. In cultured rabbit aortic smooth muscle cells, Tham et al.[24] and Peng et al.[25] found that 25-OH was one of the most potent inhibitors of HMG-CoA reductase (50% inhibition at 0.1 μg/ml, as shown in Figure 1). Using 3 μg/ml of each sterol during 2-h incubations, HMG-CoA reductase was inhibited 83% by 25-OH or 20α-OH, 78% by 7-keto, and in the range of 51 to 63% by 7α- and 7β-hydroxycholesterol, cholestane 3β,5α,6β-triol, and 5,6-epoxycholesterol. Cholesterol biosynthesis from acetate was inhibited by 80 to 90% after incubation of cultured calf aortic smooth muscle cells with 10 μg/ml 25-OH for 6 h[26,27] and human aortic smooth muscle cells showed 58% inhibition of HMG-CoA reductase after incubation with 0.5 μg/ml of 25-OH for 4 h.[28] In aortic smooth muscle cells from pigeons lacking a functional LDL receptor pathway, 25-OH still inhibited cholesterol biosynthesis, indicating that the LDL receptors are not involved in mediating the inhibitory effects of the cholesterol oxides.[29]

Lymphocytes are also quite sensitive to inhibition of cholesterol biosynthesis by cholesterol oxides. In normal guinea pig lymphocytes, and in leukemic lymphocytes, which produce cholesterol at 30 times the normal rate, 50% inhibition of sterol synthesis was noted at 10^{-7} *M* 25-OH.[30,32] Similarly, normal human lymphocytes, and cells from patients with hairy cell leukemia (a type of B-lymphocyte leukemia), which synthesize cholesterol at 5 to 6 times the normal rate, were equally sensitive to sterol biosynthesis inhibition by 25-OH in the same range as the guinea pig lymphocytes.[33] Greater than 90% inhibition of sterol synthesis in human lymphocytes stimulated by phytohemagglutin was noted after incubation for 2 h with 1.5 μg/ml of 7α-hydroxycholesterol.[34]

Inhibition of cholesterol biosynthesis and HMG-CoA reductase activity by cholesterol oxides has also been noted in Chinese hamster ovary and lung cells, nerve cells, skin cells, HeLa cells, and other cell lines as listed in Table 1.

C. RELATIONSHIP OF CHOLESTEROL OXIDE STRUCTURE TO INHIBITORY EFFECTS

Sterols with a greater molecular distance between the 3β-hydroxyl and the second oxygenated position, as in the case of 20α-, 25- and 26-hydroxycholesterol, appear to have the greatest inhibitory effects.[16,35] An intact isooctyl side chain is required for complete inhibitory activity. The presence or absence of a double bond at carbons 5 and 6 has little effect. Oxidation of the 3β-hydroxyl to a ketone decreases the inhibitor action of 7α-hydroxycholesterol, but cholestane-3,6-diones still have significant inhibitory activity. Sterols with a third hydroxyl group have only weak inhibitory activity in liver cells, although cholestane-3β,5α,6β-triol does significantly inhibit HMG-CoA reductase in arterial smooth muscle cells.[25] Cholesterol 5α,6α-epoxide was as inhibitory as the triol in liver cells but did not inhibit mouse L cell fibroblasts.[16] Conversely, cholestane-3β,5α-diol inhibited HMG-CoA reductase in L cells but was inactive in liver. Sterols with ketones at positions 3, 6, or 7 in combination with one or more double bonds were inactive.

D. RESISTANCE TO CHOLESTEROL OXIDE-INDUCED INHIBITION

Mutant strains of Chinese hamster ovary and lung cells have been developed in which HMG-CoA reductase activity is not inhibited by 25-OH.[36,37] Lung cells resistant to 25-OH were also resistant to sterols oxygenated at the 6, 7, or 15 positions of the steroid nucleus or the 20, 22, or 24 positions of the side chain.[39] This suggests that all the cholesterol oxides may inhibit cholesterol biosynthesis by the same mechanism. Since these cholesterol oxides are not metabolized by Chinese hamster lung cells, the resistance to inhibition is apparently not a result of enhanced clearance of the oxides. There are no differences in the kinetic properties or levels of activity between wild-type cells and the mutant variety. It was concluded by Cavenee et al.,[38] therefore, that the resistance was not due to increased catalytic efficiency or increased numbers of enzyme molecules and unlikely to be due to an alteration in the gene regulating the reductase. The mechanism of resistance was theorized to be a mutation of another gene normally influenced by the oxysterols which then can no longer regulate expression of the structural gene for the reductase. In other studies, Westhafer et al.[39] used DNA from 25-OH-resistant hamster cells to transfect mouse L-M cells. Several resistant colonies developed after 3 weeks, and the phenotype remained constant for 10 generations of cells. These studies indicated that resistance to 25-OH inhibition is a dominant selectable genetic trait.

TABLE 1
In Vitro Inhibition of HMG-CoA Reductase by Cholesterol Oxides

Tissue	Cholesterol oxide	Inhibition	Ref.
Liver			
Fetal mouse	7α-Hydroxycholesterol	50% at 50 μ*M*	16
	7β-Hydroxycholesterol	50% at 6.3 μ*M*	16
	7-Ketocholesterol	50% at 11.3 μ*M*	16
	Cholestane-3β,5α,6β-triol	50% at 11.5 μ*M*	16
	Cholesterol 5α,6α-epoxide	50% at 7.5 μ*M*	16
Rat	7-Ketocholesterol	50% at 1.5 μ*M*	78
	25-Hydroxycholesterol	50% at 50 μ*M*	44,81
		50% at 5 μ*M*	74
		90% at 10 μg/ml	5
Intestine			
Rat	25-Hydroxycholesterol	50% at 0.1 μg/ml	69
Rabbit	7-Ketocholesterol	50% at 0.1 m*M*	22
	25-Hydroxycholesterol	50% at 0.1 m*M*	22
Dog	7-Ketocholesterol	32% at 0.2 m*M*	23
	25-Ketocholesterol	42% at 0.2 m*M*	23
Fibroblasts			
Mouse	7α-Hydroxycholesterol	50% at 2.5—10 μ*M*	16,81-84
	7β-Hydroxycholesterol	50% at 1.9—5.7 μ*M*	16,79-82
	7-Ketocholesterol	50% at 1.7—3.0 μ*M*	16,79-82
	20(S)-Hydroxycholesterol	50% at 0.3—1.5 μ*M*	81-84
	25-Hydroxycholesterol	50% at 0.05—0.17 μ*M*	4,68,81-83 85
	Cholestane-3β,5α,6β-triol	50% at 13.3 μ*M*	16
	Cholesterol 5α,6α-epoxide	50% at >25 μ*M*	16
Rat embryo	7-Ketocholesterol	94% at 6 μ*M*	59
	25-Hydroxycholesterol	94% at 2.5 μg/ml	17,59,88
Chinese hamster	7α-Hydroxycholesterol	50% at 11.5 μ*M*	38
	7β-Hydroxycholesterol	50% at 3.1 μ*M*	38
	7-Ketocholesterol	50% at 2.0 μ*M*	38
	25-Hydroxycholesterol	50% at 0.16 μ*M*	68
	24(S),25-Epoxycholesterol	50% at 0.89 μ*M*	68
	Cholestane-3β,5α,6β-triol	50% at >15 μ*M*	38
Human	7α-Hydroxycholesterol	50% at 0.1 μg/ml	76
	7β-Hydroxycholesterol	50% at 0.1 μg/ml	76
	7-Ketocholesterol	50% at 0.1 μg/ml	19,75
	25-Hydroxycholesterol	50% at 0.8 μg/ml	19
Artery smooth muscle			
Rabbit	7α-Hydroxycholesterol	59% at 3 μg/ml	25
	7β-Hydroxycholesterol	51% at 3 μg/ml	25
	7-Ketocholesterol	78% at 3 μg/ml	25
	25-Hydroxycholesterol	50% at 0.1—0.5 μg/ml	25,28
		83% at 3 μg/ml	25
	Cholestane-3β,5α,6β-triol	63% at 3 μg/ml	25
Pigeon	25-Hydroxycholesterol	90% at 1 μg/ml	29
Monkey	25-Hydroxycholesterol	75% at 1 μg/ml	29
Human	25-Hydroxycholesterol	80% at 1 g/ml	89,90
Lymphocytes			
Mouse	25-Hydroxycholesterol	50% at 0.4 μ*M*	55
Human	7-Ketocholesterol	83% at 5 μg/ml	91
		61% at 5 μg/ml	92
		75%—98% at 25 μg/ml	34,92
	25-Hydroxycholesterol	94% at 5 μg/ml	92,93
	(25R),26-Hydroxycholesterol	96% at 5 μg/ml	93
Myeloblasts			
Chicken	25-Hydroxycholesterol	90% at 1 μg/ml	46

TABLE 1 (continued)
***In Vitro* Inhibition of HMG-CoA Reductase by Cholesterol Oxides**

Tissue	Cholesterol oxide	Inhibition	Ref.
HeLa cells	7-Ketocholesterol	58—70% at 1 μg/ml	77,94
	25-Hydroxycholesterol	79—85% at 1 μg/ml	77,94
Neural cells			
Rat astrocytoma	7-Ketocholesterol	36% at 10 μg/ml	78
C-6 glioma	25-Hydroxycholesterol	50% at 25 μ*M*	95
Ovary			
Chinese hamster	7-Ketocholesterol	50% at 1.5 μ*M*	60
	25-Hydroxycholesterol	50% at 0.1—1.0 μg/ml	97,98
		80% at 0.25 μ*M*	53,81,98,99
	26-Hydroxycholesterol	40% at 0.25 μ*M*	73,96
Macrophages			
Mouse	25-Hydroxycholesterol	50% at 0.3 μg/ml	101
J774	25-Hydroxycholesterol	50% at 12.5 μ*M*	100

E. MECHANISMS OF INHIBITION OF HMG-CoA REDUCTASE BY CHOLESTEROL OXIDES

When 25-OH, 20α-OH, 7-keto, or 22-ketocholesterol were incubated with isolated microsomes, no inhibition of HMG-CoA reductase occurred, indicating that there is no direct inhibitory effect on the enzyme.[40] Even in intact enucleated cells 25-OH has no inhibitory effect on sterol synthesis, suggesting that the action of 25-OH is mediated in some manner through nuclear control.[39] Although intact cells are required, uptake and inhibitory effects by the cholesterol oxides do not require functional LDL receptors.[18] In isolated hepatocytes, the addition of 10 μg/ml of 25-OH resulted in a loss of 50% of HMG-CoA reductase activity in 27 min and a decrease to 10% with 2 h.[5] The rapidity and magnitude of effect, therefore, suggested that inhibition was occurring by inactivation or degradation of the enzyme. Reversible inactivation of HMG-CoA reductase can occur by a protein kinase-ATP-Mg^{++}-mediated phosphorylation of the enzyme, which can then be activated by phosphatase-catalyzed dephosphorylation.[41] The question of whether or not inhibitory effects of cholesterol oxides are mediated by phosphorylation is as yet unresolved. Earlier studies by Saucier and Kandutsch[42] showed that the ratio of phosphorylated to dephosphorylated HMG-CoA reductase in mouse L cells was not altered by incubation with 25-OH. Inhibition of rat liver HMG-CoA reductase as shown by Erickson et al.[43] was not reversed by incubation with a cytosolic phosphatase activator, and no changes in the levels of this activator were induced by 25-OH. Liver cell cultures are not an optimal model for study of effects on phosphorylation-dephosphorylation, since full activation occurs during cell plating. In freshly isolated hepatocytes, however, HMG-CoA reductase is 80% phosphorylated and 20% dephosphorylated.[44] The ratio of these forms was not altered by incubation with 25-OH. Inhibition by this sterol was not compensated for by dephosphorylation of more enzyme, however, suggesting the possibility of inhibition of phosphoprotein phosphatase by 25-hydroxycholesterol.

More recently, Beg et al.[45] have proposed that phosphorylation is a short-term mechanism of regulation of HMG-CoA reductase by cholesterol oxides. After only 20 minutes of incubation of cultured human fibroblasts with 25-OH, the level of expressed (dephosphorylated) enzyme activity was decreased, but total enzyme activity, as measured after addition of phosphatase, was unchanged. Their experiments suggested that low concentrations of either 25-OH or 7-keto increased phosphorylation of the enzymes, whereas incubation with higher sterol concentrations for 60 minutes or longer depressed total enzymatic activity. They postulate, therefore, a short-term regulation by phosphorylation, and long-term regulation by decreased enzyme synthesis and/or increased degradation. Tanaka et al.,[46] however, using an immunoprecipitation method to quantitate HMG-CoA reductase labeled by

$^{32}P[PO_4]$ showed that the decreased activity induced in chicken myeloblasts by incubation with 5 μg/ml of 25-OH was not associated with any increase in ^{32}P activity of the enzyme, and therefore that phosphorylation was not involved.

There is more unanimity in published reports that increased irreversible degradation of HMG-CoA reductase is a mechanism for the inhibitory effects of cholesterol oxides. In Chinese hamster ovary cells, the enzyme is degraded with a half-life of 10 to 13 hours.[47] This is increased 3- to 4-fold by incubation with 25-OH.[47,48] Similar results were reported by Tanaka et al.,[46] who found a 350% increase in degradation rate of the enzyme in chicken myeloblasts after incubation with 25-OH. The increased rate of HMG-CoA degradation induced by 25-OH in Chinese hamster ovary cells could be abolished by addition of the protein synthesis inhibitors cycloheximide, emetine, or puromycin, suggesting that the degradative effect of 25-OH was mediated by a protein.[47] The inhibitory effect of 25-OH could be prevented by preincubation with cycloheximide for only 2 hours, indicating a rapid turnover of the mediator protein.

The region of the reductase molecule that is degraded during this process has been localized by Gil et al.[49] Chinese hamster ovary cells deficient in the reductase were transfected with a plasmid expression vector containing a reductase cDNA lacking amino acids 10 to 339. This segment is the major portion of the domain that normally anchors the reductase to the endoplasmic reticulum. The deletion of this segment caused a decrease in the degradation rate to one fifth that of the complete enzyme. In cells with this partial enzyme, 25-OH did not accelerate degradation, thus establishing the locus of degradation to the 10- to 339-peptide segment.

Inhibitory effects of cholesterol oxides on synthesis of the HMG-CoA reductase enzyme may be an important mechanism of action. In rat liver cells, 25-OH was found to decrease HMG-CoA reductase activity by blocking synthesis of three polypeptides of the enzyme.[50] Using a quantitative immunoprecipitation method, Sinensky et al.[51] found that the incorporation of [^{35}S]-methionine into the reductase of Chinese hamster ovary cells was blocked by 25-OH. There was no reduced incorporation of this label into the reductase of Chinese hamster ovary cells resistant to the inhibitory effects of 25-OH. Faust et al.[48] utilized a clone of Chinese hamster ovary cells selected to grow in the presence of the HMG-CoA inhibitor compactin. These cells exhibit a level of reductase 100 to 1000 times greater than normal, but this can also be inhibited by 25-OH. When the incorporation of [^{35}S]-methionine into the reductase was studied by quantitative immunoprecipitation, 25-OH was found to decrease the incorporation rate by greater than 98%. Although degradation was also increased 3-fold, the authors concluded that the predominant mechanism of the 25-OH inhibitor effect was decreased enzyme synthesis. A subsequent study by Tanaka et al.,[46] however, showed that in avian myeloblasts, within 20 to 30 minutes after 25-OH there was a 350% increase in rate of enzyme degradation but only a 50 to 72% decrease in rate of enzyme synthesis. Their immunoprecipitated enzyme had a molecular weight of 94,000, which has been shown to be closer to the true molecular weight of the reductase. The enzymes quantitated in the studies by Sinensky et al.,[51] and Faust et al.[48] had molecular weights in the range of 52 to 68,000, and it is therefore possible that the observed inhibition of enzyme synthesis by 25-OH may have been artifactually high due to precipitation of only proteolytic fractions of the enzyme.

It has been suggested that the inhibitory effect of 25-OH on reductase synthesis might be secondary to a block in RNA synthesis by this sterol.[52-54] When added to UT-1 cells, a clone of compactin-resistant Chinese hamster ovary cells with 100 to 1000 times the normal amount of HMG-CoA reductase protein, 25-OH decreased the level of mRNA for the reductase.[53] No effects of 7α- and 7β-cholesterol or 7-keto on RNA synthesis were noted in cultured mouse liver cells or in mouse L cell cultures.[11] In mouse thymocytes, the inhibitory effect of 25-OH on sterol synthesis from acetate was not accompanied by any effect on RNA formation.[55] The RNA synthesis blocking agent actinomycin likewise had no effects on the inhibitory action of 25-OH. Tanaka et al.[46] also found no change in RNA synthesis due to 25-OH in chicken myeloblasts, and suggested that reductase synthesis is probably decreased by translational regulation.

F. CELLULAR OXYSTEROL-BINDING PROTEINS

Cytosolic proteins which specifically bind cholesterol oxides and other oxysterols have been identified in mouse L fibroblasts,[56] fetal mouse liver,[40] mouse myeloma,[52] mouse spleen,[52] rat liver,[57] hamster liver,[58] rat embryo fibroblasts,[59] Chinese hamster ovary,[60] and human lymphocytes.[61] These proteins may act as receptors for uptake of cholesterol oxides by cells and may thereby have a role in the regulation of cholesterol biosynthesis by these sterols. The oxysterol-binding proteins are different from the cellular proteins that bind cholesterol and the steroid hormones. When the binding affinity of 47 different oxysterols to oxysterol-binding protein of mouse L fibroblasts was compared to their HMG-CoA reductase inhibiting effects, 35 of these sterols showed a positive correlation.[62] Most of those that showed higher reductase inhibition in relation to binding affinity were sterols with a ketone at the 3rd position of the steroid nucleus, and these are known to be converted in cells to 3β-hydroxysterols. Chinese hamster ovary K1 cells that are resistant to the reductase inhibitory effects of 25-OH have been found to lack binding affinity for this sterol.[63] The chemical and physical properties of these binding proteins have been studied in considerable detail. An intact 8-carbon side chain and a 3β-hydroxyl group are required for binding activity.[52] Binding is competitive between the various sterols with HMG-CoA reductase inhibitory properties. The most competitive sterols in decreasing order are 25-OH > 20α-OH > 23(S)-hydroxycholesterol > 24(R)-hydroxycholesterol > 26-hydroxycholesterol.[64] Addition of a hydroxyl group at position 7 of 25-OH causes a loss of binding activity. The dissociation constant (K) for 25-OH is in the range of 4 to 5×10^{-9}, indicating a very high affinity. Binding capacity is low, in the range of 10^{-8} *M*. When Kandutsch et al.[56] incubated cultured L-cell fibroblasts with [^{3}H]25-hydroxycholesterol and then fractionated the cytosolic proteins by sucrose density gradient centrifugation, two labeled fractions, with sedimentation coefficients of 7.5 S and 5 S were isolated. The 7.5-S fraction was identified as having the specific binding activity for the oxysterols. The bound [^{3}H]25-hydroxycholesterol on this fraction could be displaced by other oxysterols in proportion to their reductase-inhibiting capacity. This 7.5-S protein appears to have similar properties to an 8-S binding protein isolated from rat embryo fibroblasts by Beseme et al.[59] and an 8.3-S protein isolated from human lymphocytes by Defay et al.[61] The labeled 5-S fraction isolated from the mouse L cytosol by the Kandutsch group appears to be a nonspecific, low-affinity binding protein.[51] The labeled 25-OH in this fraction was not displaced by other sterols. A 5-S cytosolic protein isolated from Chinese hamster ovary cells by Sinensky et al.[60] however, did display competitive binding affinity between 25-OH and other sterols that inhibit cholesterol biosynthesis.

Most of the specific oxysterol-binding protein is located in the cytosol, although small amounts have been identified in L-cell nuclei.[52] The rates of sterol binding and dissociation of intact cells are much faster than those with purified binding protein. The concentration of 25-OH reductase in cell cultures is equivalent to the amount that would saturate only 5% of the binding sites of the specific oxysterol-binding protein. In rat embryo fibroblasts, there was a 5-fold increase in oxysterol-binding activity during the G_2M phase of the cell cycle and in exponentially growing cells.[58] This appeared to be attributable to an increased number of binding sites on the protein.

Several different forms of the binding protein have been identified in L cells. An unbound protein with a sedimentation coefficient of 7.9 S and a molecular weight of 236 kDa appears to reduce to a 7.5-S, 169-kDa protein upon binding to 25-OH.[55] It becomes less elongate and its Stokes radius is reduced from 71.6 Å to 50 Å. The 7.5-S protein can be further dissociated into a 4.2-S, 97-kDa protein by incubation in urea at pH less than 6. This process can be reversed by removal of the urea, indicating that proteolytic degradation is not occurring. Taylor and Kandutsch[52,64,65] have postulated that the specific oxysterol-binding protein consists of a 97-kDa subunit containing the binding properties, a 72-kDa unit which is associated with the other subunits at neutral pH, and a 67-kDa subunit component of the unliganded protein.[52] The subunits appear to be linked by both ionic and nonionic bonds. Further evidence for the 97-kDa unit as the active binding component has been provided by photoaffinity labeling techniques.[66] An oxysterol-binding protein recently purified 40,000-fold from hamster liver by Dawson et al.[58] appears to differ somewhat from the L cell protein. The hamster protein consists of a doublet of 101-kDa and 96-kDa peptides, both of which react with the same monoclonal antibody, and may therefore be modified or proteolyzed forms of the same molecule. These units associated into a complex of 280 kDa in the presence or absence of 25-OH. Dawson et al.[67] have also recently cloned a rabbit liver cDNA encoding an oxysterol-binding protein. The predicted amino acid sequence of the protein this cDNA would transcribe indicated 809 amino acids, with a glycine-alanine rich N_2 terminus and a 35-residue leucine zipper, which could promote the previously observed oligomerization of the oxysterol-binding protein monomers. Transfection of the rabbit cDNA into simian COS cells produced the same 96- and 101-kDa forms that were previously observed in hamster liver, and these proteins had the same affinity and specificity for sterols as did the purified hamster liver protein.

A number of similarities have been observed between the monomeric and polymeric structures and physicochemical properties of the oxysterol-binding proteins and steroid hormone receptor proteins.[52,61] Both show a dissociation from a polymeric to a monomeric form in hypertonic buffer. The steroid hormone polymeric form, however, is stabilized by molybdate, whereas the oxysterol binding protein is not. Heating changes the steroid receptor to the monomeric form but does not change the oxysterol-binding protein. Ammonium sulfate precipitation activates steroid receptors only, and steroid receptors require the presence of reduced thiol compounds. Changes in pH affect the kinetics of oxysterol binding, but not steroid hormone binding.

Taylor and Kandutsch[52] have postulated that the mechanism of action of the oxysterol-protein complex is to bind to the regions in the genome that control transcription of mRNA for HMG-CoA reductase. They have demonstrated nonspecific binding of the oxysterol-binding protein to calf thymus DNA, with a high affinity at pH 5.5, similar to steroid hormone receptor binding. The oxysterol-binding protein purified from hamster liver by Dawson et al.,[58] however, unlike a steroid hormone receptor, does not change to a DNA binding form on incubation at 37°C or in high salt solutions. They have been unable to demonstrate binding of this protein to the octanucleotide sterol regulatory element component of the promoter for the HMG-CoA reductase gene. Mutations of this sterol regulatory element are associated with reduced susceptibility to sterol-mediated reductase repression. In addition, although oxysterols have reduced mRNA synthesis in some studies,[53,54] others have demonstrated no change in mRNA synthesis,[11,46,55] and have postulated a translational rather than a transcriptional mechanism of action for the oxysterols.[46]

G. PHYSIOLOGICAL RELEVANCE OF HMG-CoA REDUCTASE INHIBITION

Although cholesterol oxides have definite inhibitory effects on HMG-CoA reductase when administered *in vivo* or when added to cultured cells *in vitro*, it is not yet established whether these compounds are natural regulators of this enzyme. The inhibitory effects of 25-OH on HMG-CoA reductase, LDL receptor activity, and other cholesterol regulatory enzymes such as acyl CoA cholesterol:acyltransferase, are very similar to the effects of LDL.[35] Cells that are resistant to the inhibitory effects of 25-OH on HMG-CoA reductase are also resistant to inhibition by LDL, suggesting that LDL may contain 25-OH or another

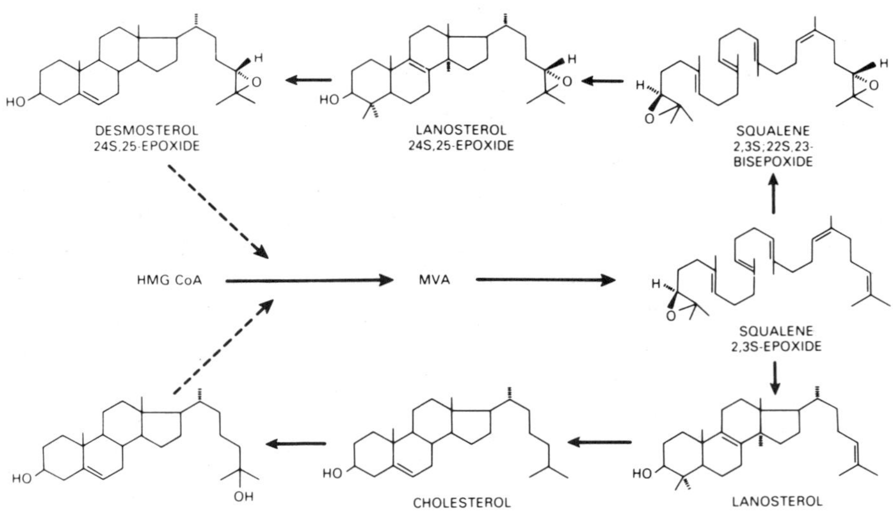

FIGURE 2. Possible mechanisms of physiological regulation of cholesterol biosynthesis by oxysterols. (Reprinted with permission from Smith, L. L. and Johnson, B. H., *Free Rad. Biol. Med.*, 7, 285, 1989. Copyright 1989, Pergamon Press PLC.)

inhibitory sterol or may acquire one after entering the cell.[36] Chinese hamster lung fibroblasts have been found to contain 1.5 fg/cell of 25-OH, which is in the range required for inhibitory effects on HMG-CoA reductase.[68] In the liver, 25-hydroxylation of cholesterol is catalyzed by a mitochondrial P-450 system.[68] In rat intestinal epithelial cells, 25-OH is formed from 25-hydroxylanosterol and 25-hydroxylanostene-3-one, both of which are also HMG-CoA reductase inhibitors.[69] Another oxysterol 24(S),25-epoxycholesterol has been identified in both Chinese hamster lung fibroblasts and in human liver.[68,70] In the hamster fibroblasts its concentration of 7.2 fg/cell was higher than that of 25-OH, but its inhibitory effects were somewhat lower, so that the net effects of the two sterols were similar, and together accounted for most of the inhibitory effects of all the endogenous cells sterols. Production of 24(S),25-epoxycholesterol occurs in liver by a branch in the cholesterol biosynthetic pathway involving conversion of squalene 2,3(S),22(S),23-dioxide to 24(S),25-oxidolanosterol and then to the epoxycholesterol. A similar conversion has been noted in rat intestinal epithelial cells,[71] and in liver, where the epoxycholesterol both inhibits HMG-CoA reductase activity and binds to the oxysterol binding protein.[70] It is slowly metabolized by cultured cells. The levels of squalene epoxidase or oxidosqualese cyclase activity, therefore, could regulate the formation of 24(S),25-epoxycholesterol, which may function as a natural regulator of cholesterol biosynthesis. The possible mechanisms for effects of this sterol and 25-OH are summarized in Figure 2. Mouse L929 fibroblasts are able to metabolize 25-OH via a side-chain hydroxylation and scission of the C20-22 bond to form pregnenolone.[72] This results in restoration of HMG-CoA reductase activity, and indicates that a mechanism does exist to reduce levels of 25-OH under conditions where increased cholesterol biosynthesis might be required.

Significant levels of 20α-hydroxycholesterol are produced in endocrine organs, and this potent inhibitory sterol with a high binding affinity for the oxysterol-binding protein could well have a physiological role in regulation of cholesterol biosynthesis. In the liver, 7α-hydroxycholesterol is produced in the course of conversion of cholesterol to bile acids. Although this sterol does inhibit HMG-CoA reductase, its affinity for the oxysterol-binding protein is relatively low,[62] making it less likely that it is a physiological regulator. Another oxysterol produced by hepatic mitochondrial enzymatic activity during the course of bile acid synthesis is 26-hydroxycholesterol. This compound has reductase-inhibiting effects similar to 25-OH[73] and also has high binding affinity for the oxysterol-binding protein, making it a likely physiological regulator.

H. INHIBITION OF CHOLESTEROL BIOSYNTHESIS AT STAGES BEFORE AND AFTER HMG-CoA REDUCTASE

Inhibition of sterol synthesis in rat liver cell cultures by 7-keto is much greater than the degree of inhibition of HMG-CoA reductase by this cholesterol oxide, indicating that inhibition at an additional step must be occurring.[101] In HeLa cells, both 7-keto and 25-OH at 1 μg/ml for 24 h inhibit the activity of HMG-CoA synthetase, but less so than the reductase.[103] The synthetase activity changed coordinately with the rate of cholesterol biosynthesis. Chang and Limanek[104] found that in Chinese hamster ovarian cells, 25-OH inhibited the activities of acetoacetyl CoA thiolase, HMG-CoA synthetase, and mevalonate kinase, as well as HMG-CoA reductase. Mutant strains of cells resistant to 25-OH showed an inhibitory response to all four of these early enzymatic steps in the cholesterol biosynthetic pathway, suggesting that these are all coordinately regulated at the level of enzyme synthesis.[104,105] The effects of 25-OH on the activities of these enzymes were very similar to the effects of LDL.

Later stages of cholesterol biosynthesis are also subject to inhibition by cholesterol oxides. Accumulation of methyl sterols has been noted in primary rat hepatocyte cultures treated with 25-OH[105] and in normal and leukemic guinea pig lymphocytes treated with either 25-OH or 7β-hydroxycholesterol.[107] Inhibition of lanosterol demethylation has been produced in cultured Morris hepatic tumor cells by 7α- and 7β-hydroxycholesterol,[108] although other investigators observed no change in rat liver methyl sterol oxidase activity with either 7α-hydroxycholesterol or 7-keto.[112] Cholestane-3β,5α,6β-triol seems to have greater effects on this stage than do other cholesterol oxides. Rat liver methyl sterol oxidase was 83% inhibited by 0.05 m*M* of the triol, whereas only 52% inhibition occurred with 0.1 m*M* of 25-OH.[109] Grinstead et al.,[110] however, observed a 48% inhibition of methyl sterol oxidase on incubation of rat liver microsomes with 50 μ*M* of 25-OH, but only a 23% inhibition with triol. In primary rat hepatocyte cultures, Gaylor and Delwiche[109] observed a 97% inhibition of sterol demethylation by 0.05 m*M* triol, and a much greater accumulation of 4,4-dimethyl monoene sterols (23%) as compared to the effect of 25-OH (2%). The possible physiological regulatory oxysterol, 24(S),25-epoxycholesterol has been shown to inhibit conversion of 24,25-dihydroxylanosterol to cholesterol when added to rat liver homogenates at 0.04 m*M*.

II. CHOLESTEROL SIDE-CHAIN CLEAVAGE (EC 1.14.15.6)

Several cholesterol oxides have been shown to inhibit hormonal steroid production from cholesterol. 7-keto added to hog adrenal mitochondria was found to inhibit side-chain cleavage of labeled cholesterol.[111] In isolated rat adrenal cells with added ACTH and glutethimide, 25-OH inhibited steroid formation from endogenous substrates.[112] The effects of 25-OH were additive to the inhibition from glutethimide, and thought to be a competitive type of inhibition. Similar inhibitory effects additive to glutethimide were produced by 22(S)-hydroxycholesterol in rat adrenal cells.[113] In isolated human placental mitochondrial enzymes, side-chain cleavage of labeled cholesterol was inhibited by 20α-hydroxycholesterol, 22(R)-hydroxycholesterol, and 20α-22R-dihydroxycholesterol, which are formed as intermediates during side-chain cleavage.[114] Inhibition of pregnenolone synthesis in bovine adrenal mitochondria was observed with both R and S enantiomers of 20,22-epoxycholesterol.[115] The effectiveness of cholesterol epoxides in inhibiting pregnenolone formation from cholesterol was greatest with those compounds having the same stereochemical configuration as cholesterol at carbon 20.

Although steroidogenesis from exogenous or endogenous cholesterol is inhibited by 20α- and 25-hydroxycholesterol, these oxides can also be utilized as substrates for steroid synthesis, and stimulatory effects on corticosterone synthesis in rat adrenal cells[116] and on total pregnenolone synthesis have been observed in isolated rat testes mitochondria[117] and in rat luteal cells.[118] The stimulation is greater than that produced by addition of cholesterol. In cultured rat granulosa cells, the 80% inhibition of steroidogenesis by incubation in lipoprotein-deficient media can be reversed by addition of 25-OH.[119] Stimulation of steroidogenesis in rat luteal cells occurs in the presence or absence of luteinizing hormone (LH).[120] The effect of 25-OH in hamster granulosa and luteal cells is in fact, more rapid than LH, with stimulation occurring within 15 minutes, as compared to 30 minutes with LH.[121] The effect of LH can be abolished by cycloheximide, whereas the effect of 25-OH is not abolished by this protein synthesis inhibitor. The stimulation by LH of steroidogenesis from cholesterol may require a labile protein to facilitate transport of cholesterol to the side-chain cleavage system on the internal mitochondrial membrane, whereas 25-OH may gain direct access. Similar effects have been observed with rat Leydig cell testosterone synthesis stimulated by 25-OH, which bypasses the cAMP or trophic hormone-induced transport of cholesterol to the P-450 side-chain cleavage system.[117] Binding of 25-OH to rat adrenal cytochrome P-450 has been documented by the specific spectral change that is produced.[120]

III. CHOLESTEROL 7α-HYDROXYLASE (EC 1.14.13.17)

The enzyme cholesterol 7α-hydroxylase is believed to be the primary rate-limiting step in the hepatic conversion of cholesterol to bile acids. In rat liver microsomes, the reaction product, 7α-hydroxycholesterol, inhibits the formation of bile acids from cholesterol.[123] The hydroxylase is also inhibited by 7β-hydroxycholesterol and 7-ketocholesterol.[124] The latter compound, which is not derived from 7α-hydroxycholesterol, appears to have a particularly strong inhibitory effect, with a K_i of 7 *M*.[125] In isolated rat hepatocytes, 25- and 26-hydroxycholesterol, as well as 7-keto, inhibit total bile acid synthesis.[126] The mechanism of inhibition by these cholesterol oxides has not been established, and may be secondary to depletion of cholesterol substrate as a result of HMG-CoA reductase inhibition or may be a direct effect on the hydroxylase enzyme.

In the isolated rat hepatocyte model, addition of 7α-hydroxycholesterol bypasses the cholesterol 7α-hydroxylase step and results in a 2- to 3-fold increase in the synthesis of cholic and chenodeoxycholic acids.[126] Cholesterol 5α,6α-epoxide, although having no effect on endogenous bile acid synthesis from cholesterol *in vivo* in rats, also produces an increased total output of bile acids, probably because it acts as a substrate for bile acid synthesis at a stage distal to cholesterol 7α-hydroxylase.[127]

IV. CHOLESTEROL ESTERIFICATION AND CHOLESTERYL ESTER HYDROLYSIS

A. LIVER

Esterification of cholesterol in liver and other tissues is catalyzed by the microsomal enzyme acyl CoA:cholesterol acyltransferase (ACAT) (EC 2.3.1.26). Studies by Drevon et al.[80] demonstrated that cholesterol esterification in cultured rat hepatocytes was stimulated 3- to 6-fold by the addition of 10 μg/ml of 25-OH. The effect occurred whether labeled oleic acid, mevalolactone, or cholesterol was used as the added substrate. The 25-OH effect was specific for cholesterol esterification, since oleic acid incorporation into phospholipids and triglycerides was not increased. Pure cholesterol at 10 μg/ml had no effect on cholesterol-esterifying activity. Maximal stimulation of cholesterol esterification was observed after 15 min of incubation with 25-OH, was reduced after 4 to 6 h, and had disappeared after 18 h. This differs from the inhibitory effect of 25-OH on HMG-CoA reductase activity, which persists for greater than 22 h of incubation. In other experiments by Drevon et al.[128] hepatic microsomes isolated from cells preincubated with 25-OH showed a 3-fold increase in ACAT activity. Direct incubation of microsomes with 25-OH also increased ACAT 2-fold. Lichtenstein and Brecher have demonstrated similar stimulation of rat liver microsomal ACAT by 25-OH.[129] In these studies the inhibitory effect of progesterone on ACAT could not be reversed by 25-OH.

The stimulatory effect of 25-OH on cholesterol esterification *in vitro* does not induce any increase in cellular cholesterol ester content in fetal mouse liver cells. Feeding of 25-OH to rats likewise did not result in any change in the concentrations of esterified or free cholesterol in liver microsomes, but 7-keto feeding did increase the proportion of cholesterol esters in the microsomes. The stimulatory effects of 25-OH on cholesterol esterification noted by Drevon et al.[128] in cultured rat hepatocytes did result in an increased amount of esterified cholesterol in VLDL (very-low-density lipoproteins) secreted by these cells.

B. INTESTINE

The effects of 25-OH on cholesterol esterification were studied by Field and Mathur[130] in both isolated rabbit jejunal microsomes and in intact intestinal absorbtive cells (enterocytes). In the microsomal preparation 25-OH increased the rate of cholesterol esterification, but the effect was diminished in microsomes with higher cholesterol content. 25-OH stimulated cholesterol esterification 6-fold in isolated enterocytes, where substrate was provided either in the form of phospholipid liposomes or bile salt micelles. The uptake of the cholesterol substrate was not influenced by 25-OH. An oleic acid substrate was not incorporated into cholesterol esters at a faster rate in the presence of 25-OH, even with large amounts of cholesterol available. Cholesterol esterification in enterocytes from rabbits previously fed cholesterol was stimulated by 25-OH only half as much as cells from rabbits not fed cholesterol. Further experiments indicated that 25-OH stimulated esterification of both existing cell membrane cholesterol and newly synthesized cholesterol. Studies with two other oxysterols, 24(S),25-epoxylanosterol and 25-hydroxylanosterol, have also shown enhancing effects on cholesterol esterification in rat intestinal epithelial cells.

C. FIBROBLASTS

In cultured human fibroblasts the rate of cholesterol esterification and ACAT activity is stimulated by 25-OH, 7-keto, or 6-keto when cells grown in lipoprotein-deficient media are used.[131] The stimulation of esterification correlated with the inhibition of HMG-CoA reductase induced by these agents. An 8-fold increase in ACAT activity was observed in cells incubated for 5 hours with 5 μg/ml of 25-OH. This was similar to the stimulatory effect of added LDL on ACAT activity. The stimulatory effect of 25-OH on ACAT in human fibroblasts could be blocked by incubation with 7-keto-20-oxacholesterol, an analog of 7-ketocholesterol.[132]

D. ARTERIAL SMOOTH MUSCLE CELLS

Studies by Morin and Peng[133] in cultured rabbit aortic smooth muscle cells showed a 4-fold increase in ACAT activity after incubation with 5 g/ml of 25-OH for 1 h. Incubation with 5 g/ml of cholestane-3β,5α,6β-triol or pure cholesterol had no significant effects. No stimulation by 25-OH was noted at 15 min, the stimulatory effect had disappeared after 5 h, and no effect was noted after 24 h. In arterial smooth muscle cells, maximal suppression of HMG-CoA reductase activity by 25-OH is reached at about one hour.[25] This could result in depletion of microsomal membrane cholesterol at this time interval, which could increase the accessibility of exogenous cholesterol to the active sites on ACAT and account for the maximal stimulation of esterification observed at one hour.

Cholesterol oxides can induce injury to arterial cells both *in vitro* and *in vivo*, and have been implicated as possible etiologic agents in atherogenesis.[134,135] ACAT activity is low in normal rabbit arteries, but is increased up to 70-fold in experimental atherosclerosis induced by cholesterol feeding.[135,136] The observed stimulation of cholesterol esterification by 25-OH could be one mechanism for accumulation of cholesterol esters in arterial smooth muscle cells, leading to foam cell formation and progression of atherosclerosis.

E. ADRENAL

ACAT activity in rat or bovine adrenal cortical microsomes or intact cells is not influenced by incubation with either 25-OH or cholesterol.[138-140] Since the cholesterol content of adrenal microsomes is higher than in other tissues (66 μg/ml protein in bovine adrenal vs. 35 μg/mg protein in rat liver),[139] the lack of 25-OH effect may reflect the fact that adrenal ACAT is more saturated with substrate than in liver, intestine, or other tissues. The inhibitory effect of the drug Sandoz 58-035 on adrenal ACAT can be partially prevented by addition of 25-OH.[138]

F. ESTERIFICATION OF 25-HYDROXYCHOLESTEROL

Human fibroblasts incubated with 5 μg/ml 25-OH for 5 hours were found to contain 5 μg/mg protein of 25-OH in the esterified form.[131] Liver microsomes were also noted to esterify 25-OH, with the esterification occurring at the 3β-hydroxyl group of 25-OH.[141] Under incubation conditions where the substrate concentration was not rate limiting, cholesterol did not affect the rate of 25-OH esterification, and vice versa, suggesting that 25-OH esterification is catalyzed by an enzyme other than ACAT.[141] Additional evidence for this is that the rat ovary, although possessing ACAT activity, does not esterify 25-OH.[141]

G. MECHANISMS FOR THE STIMULATORY EFFECT OF 25-HYDROXYCHOLESTEROL ON ACAT

In liver cells, the stimulatory effect of 25-OH on ACAT is not blocked by cycloheximide.[80] This plus the rapidity of onset of the 25-OH effect indicates that the effect is not mediated by new synthesis of enzyme protein. The rapid stimulation of ACAT by addition of 25-OH to liver microsomes also suggests that a direct effect of 25-OH is involved rather than an indirect metabolic effect of 25-OH or its conversion products. When microsomal membranes were incubated with 25-OH and re-isolated by ultracentrifugation, 25-OH was bound to these membranes.[142] Drevon et al.[80] have postulated that 25-OH may directly affect ACAT activity by altering the lipid environment of the enzyme in the microsomal membrane or by allosteric modification of the enzyme protein. The diminishing effect of 25-OH in the presence of increased microsomal cholesterol content suggests that 25-OH may act to release cholesterol from microsomal membranes or to increase substrate availability to ACAT in some other manner.[130] Alternatively, excess cholesterol may prevent binding of 25-OH to its effector site on ACAT. In Chinese hamster ovary cells there is evidence for a short-lived protein factor which may either inhibit ACAT or participate in the removal of cholesterol from the ACAT substrate pool.[143] A possible mechanism of 25-OH action may be to block the activity of this factor.

H. ARTERIAL CHOLESTERYL ESTER HYDROLYSIS

Hydrolysis of cholesteryl esters is catalyzed by an acid cholesteryl ester hydrolase (ACEH) (EC3.1.1.13) and a neutral cholesteryl ester hydrolase (NCEH) (EC 3.1.1.13). Stimulation of ACEH activity by 4-fold after incubation of acetone-butanol extracts of pig aortas with 50 μg/ml of 26-hydroxycholesterol and 9-fold stimulation with 50 g/ml cholestane-3β,5α,6β-triol has been observed.[144] These effects were attributed to a dispersing effect of the oxysterols, improving access of the enzyme to the cholesteryl ester substrate. Later studies using addition of 5 μg/ml 25-OH or cholestanetriol to cultured rabbit aortic smooth muscle cells showed no significant effects on either ACEH or NCEH.[133] The divergent results may be due to the use of intact cells rather than extracts or to the different oxysterol concentrations used. The stimulation of arterial cholesterol ACAT by 25-OH as described above, without a parallel increase in cholesteryl ester hydrolytic activity may result in atherogenic cholesteryl ester accumulation.

V. SUMMARY

A number of cholesterol oxides, including 25-hydroxycholesterol, 7-ketocholesterol, 7α- and 7β-hydroxycholesterol, cholestane-3β,5α,6β-triol, and others have potent inhibitory effects on cholesterol synthesis in a variety of tissues. This effect appears to be mediated primarily by inhibition of HMG-CoA reductase, although inhibitory effects on enzymes at other stages of cholesterol biosynthesis, such as HMG-CoA synthetase and methyl sterol oxidase have also been noted. The mechanism of inhibition of HMG-CoA reductase by cholesterol oxides appears to be primarily via an increase in enzyme degradation, although decreases in the rate of enzyme synthesis have been noted by several investigators. The effects of the cholesterol oxides may be mediated by binding to one or more cellular oxysterol-binding proteins, which may then transport the sterols to a nuclear or cytoplasmic site of action. The physiological relevance of the HMG-CoA inhibition is as yet unresolved. 25-OH and 24(S),25-epoxycholesterol are formed in tissues at levels which could have inhibitory effects, and these sterols can be metabolized by cells to reverse the inhibition. Cholesterol side-chain cleavage, hormonal steroid production from cholesterol, and cholesterol 7α-hydroxylase are also inhibited by cholesterol oxides. Esterification of cholesterol and the activity of acyl CoA:cholesterol acyltransferase in liver, fibroblasts, arterial smooth muscle cells, and other tissues is increased by 25-OH. This stimulation of ACAT, together with inhibition of HMG-CoA reductase and other cholesterol biosynthetic enzymes may function in the co-ordinate regulation of cellular cholesterol metabolism by cholesterol oxides.

REFERENCES

1. **Tomkins, G. M., Sheppard, I., and Chaikoff, I. L.,** Cholesterol synthesis by liver. IV. Suppression by steroid administration, *J. Biol. Chem.*, 203, 781, 1953.
2. **Steinberg, D. and Fredrickson, D. S.,** Inhibitors of cholesterol biosynthesis and the problem of hypercholesterolemia, *Ann. N.Y. Acad. Sci.*, 64, 579, 1956.
3. **Packie, R. M. and Kandutsch, A. A.,** Rates of sterol synthesis and hydroxymethylglutaryl coenzyme A reductase levels, and the effects of cholest-4-en-3-one on these parameters, in the livers of inbred strains of mice, *Biochem. Genet.*, 4, 203, 1970.
4. **Chen, H. W. and Kandutsch, A. A.,** Effects of cholesterol derivatives on sterol biosynthesis, in *Atherosclerosis Drug Discovery*, Day, C., Ed., Plenum Press, New York, 1976, 405.
5. **Erickson, S. K., Matsui, S. M., Shrewsbury, M. A., Cooper, A. D., and Gould, R. G.,** Effects of 25-hydroxycholesterol on rat hepatic 3-hydroxy-3-methylglutaryl coenzyme A reductase activity in vivo, in perfused liver, and in hepatocytes, *J. Biol. Chem.*, 253, 4159, 1978.
6. **Langdon, R., El-Masry, S., and Counsell, R. E.,** Induction of HMG-CoA reductase by the administration of 20,25-diazacholesterol, *J. Lipid Res.*, 18, 24, 1977.
7. **Erickson, S. K., Cooper, A. D., Matsui, S. M., and Gould, R. G.,** 7-ketocholesterol: its effects on hepatic cholesterogenesis and its hepatic metabolism in vivo and in vitro, *J. Biol. Chem.*, 252, 5186, 1977.
8. **Kandutsch, A. A., Heiniger, H.-J., and Chen, H. W.,** Effects of 25-hydroxycholesterol and 7-ketocholesterol inhibitors of sterol synthesis, administered orally to mice, *Biochim. Biophys. Acta*, 486, 260, 1977.
9. **Vargas, R. E., Allred, J. B., Biggert, M. D., and Naber, E. C.,** Effect of dietary 7-ketocholesterol, pure, or oxidized cholesterol on hepatic 3-hydroxy-3-methylglutaryl-coenzyme A reductase activity, energy balance, egg cholesterol concentration, and c-acetate incorporation into yolk lipids of laying hens, *Poultry Sci.*, 65, 1333, 1986.
10. **Kandutsch, A. A. and Packie, R. M.,** Comparison of the effects of some C_{27}-, C_{21}-, and C_{19}-steroids upon hepatic sterol synthesis and hydroxymethylglutaryl-CoA reductase activity, *Arch. Biochem. Biophys.*, 140, 122, 1970.

11. **Kandutsch, A. A. and Chen, H. W.,** Inhibition of sterol synthesis in cultured mouse cells by 7α-hydroxycholesterol, 7β-hydroxycholesterol, and 7-ketocholesterol, *J. Biol. Chem.*, 248, 8408, 1973.
12. **Kandutsch, A. A. and Chen, H. W.,** Inhibition of sterol synthesis in cultured mouse cells by cholesterol derivatives oxygenated in the side chain, *J. Biol. Chem.*, 249, 6057, 1974.
13. **Kandutsch, A. A. and Chen, H. W.,** Regulation of sterol synthesis in cultured cells by oxygenated derivatives of cholesterol, *J. Cell Physiol.*, 85, 415, 1975.
14. **Yachnin, S., Streuli, R. A., Gordon, L. I., and Hsu, R. C.,** Alteration of peripheral blood cell membrane function and morphology by oxygenated sterols; a membrane insertion hypothesis, *Curr. Top. Hematol.*, 2, 245, 1979.
15. **Spence, J. T. and Gaylor, J. L.,** Investigation of regulation of microsomal hydroxymethylglutaryl coenzyme A reductase and methyl sterol oxidase of cholesterol biosynthesis, *J. Biol. Chem.*, 252, 5852, 1977.
16. **Kandutsch, A. A. and Chen, H. W.,** Inhibition of cholesterol synthesis by oxygenated sterols, *Lipids*, 13, 704, 1978.
17. **Astruc, M., Roussillon, R., Defay, R., Descomps, B., and Crastes De Paulet, A.,** DNA and cholesterol biosynthesis in synchronized embroyonic rat fibroblasts. II. Effects of sterol biosynthesis inhibitors on cell division, *Biochim. Biophys. Acta*, 763, 11, 1983.
18. **Brown, M. S. and Goldstein, J. L.,** Suppression of 3-hydroxy-3-methylglutaryl coenzyme A reductase activity and inhibition of growth of human fibroblasts by 7-ketocholesterol, *J. Biol. Chem.*, 249, 7306, 1974.
19. **Breslow, J. L., Lothrop, D. A., Spaulding, D. R., and Kandutsch, A. A.,** Cholesterol, 7-ketocholesterol and 25-hydroxycholesterol uptake studies and effect on 3-hydroxy-3-methylglutaryl-coenzyme A reductase activity in human fibroblasts, *Biochim. Biophys. Acta*, 398, 10, 1975.
20. **Krieger, M., Goldstein, J. L., and Brown, M. S.,** Receptor-mediated uptake of low density lipoprotein reconstituted with 25-hydroxycholesteryl oleate suppresses 3-hydroxy-3-methylglutaryl-coenzyme A reductase and inhibits growth of human fibroblasts, *Proc. Natl. Acad Sci. U.S.A.*, 75, 5052, 1978.
21. **Stange, E. F., Schneider, A., Preclik, G., Alavi, M., and Ditschuneit, H.,** Regulation of 3-hydroxy-3-methylglutaryl CoA reductase by analogs of cholesterol and bile acids in cultured intestinal mucosa, *Arch. Biochem. Biophys.*, 244, 310, 1986.
22. **Gebhard, R. L. and Cooper, A. D.,** Regulation of cholesterol synthesis in cultured canine intestinal mucosa, *J. Biol. Chem.*, 253, 2790, 1978.
23. **Svirdov, D. D., Safanova, I. G., Talalaev, A. G., Repin, V. S., and Smirnov, V. N.,** Regulation of cholesterol synthesis in isolated epithelial cells of human small intestine, *Lipids*, 21, 759, 1986.
24. **Tham, P., Peng, S.-K., Taylor, C. B., and Mikkelson, B.,** Suppression of cholesterol synthesis in aortic smooth muscle cell cultures by oxidation derivatives of cholesterol, *Fed. Proc. Fed. Am. Soc. Exp. Biol.*, 37, 1381, 1978.
25. **Peng, S.-K., Tham, P., Taylor, C. B., and Mikkelson, B.,** Cytotoxicity of oxidation derivatives of cholesterol on cultured aortic smooth muscle cells and their effect on cholesterol biosynthesis, *Am. J. Clin. Nutr.*, 32, 1033, 1979.
26. **Adamany, A. M. and Tugendhaft Mills, J.,** Regulation of glycoprotein synthesis: interdependence of cholesterol synthesis and glycoprotein assembly, *Fed. Proc. Fed. Am. Soc. Exp. Biol.*, 39, 1830, 1980.
27. **Mills, J. and Adamany, A. M.,** Impairment of dolichyl saccharide synthesis and dolichol-mediated glycoprotein assembly in the aortic smooth muscle cell in culture by inhibitors of cholesterol biosynthesis, *J. Biol. Chem.*, 253, 5270, 1978.
28. **Naseem, S. M. and Heald, F. P.,** Cytotoxicity of cholesterol oxides and their effects on cholesterol metabolism in cultured human aortic smooth muscle cells, *Biochem. Int.*, 14, 71, 1987.
29. **Randolph, R. K., Smith, B. P., and St. Clair, R. W.,** Cholesterol metabolism in pigeon smooth muscle cells lacking a functional low density lipoprotein receptor pathway, *J. Lipid Res.*, 25, 903, 1984.
30. **Philippot, J. R., Cooper, A. G., and Hoelzl Wallach, D. F.,** A nitroxide-sterol derivative potently modifies cholesterol biosynthesis by normal and neoplastic guinea pig lymphocytes, *Biochim. Biophys. Acta.*, 406, 161, 1975.
31. **Philippot, J. R., Cooper, A. G., and Hoelzl Wallach, D. F.,** 25-hydroxycholecalciferol and 1,25-dihydroxycholecalciferol are potent inhibitors of cholesterol biosynthesis by normal and leukemic (L_2C) guinea pig lymphocytes, *Biochem. Biophys. Res. Commun.*, 72, 1035, 1976.
32. **Philippot, J. R., Cooper, A. G., and Wallach, D. F. H.,** Regulation of cholesterol biosynthesis by normal and leukemic (L_2C) guinea pig lymphocytes, *Proc. Natl. Acad Sci. U.S.A.*, 74, 956, 1977.
33. **Yachnin, S., Golomb, H. M., West, E. J., and Staffold, C.,** Increased cholesterol biosynthesis in leukemic cells from patients with hairy cell leukemia, *Blood*, 61, 50, 1983.
34. **Astruc, M., Laporte, M., Tabacik, C., and Crastes de Paulet, A.,** Effect of oxygenated sterols on 3-hydroxy-3-methylglutaryl coenzyme a reductase and DNA synthesis in phytohemagglutinin-stimulated human lymphocytes, *Biochem. Biophys. Res. Commun.*, 85, 691, 1978.
35. **Gibbons, G. F.,** The role of oxysterols in the regulation of cholesterol biosynthesis, *Biochem. Soc. Trans.*, 11, 649, 1983.

36. **Sinenksy, M.,** Isolation of a mammalian cell mutant resistant to 25-hydroxy cholesterol, *Biochem. Biophys. Res. Commun.*, 78, 863, 1977.
37. **Chen, H. W., Cavenee, W. K., and Kandutsch, A. A.,** Sterol synthesis in variant chinese hamster lung cells selected for resistance to 25-hydroxycholesterol, *J. Biol. Chem.*, 254, 715, 1979.
38. **Cavenee, W. K., Gibbons, G. F., Chen, H. W., and Kandutsch, A. A.,** Effects of various oxygenated sterols on cellular sterol biosynthesis in chinese hamster lung cells resistant to 25-hydroxycholesterol, *Biochim. Biophys. Acta*, 575, 255, 1979.
39. **Westhafer, M. A., Hall, S. L., and Nordstrom, J. L.,** Resistance to 25-hydroxy cholesterol is a dominant selectable genetic trait for mammalian cells, *Fed. Proc. Fed. Am. Soc. Exp. Biol.*, 42, 2223, 1983.
40. **Schroepfer, G. J.,** Sterol biosynthesis, *Annu. Rev. Biochem.*, 50, 585, 1981.
41. **Rudney, H. and Sexton, R. C.,** Regulation of cholesterol biosynthesis. A, *Rev. Nutr.*, 6, 245, 1986.
42. **Saucier, S. E. and Kandutsch, A. A.,** Inactive 3-hydroxy-3-methylglutaryl-coenzyme A reductase in broken cell preparations of various mammalian tissues and cell cultures, *Biochim. Biophys. Acta*, 572, 541, 1979.
43. **Erickson, S. K., Shrewsbury, M. A., Gould, R. G., and Cooper, A. D.,** Studies on the mechanisms of the rapid modulation of 3-hydroxy-3-methylglutaryl coenzyme a reductase in intact liver by mevalonolactone and 25-hydroxycholesterol, *Biochim. Biophys. Acta*, 70, 620, 1980.
44. **Cighetti, G., Galli, G., and Galli Kienle, M.,** A simple model for studies on the regulation of cholesterol synthesis using freshly isolated hepatocytes, *Eur. J. Biochem.*, 133, 573, 1983.
45. **Beg, Z. H., Renzikov, D. C., and Avigan, J.,** Regulation of 3-hydroxy-3-methylglutaryl coenzyme A reductase activity in human fibroblasts by reversible phosphorylation: modulation of enzymatic activity by low density lipoprotein, sterols, and mevalonolactone, *Arch. Biochem. Biophys.*, 244, 310, 1986.
46. **Tanaka, R. D., Edwards, P. A., Lan, S.-F., and Fogelman, A. M.,** Regulation of 3-hydroxy-3-methylglutaryl coenzyme A reductase activity in avian myeloblasts, *J. Biol. Chem.*, 258, 13331, 1983.
47. **Chang, T. Y., Limanek, J. S., and Chang, C. C. Y.,** Evidence indicating that inactivation of 3-hydroxy-3-methylglutaryl coenzyme A reductase by low density lipoprotein or by 25-hydroxycholesterol requires mediator protein(s) with rapid turnover rate, *J. Biol. Chem.*, 256, 6174, 1981.
48. **Faust, J. R., Luskey, K. L., Chin, D. J., Goldstein, J. L., and Brown, M. S.,** Regulation of synthesis and degradation of 3-hydroxy-3-methylglutaryl-coenzyme A reductase by low density lipoprotein and 25-hydroxycholesterol in UT-1 cells, *Proc. Natl. Acad. Sci. U.S.A.*, 79, 5205, 1982.
49. **Gil, G., Faust, J. R., Chin, D. J., Goldstein, J. L., and Brown, M. S.,** Membrane-bound domain of HMG CoA reductase is required for sterol-enhanced degradation of the enzyme, *Cell*, 41, 249, 1985.
50. **Edwards, P. A. and Fogelman, A. M.,** Effect of inhibitors of the activity of 3-hydroxy-3-methyl-glutaryl coenzyme A reductase on enzyme synthesis and degradation, *Fed. Proc. Fed. Am. Soc. Exp. Biol.*, 41, 1388, 1982.
51. **Sinensky, M., Torget, R., and Edwards, P. A.,** Radioimmune precipitation of 3-hydroxy-3-methylglutaryl coenzyme a reductase from chinese hamster fibroblasts, *J. Biol. Chem.*, 256, 11774, 1981.
52. **Taylor, F. R. and Kandutsch, A. A.,** The oxysterol receptor, in *Steroid and Sterol Hormone Action*, Spelsberg, T. C. and Kumar, R., Eds., Martinus Nijhoff, Boston, 1987, chap. 21.
53. **Sinenksy, M., Torget, R., and Edward, P. A.,** Radioimmune precipitation of 3-hydroxy-3 methylglutaryl coenzyme A reductase from chinese hamster fibroblasts, *J. Biol. Chem.*, 256, 11774, 1981.
54. **Lusky, K. L., Faust, J. R., Chin, D. J., Brown, M. S., and Goldstein, J. L.,** Amplification of the gene for 3-hydroxy-3-methylglutaryl coenzyme A reductase, but not for the 53-kDa protein in UT-1 cells, *J. Biol. Chem.*, 258, 8462, 1983.
55. **Picard, F., Duval, D., and Homo, F.,** Effect of 25-hydroxycholesterol and dexamethasone on sterol synthesis in isolated mouse thymocytes, *J. Steroid Biochem.*, 15, 461, 1981.
56. **Kandutsch, A. A., Taylor, F. R., and Shown, E. P.,** Different forms of the oxysterol-binding protein, *J. Biol. Chem.*, 259, 12388, 1984.
57. **Erickson, S. K., Meyer, D. J., and Gould, R. G.,** Purification and characterization of a new cholesterol-binding protein from rat liver cytosol, *J. Biol. Chem.*, 253, 1817, 1978.
58. **Dawson, P. A., Van Der Westhuyzen, D. R., Goldstein, R. L., and Brown, M. S.,** Purification of oxysterol binding protein from hamster liver cytosol, *J. Biol. Chem.*, 264, 9046, 1989.
59. **Beseme, F., Astruc, M. E., Defay, R., Descomps, B., and Crastes de Paulet, A.,** Characterization of oxysterol-binding protein in rat embryo fibroblasts and variations as a function of the cell cycle, *Biochim. Biophys. Acta*, 886, 96, 1986.
60. **Sinensky, M. and Mueller, G.,** Cytosolic 25-hydroxycholesterol binding activity of chinese hamster ovary cells, *Arch. Biochem. Biophys.*, 209, 314, 1981.
61. **Defay, R., Astruc, M., Beseme, F., Descomps, B., and Crastes de Paulet, A.,** Physico-chemical properties of the hydroxysterol binding protein of human lymphocyte cytosol, *FEBS Lett.*, 173, 1682, 1984.
62. **Taylor, F. R., Saucier, S. E., Shown, E. P., Parish, E. J., and Kandutsch, A. A.,** Correlation between oxysterol binding to a cytosolic binding protein and potency in the repression of hydroxymethylglutaryl coenzyme A reductase, *J. Biol. Chem.*, 259, 12382, 1984.

63. **Sinenksy, M., Loge, J., and Torget, R.,** A somatic cell mutant missing 25-hydroxycholesterol binding activity, *Fed. Proc. Fed. Am. Soc. Exp. Biol.*, 41, 910, 1982.
64. **Kandutsch, A. A.,** A model for the regulation of HMG-CoA reductase by oxygenated sterols, *Fed. Proc. Fed. Am. Soc. Exp. Biol.*, 41, 1387, 1982.
65. **Taylor, F. R. and Kandutsch, A. A.,** Oxysterol binding protein, *Chem. Phys. Lipids,* 8, 187, 1985.
66. **Taylor, F. R., Kandutsch, A. A., Anzalone, L., Phirwa, S., and Spencer, T. A.,** Photoaffinity labeling of the oxysterol receptor, *J. Biol. Chem.*, 263, 2264, 1988.
67. **Dawson, P. A., Ridgway, N. D., Slaughter, C. A., Brown, M. S., and Goldstein, J. L.,** cDNA cloning and expression of oxysterol-binding protein, an oligomer with a potential leucine zipper, *J. Biol. Chem.*, 264 (28), 16798, 1989.
68. **Saucier, S. E., Kandutsch, A. A., Taylor, F. R., Spencer, T. A., Phirwa, S., and Gayen, A. K.,** Identification of regulatory oxysterols, 24(S),25-epoxycholesterol and 25-hydroxycholesterol, in cultured fibroblasts, *J. Biol. Chem.*, 260, 14571, 1985.
69. **Panini, S. R., Sexton, R. C., Gupta, A. K., Parish, E. J., Chitrakorn, S., and Rudney, H.,** Regulation of 3-hydroxy-3-methylglutaryl coenzyme A reductase activity and cholesterol biosynthesis by oxylanosterols, *J. Lipid Res.*, 27, 1190, 1986.
70. **Spencer, T. A., Gayen, A. K., Phirwa, S., Nelson, J. A., Taylor, F. R., Kandutsch, A. A., and Erickson, S. R.,** 24(S),25-epoxycholesterol, evidence consistent with a role in the regulation of hepatic cholesterogenesis, *J. Biol. Chem.*, 260, 13391, 1985.
71. **Panini, S. R., Sexton, R. C., and Rudney, H.,** Regulation of 3-hydroxy-3-methylglutaryl coenzyme A reductase by oxysterol by-products of cholesterol biosynthesis, *J. Biol. Chem.*, 259, 7767, 1984.
72. **Taylor, F. R. and Kandutsch, A. A.,** Metabolism of 25-hydroxycholesterol in mammalian cell cultures. Side-chain scission to pregnenolone in mouse L929 fibroblasts, *J. Lipid Res.*, 30, 899, 1989.
73. **Esterman, A. L., Baum, H., Javitt, N. B., and Darlington, G. J.,** 26-hydroxycholesterol: regulation of hydroxymethylglutaryl-CoA reductase activity in chinese hamster ovary cell culture, *J. Lipid Res.*, 24, 1304, 1983.
74. **Gould, R. G.,** Some aspects of the control of hepatic cholesterol biosynthesis, in *Cholesterol Metabolism and Lipolytic Enzymes,* Polonovski, J., Ed., Masson, New York, 1977, 13.
75. **Brown, M. S., Brannan, P. G., Bohmfalk, H. A., Brunschede, G. Y., Dana, S. E., Helgeson, J., and Goldstein, J. L.,** Use of mutant fibroblasts in the analysis of the regulation of cholesterol metabolism in human cells, *J. Cell Physiol.*, 85, 425, 1975.
76. **Goldstein, J. L., Faust, J. R., Brunschede, G. Y., and Brown, M. S.,** Steroid requirements for suppression of HMG CoA reductase activity in cultured human fibroblasts, *Adv. Exptl. Med. Biol.*, 63, 77, 1975.
77. **Cavenee, W. K., Johnston, D., and Melnykovych, G.,** Regulation of cholesterol biosynthesis in HeLa S3G cells by serum lipoproteins: dexamethasone-mediated interference with suppression of 3-hydroxy-3-methylglutaryl coenzyme A reductase, *Proc. Natl. Acad. Sci. U.S.A.*, 75, 2103, 1978.
78. **Volpe, J. J. and Hennessy, S. W.,** Cholesterol biosynthesis and 3-hydroxy-3-methylglutaryl coenzyme A reductase in cultured glial and neuronal cells, *Biochim. Biophys. Acta,* 486, 408, 1977.
79. **Martin, G. G.,** Regulation of rat liver 3-hydroxy-3-methylglutaryl coenzyme A reductase by methoxy-polyoxyethylated cholesterol, *Biochim. Biophys. Acta,* 837, 163, 1985.
80. **Drevon, C. A., Weinstein, D. B., and Steinberg, D.,** Regulation of cholesterol esterification and biosynthesis in monolayer cultures of normal adult hepatocytes, *J. Biol. Chem.*, 255, 9128, 1980.
81. **Chang, T. Y., Limanek, J. S., and Chang, C. C. Y.,** Evidence indicating that inactivation of 3-hydroxy-3-methylglutaryl coenzyme A reductase by low density lipoprotein or by 25-hydroxycholesterol requires mediator protein(s) with rapid turnover rate, *J. Biol. Chem.*, 256, 6174, 1981.
82. **Taylor, F. R., Saucier, S. E., Shown, E. P., Parish, E. J., and Kandutsch, A. A.,** Correlation between oxysterol binding to a cytosolic binding protein and potency in the repression of hydroxymethylglutaryl coenzyme A reductase, *J. Biol. Chem.*, 259, 12382, 1984.
83. **Parish, E. J., Chitrakorn, S., Taylor, F. R., and Saucier, S. E.,** Chemical synthesis of 4,4′-dimethyl-7-oxygenated sterols. Inhibitors of 3-hydroxy-3-methylglutaryl reductase, *Chem. Phys. Lipids,* 36, 179, 1984.
84. **Parish, E. J., Nanduri, V. B. B., Kohl, H. H., and Taylor, F. R.,** Oxysterols: chemical synthesis, biosynthesis and biological activities, *Lipids,* 21, 27, 1986.
85. **Kandutsch, A. A. and Thompson, E. B.,** Cytosolic proteins that bind oxygenated sterols. Cellular distribution, specificity, and some properties, *J. Biol. Chem.*, 255, 10813, 1980.
86. **Chen, H. W. and Leonard, D. A.,** Chloroquine inhibits cyclization of squalene oxide to lanosterol in mammalian cells, *J. Biol. Chem.*, 259, 8156, 1984.
87. **Taylor, F. R., Kandutsch, A. A., Gayen, A. K., Nelson, J. A., Steckbeck Nelson, S., Phirwa, S., and Spencer, T. A.,** 24,25-Epoxysterol metabolism in cultured mammalian cells and repression of 3-hydroxy-3-methylglutaryl-CoA reductase, *J. Biol. Chem.*, 261, 15039, 1986.

88. **Astruc, M., Roussillon, S., Defay, R., Descomps, B., and Crastes de Paulet, A.,** DNA and cholesterol biosynthesis in synchronized embryonic rat fibroblasts. II. Effects of sterol biosynthesis inhibitors on cell division, *Biochim. Biophys. Acta,* 763, 11, 1983.
89. **Hill, J. C.,** Effects of cytotoxic cholesterol autoxidation derivatives on membrane function, *Diss. Abstr. Int. B.,* 45, 1158, 1984.
90. **Habenich, A. J. R., Glomset, J. A., and Ross, R.,** Relation of cholesterol and mevalonic acid to the cell cycle in smooth muscle and Swiss 3T3 cells stimulated to divide by platelet derived growth factor, *J. Biol. Chem.,* 255, 5134, 1980.
91. **Kayden, H. J., Hatam, L., and Beratis, N. G.,** Regulation of 3-hydroxy-3-methylglutaryl coenzyme A reductase activity and the esterification of cholesterol in human long term lymphoid cell lines, *Biochemistry,* 15, 521, 1976.
92. **Defay, R. E., Astruc, M. E., Roussillon, S., Descomps, B., and Crastes de Paulet, A.,** A specific hydroxysterol binding protein in human lymphocyte cytosol, *Biochimie,* 64, 331, 1982.
93. **Defay, R., Astruc, M. E., Roussillon, S., Descomps, B., and Crastes de Paulet, A.,** DNA synthesis and 3-hydroxy-3-methylglutaryl CoA reductase activity in PHA stimulated human lymphocytes: a comparative study of the inhibitory effects of some oxysterols with special reference to side chain hydroxylated derivatives, *Biochem. Biophys. Res. Commun.,* 106, 362, 1982.
94. **Ramachandran, C. K., Gray, S. L., and Melnykovych, G.,** Cytoplasmic 3-hydroxy-3-methylglutaryl coenzyme A synthase and the regulation of sterol synthesis in tissue culture cells, *Biochim. Biophys. Acta,* 618, 439, 1980.
95. **Maltese, W. A., Reitz, B. A., and Volpe, J. J.,** Selective decrease of the viability and the sterol content of proliferating versus quiescent glioma cells exposed to 25-hydroxycholesterol, *Cancer Res.,* 41, 3448, 1981.
96. **Javitt, N. B., Kok, E., Burstein, S., Cohen, B., and Kutscher, J.,** 26-hydroxycholesterol. Identification and quantitation in human serum, *J. Biol. Chem.,* 256, 12644, 1981.
97. **Chin, J. and Chang, T.-Y.,** Evidence for coordinate expression of 3-hydroxy-3-methylglutaryl coenzyme A reductase and low density lipoprotein binding activity, *J. Biol. Chem.,* 256, 6304, 1981.
98. **Luskey, K. L. and Stevens, B.,** Human 3-hydroxy-3-methylglutaryl coenzyme A reductase. Conserved domains responsible for catalytic activity and sterol-regulated degradation, *J. Biol. Chem.,* 260, 10271, 1985.
99. **Esterman, A. L., Baum, H., Javitt, N. B., and Darlington, G. J.,** 26-Hydroxycholesterol: regulation of hydroxymethylglutaryl-CoA reductase activity in Chinese hamster ovary cell culture, *J. Lipid Res.,* 24, 1304, 1983.
100. **Tabas, I., Weilgand, D. A., and Tall, R. R.,** Unmodified low density lipoprotein causes cholesteryl ester accumulation in J774 macrophages, *Proc. Natl. Acad. Sci. U.S.A.,* 82, 416, 1985.
101. **Miller, S. C. and Melynkovych, G.,** Regulation of cholesterol biosynthesis and esterification by 25-hydroxycholesterol in a macrophage-like cell line: uncoupling by progesterone, *J. Lipid Res.,* 25, 991, 1984.
102. **Rintoul, D. A. and Simoni, R. D.,** Isolation and preliminary characterization of rat liver cells resistant to 7-ketocholesterol, *Biochim. Biophys. Acta,* 531, 322, 1978.
103. **Ramachandran, C. K., Gray, S. L., and Melynkovych, G.,** Cytoplasmic 3-hydroxy-3-methylglutaryl coenzyme A synthase and the regulation of sterol synthesis in tissue culture cells, *Biochim. Biophys. Acta,* 618, 439, 1980.
104. **Chang, T.-Y. and Limanek, J. S.,** Regulation of cytosolic acetoacetyl coenzyme A thiolase, 3-hydroxy-3-methylglutaryl coenzyme A synthase, 3-hydroxy-3-methylglutaryl coenzyme A reductase, and mevalonate kinase by low density lipoprotein and by 25-hydroxycholesterol in Chinese hamster ovary cells, *J. Biol. Chem.,* 255, 7787, 1980.
105. **Schnitzer-Polokoff, R. L. and Sinensky, M.,** Coordinate regulation of 3-hydroxy-3-methyl glutaryl coenzyme A synthase and 3-hydroxy-3-methyl glutaryl coenzyme A reductase activities in Chinese hamster fibroblasts, *Fed. Proc. Fed. Am. Soc. Exp. Biol.,* 41, 667, 1982.
106. **Have, C., Hansbury, E., Scallen, T. J., and Watson, J. A.,** Regulation of cholesterol synthesis in primary rat hepatocyte culture cells, *J. Biol. Chem.,* 254, 9573, 1979.
107. **Tabacik, C., Alianu, S., and Sanite-Marie, J.,** Regulation of cholesterol biosynthesis at a stage after the 3-hydroxy-3-methylglutaryl-CoA reductase step, in normal and leukemic (L_2C) guinea pig lymphocytes, *Biochim. Biophys. Acta,* 921, 504, 1987.
108. **Ortiz de Montellano, P. R., Beck, J. P., and Ourisson, G.,** Regulation of sterol biosynthesis and lysis of cultured hepatoma cells: inhibition of lanosterol demethylation by hydroxysterols, *Biochem. Biophys. Res. Commun.,* 90, 897, 1979.
109. **Gaylor, J. L. and Delwiche, C. V.,** Purification of a soluble rat liver protein that stimulates microsomal 4-methyl sterol oxidase activity, *J. Biol. Chem.,* 251, 6638, 1976.
110. **Grinstead, G. F., Trzaskos, J. M., Billheimer, F. T., and Gaylor, J. L.,** Cytosolic modulators of activities of microsomal enzymes of cholesterol biosynthesis. Effects of Acyl-CoA inhibition and cytosolic Z-protein, *Biochim. Biophys. Acta,* 751, 41, 1983.

111. **Kobayashi, S. and Ighi, S.,** Inhibition of cholesterol analogues of the side-chain cleavage of cholesterol and 20α-hydroxycholesterol in a preparation of hog adenocortical mitochondria, *J. Biochem.*, 66, 51, 1969.
112. **Falke, H. E., Degenhart, H. J., Abeln, G. J. A., and Visser, H. K. A.,** Effects of 25-hydroxycholesterol and aminoglutethmide in isolated rat adrenal cells. A model for congenital lipoid adrenal hyperplasia, *Mol. Cell. Endocrinol.*, 4, 107, 1976.
113. **Hujimans, J. G., Degenhart, H. J., and Korleve, D. J.,** Effects of hydroxylated sterols on isolated rat adrenal cells, *Proc. Soc. Endocrinol.*, 3P, 1977.
114. **Rabe, T., Kiesel, L., and Runnebaum, B.,** Regulation of human placental progesterone synthesis in vitro by naturally occurring steroids, *J. Steroid Biochem.*, 22, 657, 1985.
115. **Teicher, B. A., Koizumi, N., Koreeda, M., Shikita, M., and Talalay, P.,** Biosynthesis of pregnenolone from cholesterol by mitochondrial enzymes of bovine adrenal cortex, *Eur. J. Biochem.*, 91, 11, 1978.
116. **Falke, H. E., Degenhart, H. J., Abeln, G. J. A., and Visser, H. K. A.,** Studies on isolated rat adrenal cells metabolism of hydroxylated sterols, *Mol. Cell. Endocrinol.*, 3, 375, 1975.
117. **Georgiou, M. and Payne, A. H.,** 25-Hydroxycholesterol-supported and 8-bromo-adenosine 3′, 5′-monophosphate-stimulated testosterone production by primary cultures of two populations of rat Leydig cells, *Endocrinology,* 117, 1184, 1985.
118. **Kim, I. and Greenwald, G. S.,** Steroidogenic effects of lipoproteins and 25-hydroxycholesterol on luteal and ovarian cells: a comparison of two pseudopregnant rat models (42248), *Proc. Soc. Exp. Biol. Med.*, 181, 242, 1986.
119. **Rosenblum, M. F., Huttler, C. R., and Strauss, J. F.,** Control of sterol metabolism in cultured rat granulosa cells, *Endocrinology,* 109, 1518, 1981.
120. **Kim, I. and Greenwald, G. S.,** Effect of lipoproteins (LPS) and 25-OH cholesterol (25-OH Chol) on luteal production of progesterone (P_4) in the pregnant (PREG), pseudopregnant (PSP) or hypophysectomized rat and hamster, *J. Steroid Biochem.*, 20, 1517, 1984.
121. **Lilavin, S. L. and Strauss, J. F.,** Progesterone production by hamster granulosa and luteal cells during short-term incubation. Effects of lipoproteins, compactin and 25-hydroxycholesterol, *Biol. Reprod.*, 29, 1163, 1983.
122. **Jefcoate, C. R., Simpson, E. R., and Boyd, G. S.,** Spectral properties of rat adrenal-mitochondrial cytochrome P-450, *Eur. J. Biochem.*, 42, 539, 1974.
123. **Boyd, G. S., Grimwade, A. M., and Lawson, M. E.,** Studies on rat-liver microsomal cholesterol 7α-hydroxylase, *Eur. J. Biochem.*, 37, 334, 1973.
124. **Shefer, S., Hauser, S., and Mosbach, E. H.,** 7α-hydroxylation of cholestanol by rat liver microsomes, *J. Lipid Res.*, 9, 328, 1968.
125. **Van Cantfort, J.,** The in-vitro formation and the strong inhibitory action of 7-keto-cholesterol on cholesterol-7-hydroxylase activity, *Life Sci.*, 11, 773, 1972.
126. **Botham, K. M., Beckett, G. J., Percy-Robb, I. W., and Boyd, G. S.,** Bile acid synthesis in isolated rat liver cells, *Eur. J. Biochem.*, 103, 299, 1980.
127. **Raicht, R. F. and Cohen, B. I.,** Effect of cholesterol and cholesterol-epoxide on sterol metabolism in rats fed liquid diets, *Biochim. Biophys. Acta,* 666, 455, 1981.
128. **Drevon, C. A., Englehorn, S. C., and Steinberg, D.,** Secretion of very low density lipoproteins enriched in cholesteryl esters by cultured rat hepatocytes during simulation of intracellular cholesterol esterification, *J. Lipid Res.*, 21, 1065, 1980.
129. **Lichtenstein, A. H. and Brecher, P.,** Properties of acyl-CoA:cholesterol acyltransferase in rat liver microsomes, *J. Biol. Chem.*, 255, 9098, 1980.
130. **Field, F. J. and Mathur, S. N.,** Regulation of acyl CoA:cholesterol acyltransferase by 25-hydroxycholesterol in rabbit intestinal microsomes and absorptive cells, *J. Lipid Res.*, 24, 1049, 1983.
131. **Brown, M. S., Dana, S. E., and Goldstein, J. L.,** Cholesterol ester formation in cultured human fibroblasts, *J. Biol. Chem.*, 250, 4025, 1975.
132. **Goldstein, J. L., Faust, J. R., Dygos, J. H., Chorvat, R. J., and Brown, M. S.,** Inhibition of cholesteryl ester formation in human fibroblasts by an analogue of 7-ketocholesterol and by progesterone, *Proc. Natl. Acad. Sci. U.S.A.*, 75, 1877, 1978.
133. **Morin, R. J. and Peng, S.-K.,** Effects of cholesterol oxidation derivatives on cholesterol esterifying and cholesteryl ester hydrolytic enzyme activity of cultured rabbit aortic smooth muscle cells, *Lipids,* 24, 217, 1989.
134. **Peng, S.-K., Taylor, B., Tham, P., Werthessen, N. T., and Mikkelson, B.,** Effect of auto-oxidation products from cholesterol on aortic smooth muscle cells, *Arch. Pathol. Lab. Med.*, 102, 57, 1978.
135. **Peng, S.-K., Taylor, C. B., Hill, J. C., and Morin, R. J.,** Cholesterol oxidation derivatives and arterial endothelial damage, *Atherosclerosis,* 54, 121, 1985.
136. **Hashimoto, S., Dayton, S., Alfin-Slater, R. B., Bui, P. T., Baker, N., and Wilson, L.,** Characteristics of the cholesterol-esterifying activity in normal and atherosclerotic rabbit aortas, *Circ. Res.*, 34, 176, 1974.
137. **Brecher, P. and Chan, C. T.,** Properties of acyl-CoA: cholesterol-0-acyltransferase in aortic microsomes from atherosclerotic rabbits, *Biochim. Biophys. Acta,* 617, 458, 1980.

138. **Beins, D. M., Vinning, R., and Balasubramaniam, S.,** Regulation of neutral cholesterol esterase and acyl-CoA: cholesterol acyltransferase in the rat adrenal gland, *Biochem. J.*, 202, 631, 1982.
139. **Suckling, K. E.,** Regulation of acyl-CoA: cholesterol acyltransferase, *Biochem. Soc. Trans.*, 651, 1983.
140. **Suckling, K. E., Tocher, D. R., Smellie, C. G., and Boyd, G. S.,** In vitro regulation of bovine adrenal cortical acyl-CoA: cholesterol acyltransferase and comparison with the rat liver enzyme, *Biochim. Biophys. Acta*, 753, 422, 1983.
141. **Tavani, D., Avart, S., and Billheimer, J. T.,** The esterification of 25-hydroxycholesterol and the effect of 25-hydroxycholesterol on acyl-coenzyme A: cholesterol acyltransferase, *Fed. Proc. Fed. Am. Soc. Exp. Biol.*, 44, 1601, 1985.
142. **Lichtenstein, A. H. and Brecher, P.,** Esterification of cholesterol and 25-hydroxycholesterol by rat liver microsomes, *Biochim. Biophys. Acta*, 751, 340, 1983.
143. **Chang, C. C. Y., Doolittle, G. M., and Chang, T. Y.,** Cycloheximide sensitivity in regulation of ayl coenzyme A: cholesterol acyltransferase activity in Chinese hamster ovary cells. I. Effect of exogenous sterols, *Biochemistry*, 25, 1693, 1986.
144. **Smith, A. G., Brooks, C. J. W., and Harland, W. A.,** Acid cholesterol ester hydrolase in pig and human aortas, *Steroids Lipids Res.*, 5, 150, 1974.

Chapter 7

EFFECTS OF CHOLESTEROL OXIDES ON CELL MEMBRANES

Shi-Kaung Peng and Robert J. Morin

TABLE OF CONTENTS

I. NORMAL MEMBRANE STRUCTURE

The most widely accepted model of cell membrane structure is the fluid mosaic model of Singer and Nicolson.[1] According to this model, biological membranes are best described as fluid lipid bilayers which allow lateral movement of membrane components in the plane of the membrane. The cell's limiting membrane, the plasma membrane, may be a mosaic of patches of lipid, protein, or carbohydrate. The protein component of the membrane is divided into two categories: integral proteins, which may span the entire membrane, and peripheral proteins, which are loosely associated with the membrane. Carbohydrates found in biological membranes are linked to either proteins or lipids and are usually found on the extracellular surface of the plasma membrane. The lipid components of membranes include phospholipids, glycolipids, and neutral lipids.

Integral proteins are firmly embedded in the lipid layer. Extraction of these proteins requires solubilization of the membrane with organic solvents or detergents.[2] Once isolated, these proteins exist as lipoprotein complexes, and further removal of lipid may result in loss of the protein's function. Most studies have shown that membrane proteins have an α-helical structure and most are not evenly distributed in a particular membrane. The asymmetric distribution of membrane proteins was derived from freeze fracture electron microscopy studies and functional studies, such as enzyme assays.[3] Asymmetry of membrane proteins plays an important role in cellular metabolism by providing direction for specific processes. Many of the integral proteins are free to laterally diffuse in the plane of the membrane. Lateral diffusion of proteins was initially studied in virally induced mouse and human heterokaryons.[4] The hybrid cells have half of their plasma membrane from mouse cells and the other from human cells. Specific plasma membrane markers were followed after fusion, and by one hour these markers were evenly distributed throughout the whole membrane. Peripheral proteins are loosely associated with the membrane. These proteins can be removed from the membrane by mild extraction procedures, such as chelating agents or addition of salts.[2] These proteins may be associated with either side of the membrane.

Membrane phospholipids are composed of phosphoglycerides and sphingolipids. The phosphoglycerides are derived from the glycerol molecule, where two of the carbons contain esters of long-chain fatty acids, and the remaining carbon is esterified to a polar group by a phosphodiester bond. The fatty acid at position two (middle carbon) usually contains unsaturated bonds. Phospholipids are defined by the polar group, all of which contain a phosphate moiety. The polar groups include choline, serine, ethanolamine, etc., and various modifications of the glycerol molecule.[5] Sphingolipids are based on sphingosine, a long-chain amino alcohol. A fatty acid residue is linked to sphingosine through an amide bond. The polar group may be a phosphocholine (sphingomyelin) or a free hydroxyl residue (ceramides). Attachment of sugars to the primary hydroxyl of sphingosine defines the glycolipids. The cerebrosides and gangliosides have a single glucose and complex polysaccharides respectively, on the primary hydroxyl of sphingosine. Neutral lipids found in membranes include cholesterol and free fatty acids.

The amphipatic nature of membrane lipids is essential to membrane structure. Phospholipids are aligned in a tuning fork fashion, where the hydrophilic polar group faces the aqueous environment and the hydrophobic acyl chains extend into the core of the membrane. Membrane lipids are also asymmetrically distributed across the lipid bilayer. Turner[6] was the first to show the asymmetry of lipids in mammalian cells. By comparing the lytic action of phospholipase on ruminant red blood cells, he demonstrated that sphingomyelin was the predominant lipid in the outer half of the membrane bilayer. Chemical labeling methods used to locate amino lipids have also provided evidence for the asymmetry of membrane lipids.[7]

Lipid molecules are free to laterally diffuse in the plane of the membrane. In artificial phospholipid dispersions, it was estimated that each molecule could exchange with its neighbor at the rate of 10^6 times per second.[8] With the use of spin labels, rapid lipid diffusion was also demonstrated in biological membranes.[9] This rapid motion of lipids does not occur in the transbilayer (flip-flop) motion of lipids. The lowest free energy state for the membrane bilayer exists when the polar groups of the lipids are in contact with the aqueous environment. Transbilayer diffusion would necessitate the movement of the polar group through the hydrophobic interior of the membrane.[5] Nevertheless, transmembrane movement is facilitated by the presence of membrane proteins[10] or through structural transitions of phospholipids to nonbilayer membrane configurations.[11]

Cholesterol is the most abundant sterol in animal tissues and is present in blood, bile, and cellular membranes. It may be found as free cholesterol in membranes or esterified to long-chain fatty acids at other sites, particularly in the lipoprotein complexes in blood. In mammalian cells, cholesterol is present in molar ratios of 0.4:1.0 (cholesterol:phospholipid) in the plasma membranes.[12] Intracellular membranes, such as mitochondria and endoplasmic reticulum have a lower proportion of cholesterol. Within a given membrane, cholesterol has been shown to have an unequal distribution in the membrane bilayer. Using electron microscopic autoradiography, it was demonstrated that cholesterol was present in specific regions in red blood cell membranes.[13] Another method, which employed filipin, a polyene antibiotic that specifically binds cholesterol, demonstrated cholesterol-rich domains in rat intestinal smooth muscle cell plasma membrane.[14]

Biological membranes are considered to be in a fluid liquid-crystalline state. Fluidity is a measure of the viscosity of the membrane. It is a function of the molecular area of the membrane components, the structural arrangement and interaction of these components, and the dynamics of their motion. Fluidity is measured in terms of microviscosity and is often expressed in absolute macroscopic units (poise) to relate it to other liquids.[15] Membrane fluidity is frequently measured by analysis of probe behavior in the lipid bilayer. The motion of probes can be detected by nuclear magnetic resonance, electron spin resonance, or fluorescence.

The predominant physical effect of cholesterol in lipid membranes is on the structural component of their fluidity. This effect was first studied in model systems of membranes. It was shown that addition of cholesterol to phospholipid, which was spread at an air-water interface, resulted in a reduction of the mean molecular area of the phospholipids.[16] Cholesterol also affects the molecular motion of phospholipids. The fatty acyl chains of phospholipids which are predominantly in the disordered liquid-crystalline state in biological membranes, become less mobile because of steric and hydrophobic interactions between cholesterol and phospholipid (condensing effect of cholesterol). In contrast, the fatty acyl chains of phospholipids that are in the ordered gel state become more mobile in the presence of cholesterol (liquifying effect of cholesterol). In effect, cholesterol modifies the motional freedom of fatty acyl chains in membranes, enhancing the movement of gel-state chains, decreasing that of fluid-state chains, and promoting an overall structural and dynamic homogeneity to the membrane intermediate fluid state.[17,18] Through studies using synthetic model membranes, mainly with liposomes, and to a lesser extent with monolayers, the parameters found to be central to this interaction with phospholipids are a 3β-hydroxyl group, a planar steroid-ring nucleus, and an isoprenoid side chain of precise size and saturation.[19]

The molecular interaction between cholesterol and membrane phospholipids remain unsettled. Demel and De Kruyff[20] proposed that hydrogen bonding occurs between phospholipid-bound water molecules and the 3β-hydroxyl of cholesterol. An alternative mechanism was proposed by Huang,[21] who suggested that the 3β-hydroxyl of cholesterol was directly engaged in hydrogen bonding with the carbonyl oxygen of the fatty acyl residues in phospholipids. He also proposed that the rigid sterol ring of cholesterol participates in Van der Waals attractive interactions with the unsaturated fatty acyl chain in phosphoglycerides. Most studies have shown a 1:1 stoichiometry between cholesterol and phospholipid.[18]

II. EFFECTS OF CHOLESTEROL OXIDES ON MEMBRANE STRUCTURE

Insertion of cholesterol oxides into cell membranes may alter membrane fluidity and affect membrane protein and phospholipid structure. Incorporation of 7α-hydroxycholesterol into human erythrocyte membranes resulted in an increase in protein helical structure, whereas the 20α-hydroxycholesterol caused increased immobilization of fatty acyl chains within the cell membrane. Both synergism and antagonism between cholesterol oxides and cholesterol may occur in control of membrane protein and phospholipid fatty acyl chain stereochemistry.[22] Many biophysical techniques have been applied to the study of these effects, including X-ray diffraction, differential scanning calorimetry, and a variety of spectroscopic techniques. Two aspects of the interaction of phospholipids with cholesterol oxides can be examined by differential scanning calorimetry; one is the effects on the gel to liquid-crystalline phase transition and the other is on the bilayer to hexagonal phase transition as a determination of effects on membrane lipid structure. Calorimetric studies on the influence of cholesterol oxides on the phase transition behavior of red cell ghost membranes provides some important information about their thermodynamic properties. Since biological membranes by virtue of their great variety of different lipids and proteins are highly complex, investigations of liposome model membranes containing one chemically pure phospholipid have been used to provide basic information concerning the influence of cholesterol oxides which could then be extended to more complex systems. Using dipalmitolylphosphatidycholine (DPPC) liposomes as a membrane model, a marked influence of cholesterol oxides on the reversible gel to liquid-crystalline phase transition behavior of cholesterol-free and cholesterol-containing model membranes was evidenced.[23,24] The influence of each cholesterol oxide on the phase transition was expressed in terms of the temperature range, the corresponding enthalpy, and the peak shape of calometric curves. The major cholesterol autoxidation products, including 7α-hydroxycholesterol, 7β-hydroxycholesterol, 7-ketocholesterol, and 25-hydroxycholesterol, all reduced the temperature, the enthalpy, and the width at half peak height for the gel to liquid-crystalline phase transition, which may result in decreased membrane fluidity with attendant altered membrane permeability and decreased stability. 7β-Hydroxycholesterol was found to be the most potent of the sterols examined with regard to its ability to lower the onset temperature of phase transition. The influence of side-chain cholesterol oxides on the gel to liquid-crystalline phase transition of DPPC liposomes was also studied by the differential scanning calorimetry technique[25,26] and the fluorescence anisotropy technique using 1,6-diphenyl,1,3,5-hexatriene as a probe.[27] The majority of these compounds, including 20α-hydroxycholesterol, 22S-hydroxycholesterol, the epimeric 23-hydroxycholesterols, and 22-ketocholesterol appeared to interact like cholesterol with phospholipid in model systems to form a stable membrane system; only 22R-hydroxycholesterol did not interact, presumably because of the altered stereochemistry of the side chain imparted by the 22R-hydroxyl configuration. Studies on the effects of 25-hydroxycholesterol on the physical properties of membrane lipids using a spin label technique showed that low concentrations of 25-hydroxycholesterol generally behaved similarly to cholesterol. At higher concentrations, however, 25-hydroxycholesterol differed markedly from cholesterol in not having the ability to increase molecular order with increasing concentration.[28] Thus it appears that the hydroxyl group in the 25 position limits the interaction of the steroid nucleus with phospholipid fatty acyl chains.

The membrane structural changes could be best studied in liposomes composed of dielaidoylphosphatidylethanolamine (DEPE) liposome because of its property of undergoing transition to the hexagonal phase. Compared with cholesterol, 7α-hydroxycholesterol had a greater effect in decreasing the bilayer to hexagonal phase transition temperature and broadening this transition.[29] In contrast, cholestane-3β,5α,6β-triol raised the bilayer to the hexagonal phase transition temperature of DEPE. This is due to its larger and more hydrophilic head group. The side-chain cholesterol oxides at the C20 and C22 positions all induced a downward shift of the melting temperature of DEPE to a lower temperature as was the case with cholesterol.[26] 20-Hydroxycholesterol had the strongest effect, and 22R-hydroxycholesterol the weakest effect. The molecular interfacial properties of cholesterol oxides were studied by a monomolecular film technique. The monolayer studies showed that cholesterol oxides, including 7-oxygenated cholesterol oxides and side-chain cholesterol oxides, exhibit a larger molecular area than cholesterol, indicating that these molecules are not completely perpendicular at the air/water interface as is cholesterol.[24] 7α-Hydroxycholesterol and 7β-hydroxycholesterol are tilted at low surface pressure. 25-Hydroxycholesterol with hydroxyl groups at both ends of the molecule has a perpendicular orientation and was not folded. 22R-Hydroxycholesterol has, however, been shown to initially orient parallel to the interface. In mixed monolayers with dioleoylphosphatidylcholine (DOPC), 7-ketocholesterol, 7β-hydroxycholesterol, and 7α-hydroxycholesterol show a condensing effect in this order, although to a lesser extent than that observed for cholesterol, whereas 25-hydroxycholesterol and other side-chain cholesterol oxides showed no significant condensing effect.[23] By contrast, in phospholipid liposomes, the 7-oxygenated cholesterol oxides have a diminished capacity to condense phospholipid fatty acyl chains, but 20α-hydroxycholesterol, 22S-hydroxycholesterol, the epimeric 23-hydroxycholesterols, and 25-hydroxycholesterol condense these phospholipid bilayers in the manner of cholesterol.[30]

Displacement of cholesterol by cholesterol oxides could affect the surface shape of several cell types. The red blood cell has been utilized in these studies because the cholesterol content of its membrane can be easily manipulated by incubation with cholesterol-rich or cholesterol-depleted or cholesterol oxide liposomes. Cholesterol is taken up or lost from the membrane by a diffusion-mediated exchange process. Cooper[18] showed that an increase in red blood cell cholesterol resulted in a striking change in cell contour characterized by folding and scalloping of cell margins. When human red blood cells were incubated with certain cholesterol oxides at 1.25×10^{-5} to 5×10^{-5} *M* concentration for one hour in lipoprotein-depleted medium, the red blood cells were transformed into spiny-shaped cells known as echinocytes.[31] Cholestane-3β,5α,6β-triol produced 100% echinocytic transformation of red blood cells at 2.5×10^{-5} *M* concentration. Cholestan-3β-5α-diol-6-one, 7β-hydroxycholesterol, 6-ketocholestanol, and 6β-hydroxycholestanol were also capable of transforming the shape of virtually the entire population. Cholesterol itself was without effect. The effects of cholesterol oxides on red blood cells appear to be a consequence of cholesterol oxide insertion into the red cell membranes. The proportion of cholesterol oxides in the incubation mixture inserted into red cell membranes was found to range from 24% for 25-hydroxycholesterol to 36% for 7β-hydroxycholesterol. When red blood cells were incubated with 7β-hydroxycholesterol, the amount of cholesterol oxide inserted in red cell membranes increased in proportion to the concentration originally present in the incubation medium, and the degree of echinocyte transformation also increased. Echinocyte formation was apparent within 2 minutes of exposure of red blood cells to cholesterol oxides. The early formation of echinocytes by membrane-inserted cholesterol oxides is a consequence of asymmetric expansion of the outer lamella of the red cell membrane lipid bilayer due to the presence of bulky hydrophilic group on sterol molecules inserted within the hydrophobic environment of the membrane fatty acyl chains. Both cholesterol oxide membrane insertion and red cell echinocyte transformation were impeded but not abolished by the presence of lipoproteins, since lipoprotein can act as acceptors of cholesterol oxides previously inserted into red cell membranes.[32]

Another morphological change observed by light and electron microscopy is the loss of cell surface microvilli, when human lymphocytes were depleted of plasma membrane cholesterol.[33] Treatment of mouse L cells with 25-hydroxycholesterol inhibited the synthesis of cellular cholesterol, followed by cholesterol depletion in membranes and diminished the number of microvilli on the cell surface which could be demonstrated by scanning electron microscopy.[34] Simultaneous addition of mevalonic acid or cholesterol to cells during the treatment with inhibitor prevented the loss of microvilli. However, rat Morris hepatoma HTC cells treated with 7β-hydroxycholesterol, in contrast, developed microvilli.[35] Deformation of bovine platelets by 7β-hydroxycholesterol or 22R-hydroxycholesterol was also observed.[36]

Membrane fluidity may also affect the growth rate and hydroxymethylglutaryl coenzyme A (HMG-CoA) reductase activity. Cell membrane fluidity measured by fluorescence polarization using 1,6-diphenyl,1,3,5-hexatriene has been examined in HTC hepatoma cells, in relation to both growth rate and response to treatment with cholesterol oxides.[37] A high level of HMG-CoA reductase activity was observed in cells at the log phase of growth, which progressively decreased to reach a sustained low level at the stationary phase. Similarly, membrane fluidity markedly decreased in relation to growth rate. 7β-Hydroxycholesterol or 25-hydroxycholesterol strongly inhibited HMG-CoA reductase activity and DNA biosynthesis and concomitantly also strongly decreased membrane fluidity, whereas the water-soluble 7β-hydroxycholesterol-7β-dihemisuccinate did not alter HMG-CoA reductase activity nor membrane fluidity but did decrease DNA biosynthesis. This suggests that HMG-CoA reductase activity does depend on membrane fluidity but DNA biosynthesis does not.

III. EFFECTS OF CHOLESTEROL OXIDES ON MEMBRANE LIPID COMPOSITION

The cholesterol content in the cell membrane can be influenced by cholesterol oxides, since many of the cholesterol oxides have been shown to be potent inhibitors of cholesterol biosynthesis, resulting in decreases in cellular cholesterol levels with depletion of the cholesterol available for membrane synthesis. This would cause a decline in the molar ratio of membrane cholesterol to phospholipd. 25-Hydroxycholesterol is one of the most potent inhibitors of cholesterol biosynthesis.[38] The coordination of the cholesterol biosynthesis with the regulation of the membrane cholesterol:phospholipid ratio and cell cycle control was investigated in proliferating L_6 myoblasts and fibroblasts (WI-38 and CEF).[39] Cells cultured in lipid-depleted medium containing 25-hydroxycholesterol or compactin, another inhibitor of HMG-CoA reductase, displayed a rapid, dose-dependent inhibition of cholesterol biosynthesis. Twenty-four hours after the addition of the inhibitor, the synthesis of DNA, RNA, and protein also declined. The inhibition of cholesterol biosynthesis led to a threefold reduction in the sterol:phospholipid ratio that paralleled the development of proliferative and G_1 cell cycle arrests and alterations in cellular morphology. All of these responses were reversed upon reinitiation of cholesterol biosynthesis or addition of exogenous cholesterol. A comparison of the timing of these responses with respect to the development of the G_1 arrest indicates that the primary factor limiting cell cycling is the availability of cholesterol provided either from an exogenous source or by *de novo* synthesis. The G_1 arrest appears to be responsible for the general inhibition of macromolecule synthesis in proliferating cells treated with 25-hydroxycholesterol.

The molar ratio of membrane cholesterol to phospholipid could also be affected by direct insertion of cholesterol oxides into cell membranes due to their similar molecular structure to cholesterol. Many membrane-associated changes, such as echinocytic transformation of red blood cells, could only be partially reversed by the addition of cholesterol to the medium and could not be reversed by the addition of mevalonic acid.[40] The echinocytic effect of cholesterol oxides on erythrocyte membranes could not be due to the inhibition of cholesterol biosynthesis, since mature red blood cells do not synthesize cholesterol. When human red blood cells were incubated with 7β-hydroxycholesterol in lipoprotein-depleted medium, the amount of the cholesterol oxide incorporated into red blood cell membranes increased in proportion to the concentration originally present in the incubation medium, and the degree of echinocyte transformation also increased.[32]

Although cholesterol can be exchanged by a simple physicochemical exchange process, the sterol is virtually insoluble in water. Cholesterol is solubilized only in lipoproteins and in membranes by phospholipids. Both cholesterol oxide membrane insertion and red cell echinocyte transformation could be impeded, but not abolished, by the presence of lipoproteins in the cholesterol oxide-red blood cell incubation medium. All density classes of serum lipoproteins can bind cholesterol oxides and lipoproteins could act as acceptors of cholesterol oxides previously inserted into red blood cells. Incorporation of 7-ketocholesterol by plasma membranes has also been shown in porcine aortic endothelial cells.[41] 7-Ketocholesterol was associated with the cell plasma membrane in substantial quantities when endothelial cells were incubated with this sterol. On the average, the molar ratio of 7-ketocholesterol to cholesterol in the isolated cell plasma membrane was 0.42 after a 2-hour incubation with endothelial cells containing 7-ketocholesterol-enriched plasma (300 mg of 7-ketocholesterol suspended in 600 ml of plasma). However, no significant difference in the cholesterol to phospholipid ratio was obtained with the 7-ketocholesterol incubation; instead a significant increase in the total sterols to phospholipid ratio to about 30% was observed. Since no significant changes either in the absolute value for cholesterol, phospholipid, or cholesterol to phospholipid ratio was seen, it is likely that the increase in total sterols to phospholipid was due primarily to the insertion of 7-ketocholesterol into the lipid bilayer of the cell membranes and was not due to substitution for the native membrane cholesterol. Human red blood cells were also shown to be able to absorb 7-ketocholesterol, cholesta-4,6-dien-3-one, or 1-methyl-19-norcholesta-1,3,5(10)-trien-3β-ol from phospholipid liposomes,[42] and 7α- and 7β-hydroxycholesterol, 20α-hydroxycholesterol, and 22-ketocholesterol from lipoprotein-depleted medium.[22] Sterol distribution within mouse 3T3 cells following administration of 25-hydroxycholesterol was readily observed.[43]

The subcellular distribution of incorporated cholesterol oxides has also been investigated.[44] To determine whether cholestane-3β,5α,6β-triol and 25-hydroxycholesterol specifically localize in particular membrane fractions, these sterols were incubated with cultured rabbit aortic smooth muscle cells for 24 hours prior to lipid analysis of specific cell fractions. The plasma membrane-enriched fraction, mitochondrial and microsomal fraction, and nuclear pellet were analyzed separately for sterol content. When the amounts of cholestane-3β,5α,6β-triol in each fraction were compared on the basis of unit protein, it was found that the plasma membrane-enriched fraction had the highest percentage of that sterol, 5.5 times higher than in other subcellular fractions. In contrast, 25-hydroxycholesterol was more evenly distributed throughout the various subcellular fractions. The results indicate that cholestane-3β,5α,6β-triol appears to be more readily incorporated into the plasma membrane than 25-hydroxycholesterol. There was no significant influence of the cholesterol oxides on membrane fatty acid composition or phospholipid composition.[45] The modification in cholesterol content by cholesterol oxides is extensive enough to alter membrane fluidity and affect a number of membrane functions, including carrier-mediated transport, the properties of certain membrane-bound enzymes, phagocytosis, pinocytosis receptor-mediated endocytosis, and cell proliferation.

IV. EFFECTS OF CHOLESTEROL OXIDES ON MEMBRANE FUNCTIONS

A. EFFECTS ON CARRIER-MEDIATED TRANSPORT

The glucose analog 2-deoxy-D-glucose, since it moves unidirectionally, is utilized for the determination of carrier-mediated hexose transport. This compound is taken up by the

TABLE 1.
Effect of Cholesterol Autoxidation Derivatives on 2-Deoxy-D-Glucose Transport

Sterol	% Changes compared to control	Transport of 2-deoxy-D-glucose (nmol/10 min/mg)
Control (vehicle only)[a]	—	225.29 ± 45.78[b]
25-Hydroxycholesterol[c]	−15.51	190.34 ± 47.30
20α-Hydroxycholesterol	−3.86	216.58 ± 29.07
5α,6α-Epoxy-cholesterol	−1.87	221.08 ± 21.80
Cholest-3β,5α-diol-6-one	−27.25[d]	163.89 ± 25.47
7-Ketocholesterol	−2.67	219.26 ± 26.50
4-Cholesten-3-one	+1.24	228.09 ± 25.82
Cholestane-3β,5α,6β-triol	−30.56[d]	156.44 ± 38.00

[a] Vehicle was ethanol which was less than 0.8% of final volume of media.
[b] Values represent mean ± S.E.M. of at least two experiments.
[c] Sterols were dispensed in basal culture medium supplemented with 10% fetal bovine serum at a final concentration of 25 μg/ml. All cells were incubated with 2 ml of medium containing a test sterol for 1.0 h prior to transport assay described in the methods.
[d] $p < .01$ compared to control value.

Reprinted from Hill, J. C., Peng, S. K., Morin, R. J., and Taylor, C. B., *Exp. Mol. Pathol.*, 41, 249, 1984. With permission.

same transport mechanism as D-glucose, but once inside the cell, it is phosphorylated and not further metabolized, and this prevents reverse transport. The effects of various cholesterol oxides on hexose transport were studied using cultured rabbit aortic smooth muscle cells.[46] Cultured aortic smooth muscle cells have a saturable carrier transport mechanism for the uptake of 2-deoxy-D-glucose. At 10 m*M* external 2-deoxy-D-glucose concentration, the rate of uptake of the hexose was saturated. The value of the Km (1.75 m*M*) for 2-deoxy-D-glucose corresponded well with that obtained in transport studies with cultured human fibroblasts.[47] The 90% inhibition of hexose uptake by cytochalasin B provided additional evidence for a glucose carrier transport system in cultured aortic smooth muscle cells. It can, therefore, be concluded that the major mechanism involved in glucose uptake by these cells was carrier-mediated transport. The effects of various cholesterol oxides on hexose transport in cultured aortic smooth muscle cells are shown in Table 1. The most potent inhibitor of hexose transport was cholestane-3β,5α,6β-triol. The inhibitory effect of the cholestanetriol appeared to be dose dependent. Increasing concentrations of the cholestanetriol resulted in progressive inhibition of hexose transport, with a 60% decrease in hexose transport at the highest level (50 μg/ml) tested (Table 2). The inhibitory effect was not the result of cell toxicity, as the total protein content of the cultured cells did not vary and dye exclusion studies showed the cultured cells to be viable during the one-hour incubation period. Furthermore, the inhibitory effect of the cholestanetriol on hexose transport was completely reversible. Cells incubated with cholestanetriol at 50 μg/ml for 30 minutes showed an approximately 50% decrease in the rate of hexose transport (Figure 1). When duplicate cultures were transferred to control nutrient media (triol-free), however, the rate of hexose transport returned to control values by 1.5 hours. There appeared to be a slight inhibitory effect of 25-hydroxycholesterol on hexose transport, but this did not reach statistical significance.

B. EFFECTS ON MEMBRANE-BOUND ENZYMES

5′-Nucleotidase has been used extensively as a membrane marker for the plasma membrane in cell fractionation studies.[48] The physiological significance of 5′-nucleotidase has

TABLE 2
Effect of Cholestane-3β,5α,6β-Triol and Cholesterol on 2-Deoxy-D-Glucose Transport

Sterol	Concentration (μg/ml)	Protein (mg/ml)	Uptake of 2-deoxy-D-glucose (nmol/10 min/mg)	% Change compared to control
Control (vehicle only)[a]	—	0.137 ± 0.011	225.29 ± 45.78[b]	—
Cholesterol (purified)[c]	50	—	243.37 ± 36.92	+8.03
Cholestane-3β,5α,6β-triol	5	0.139 ± 0.008	168.04 ± 30.19	−25.41[d]
	10	0.132 ± 0.022	146.71 ± 18.96	−34.87[e]
	25	0.129 ± 0.018	156.44 ± 38.00	−30.56[e]
	35	0.140 ± 0.024	149.68 ± 22.83	−33.56[e]
	50	0.134 ± 0.022	93.86 ± 13.53	−58.34[e]

[a] Vehicle was ethanol which was less than 0.8% of final volume of media.
[b] Values represent mean ± S.E.M. of at least two experiments.
[c] Sterols were dispensed in basal culture medium (BME) supplemented with 10% fetal bovine serum at final concentrations indicated. Cells were incubated with 2 ml of medium containing test sterol for 1.0 h prior to transport assay described in the methods.
[d] $p < .05$ compared to control value.
[e] $p < .01$ compared to control value.

Reprinted from Hill, J. C., Peng, S. K., Morin, R. J., and Taylor, C. B., *Exp. Mol. Pathol.*, 41, 249, 1984. With permission.

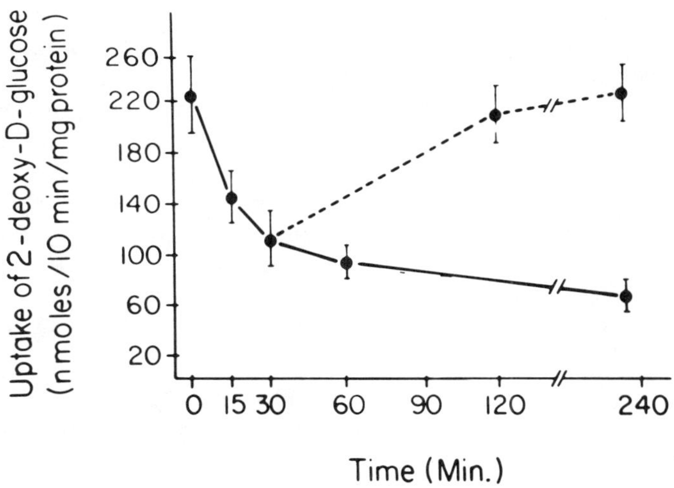

FIGURE 1. Reversibility of inhibition of hexose transport by cholestane-3β,5α,6β-triol. Cells were incubated with triol at a concentration of 50 μg/ml at time periods indicated. Cells were transferred to control medium (- -- -) and recovery of hexose transport was evident by 1.5 h. (Reprinted from Hill, J. C., Peng, S. K., Morin, R. J., and Taylor, C. B., *Exp. Mol. Pathol.*, 41, 249, 1984. With permission.)

not been established. Recent theories regarding its function include regulation of local blood flow by providing adenosine, a potent vasodilator,[49] and salvaging extracellular adenosine monophosphate (AMP) released by glycosyl transferase enzymes on the cell surface.[50] In addition to 5′-nucleotidase, the plasma membrane of a variety of cell types contain enzymes with ATPase activity. The best-characterized ATPase in plasma membranes is the

TABLE 3
Effect of Cholestane-3β,5α,6β-Triol and 25-Hydroxycholesterol on 5′-Nucleotidase Activity in Cultured Aortic Smooth Muscle Cells

Sterol	Time (h)	5′-Nucleotidase activity (μM/min/mg × 10^{-2})	% Change compared to control
Control (vehicle only)[a]	—	30.76 ± 6.75[b]	—
25-Hydroxycholesterol	24	18.11 ± 5.41[c]	−41.12
	48	22.60 ± 3.47[c]	−26.52
Cholestane-3β,5α,6β-triol	24	18.82 ± 5.89[c]	−38.81
	48	15.83 ± 3.09[c]	−48.53

[a] Control values represent the average values obtained from zero time, 24-, and 48-h incubation with vehicle (ethanol).
[b] Values represent the mean ± S.E.M. of at least 6 experiments.
[c] $p < .01$ compared to control value.

Reprinted from Peng, S. K., Hill, J. C., Morin, R. J., and Taylor, C. B., *Proc. Soc. Exp. Biol. Med.*, 180, 126, 1985. © by The Society for Experimental Biology and Medicine. With permission.

Na^+,K^+-ATPase, which has been shown to span the entire plasma membrane bilayer, having extracellular sodium-binding sites and intracellular potassium- and ATP (adenosine triphosphate)-binding sites.[51] There is wide agreement that the physiological role of this enzyme is the maintenance of intracellular ion concentration. These membrane-bound enzyme activities were studied using cultured rabbit aortic smooth muscle cells which were incubated for 24 to 48 hours with 10 μg/ml of either cholestane-3β,5α,6β-triol or 25-hydroxycholesterol. Membrane-bound enzymes 5′-nucleotidase and Na^+,K^+-ATPase were measured both cytochemically by electron microscopy[52,53] and biochemically.[44,54] There was a decrease in the deposition of reaction product on the plasma membrane of both 5′-nucleotidase and Na^+,K^+-ATPase after 24-hour incubation with both sterols and no reaction product visible on the plasma membrane after 48 hours. For biochemical determination the cells were homogenized, after which fractionation of the crude membranes was done. The plasma membrane-enriched fraction was separated by ultracentrifugation using a discontinuous sucrose density gradient. The purity of plasma membrane-enriched fraction was confirmed by assay of the membrane enzyme markers 5′-nucleotidase and Na^+,K^+-ATPase. The effect of cholestane-3β,5α,6β-triol and 25-hydroxycholesterol on 5′-nucleotidase activity is shown in Table 3 and on Na^+,K^+-ATPase activity in Table 4. The activity of 5′-nucleotidase was markedly decreased in crude membranes when cells were exposed to 10 μg/ml of either cholestane-3β,5α,6β-triol or 25-hydroxycholesterol. It was further demonstrated that 5′-nucleotidase activity was also markedly decreased in plasma membrane-enriched fractions when cells were incubated with cholestane-3β,5α,6β-triol. In contrast to 5′-nucleotidase, Na^+,K^+-ATPase activity in crude homogenates from cells incubated for 24 h with either cholesterol oxide was not significantly different from control values. There was a significant decrease (32%) in Na^+,K^+-ATPase activity in cells incubated with cholestane-3β,5α,6β-triol for 48 h. Chen et al.[55] also demonstrated that in 25-hydroxycholesterol-treated mouse fibroblast L-cells, the level of Na^+,K^+-activated ATPase activity measured in homogenates and plasma membrane preparations was not significantly different from the control values. Ouabain-sensitive uptake of $^{86}Rb^+$ (an analog of K^+), however, was enhanced in the cells treated with 25-hydroxycholesterol and 7-ketocholesterol. Ouabain-insensitive Rb^+ efflux increased in the sterol-depleted cells and the intracellular concentration of K^+ diminished while the concentration of Na^+ increased. All of these effects of 25-hydroxycholesterol were counteracted by the addition of mevalonate to the culture medium. Another study showed that human erythrocyte membranes could be manipulated by treatment with cholesterol

TABLE 4
Effect of 25-Hydroxycholesterol and Cholestane-3β,5α,6β-Triol on ATPase Activity in Cultured Aortic Smooth Muscle Cells

Sterol	Time (h)	ATPase (μmol/min/mg × 10^{-2})	% Change compared to control
Control (vehicle only)[a]	—	22.96 ± 4.18[b]	—
25-Hydroxycholesterol	24	29.10 ± 8.44	+26.74
	48	21.09 ± 5.58	−8.14
Cholestane-3β,5α,6β-triol	24	23.22 ± 1.33	+1.13
	48	15.66 ± 3.39[c]	−31.79

[a] Control values represent the average values obtained from zero time, 24-, and 48-h incubation with vehicle (ethanol).
[b] Values represent mean ± S.E.M. of at least 6 experiments.
[c] $p < .01$ compared to control value.

Reprinted from Peng, S. K., Hill, J. C., Morin, R. J., and Taylor, C. B., *Proc. Soc. Exp. Biol. Med.*, 180, 126, 1985. © by The Society for Experimental Biology and Medicine. With permission.

oxidase.[56] This enzyme converted up to 76% of cholesterol to cholest-4-en-3-one *in situ* and resulted in a decrease in Na^+,K^+-ATPase activity that was directly proportional to the extent of oxidation. The opposite effect was demonstrated with 24-dehydrocholesterol incorporated in rat brain microsomes.[57] Substitution of 25% of cholesterol by 24-dehydrocholesterol resulted in an approximately 25% increase in Na^+,K^+-ATPase activity. The effect was reversible and not due to an increase in enzyme active sites.

Only a few other membrane-bound enzymes have been studied. Upon treatment with 25-hydroxycholesterol, aortic smooth muscle cells cultured under steady-state conditions exhibited a coordinate loss of synthesis (80 to 90%) of cholesterol, dioleoyl phosphates, and glycoproteins and an accompanying decrease (40%) in cellular cholesterol content.[58] Analyses of enzyme activities showed that 5′-nucleotidase and alkaline phosphatase were conserved, α-mannosidases, galactose-transferase, and cholesterol esterase were diminished, while neutral α-glucosidase and mannose-transferase redistributed in 25-hydroxycholesterol-treated cells.

Cholesterol oxides may affect membrane-bound enzymes by one or more of the following mechanisms: (1) cholesterol oxides, due to their effect on membrane microviscosity, may alter the potential volume available to the enzyme and substrate; (2) cholesterol oxides may directly interact with the enzymes; (3) the positioning of the enzyme may be affected by cholesterol oxide-mediated changes in membrane fluidity. Borochov and Shinitzky[59] have shown that decreased membrane fluidity, via cholesterol depletion in human erythrocytes, was followed by a masking of membrane proteins. The opposite effect occurred after cholesterol enrichment, suggesting that membrane fluidity may affect vertical placement of membrane proteins.

C. EFFECTS ON CALCIUM CHANNELS

The loss of the membrane calcium barrier, and accumulation of intracellular calcium is a mechanism by which many compounds produce cell death.[60] Sevanian[61] showed that the toxicity of cholestane-3β,5α,6β-triol, isomeric cholesterol 5,6-epoxides, and 7-ketocholesterol appeared to be directly towards the cell membrane and included marked perturbations of Ca^{2+} homeostasis by being incorporated into cell membranes in cultured V79 lung fibroblasts. The toxicity of each of the cholesterol oxides was reduced by approximately 50% when cells were treated in Ca^{2+}-deficient media. The effect of the cholesterol oxides on calcium transport was examined further by measurement of $^{45}Ca^{2+}$ influx and exchange-efflux using bovine endothelial cells.[62] Verapamil is known to inhibit primarily rapid voltage-gated calcium channels. Cholesterol 5α,6α-epoxide (30 μ*M*), cholesterol 5β,6β-epoxide (25 μ*M*), and cholestane-3β,5α,6β-triol (20 μ*M*) significantly inhibited the early phase of calcium influx, and verapamil did not alter this effect. The inhibitory effect of α-epoxide was found to continue during the late phase of calcium influx, while the β-isomer produced no prolonged inhibition. In the presence of cholestane-triol, by contrast, verapamil increased the rate of late-phase calcium influx. These cholesterol oxides also inhibited the rate of calcium efflux to varying degrees. The most potent inhibition was produced by α-epoxide but was confined to early phase. A similar effect was seen for β-epoxide although this compound was considerably less potent an inhibitor than the α-isomer and was ineffective during the late phase of calcium efflux. Cholestane-3β,5α,6β-triol produced an early-phase inhibition which was reversed by verapamil. The consequences of membrane perturbation found in the endothelial cells treated with cholesterol oxides included the rapid coagulation of cytoplasmic contents, retraction of cells, and development of picnotic organelles and nuclei.[59]

In human red blood cells, the Ca^{2+}-ejecting ATPase was inhibited by vanadate and the nitrendipine-inhibitable part of Ca^{2+} influx was measured.[63] This portion of influx exhibited all the properties of a receptor-dependent calcium channel. There was a linear relationship between the cholesterol content of the membrane and Ca^{2+} influx, with a ratio of membrane cholesterol to phospholipid above 1.4 giving increased Ca^{2+} influx, and a ratio below 0.75 causing a decreased influx. Ca^{2+} influx was increased by cholestane-3β,5α,6β-triol, 22S-hydroxycholesterol, cholestane-3β,5α-diol-6-one, and 7-ketocholestanol, but decreased by 25-hydroxycholesterol, 20α-hydroxycholesterol, 7β-hydroxycholesterol, and 7-ketocholesterol. In rat hepatocytes and platelets, 7-ketocholesterol, cholestane-3β,5α,6β-triol, and 26-hydroxycholesterol stimulated Ca^{2+} influx, but Ca^{2+} efflux was increased by isomeric cholesterol 5,6-epoxides.[64] 25-Hydroxycholesterol increased Ca^{2+} uptake into cultured murine P815 mastocytoma cells.[65] These effects appeared to result from incorporation of cholesterol oxide into the plasma membrane, thereby affecting membrane cholesterol content, membrane fluidity, and the associated Ca^{2+} channels in a highly stereospecific manner.[66]

In the model membranes of phospholipid liposomes, similar results were also observed.[67] 25-Hydroxycholesterol, 26-hydroxycholesterol, and 25-hydroxyvitamin D_3 all increased Ca^{2+} uptake into phospholipid liposomes. Uptake of other divalent cations such as Mn^{2+}, Mg^{2+}, Sr^{2+}, and Ba^{2+} was also increased by 25-hydroxycholesterol.

Another mechanism by which cholesterol oxides may affect Ca^{2+} metabolism is via inhibition of the protein calmodulin which has been implicated in a variety of Ca^{2+}-dependent events.[68] This inhibition would have the effect of suppressing actions of Ca^{2+} on intracellular processes, including a Ca^{2+}-calmodulin kinase inhibiting effect on HMG-CoA reductase, thereby on *de novo* sterol biosynthesis.[69]

D. EFFECTS ON OSMOTIC FRAGILITY AND PERMEABILITY

The exchange of cholesterol between erythrocytes and plasma often reaches completion within a few hours. It was shown that large proportions of the membrane cholesterol could be replaced by other lecithin-solubilized steroids using an exchange process.[42] The osmotic fragility of erythrocytes was considerably increased after removal of part of the cholesterol complement with pure lecithin dispersions. *In vitro* uptake of cholesterol oxides into the erythrocyte plasma membranes caused an expansion of cell surface area without attendant cell volume changes, and consequently a diminution of osmotic fragility of the erythrocyte. It was shown that the osmotic fragility was decreased by 25 μ*M* concentration of 7α-hydroxycholesterol, 7β-hydroxycholesterol, 7-ketocholesterol, 20α-hydroxycholesterol, 4β-hydroxycholesterol, 6β-hydroxycholesterol, 22-ketocholesterol, 6-ketocholestanol, and cholestan-3β,5α-diol-6-one. There was an inverse linear relationship between insertion of cholesterol oxides into the membranes and the osmotic fragility of the membranes.[70] The effect could be counteracted by serum proteins, possibly because lipoproteins can act as acceptors of cholesterol oxides previously inserted into membranes.[32] Both echinocyte-forming and non-echinocyte-forming cholesterol oxides were capable of protecting red blood cells against osmotic lysis. Erythrocytes from patient with hereditary spherocytosis also respond *in vitro* to the stabilizing influence of cholesterol oxides against osmotic fragility, suggesting possible future directions for pharmacologic management of this blood disorder.[71] There was a slight increase in osmotic fragility of erythrocytes incubated with cholest-5-en-3-one, cholest-4-en-3-one, and 5α-cholestan-3-one.[72]

Significant differences in the glycerol permeabilities of the modified human and pig erythrocytes were also demonstrated. In the presence of 3-ketosteroids, the rates of penetration of glycerol were significantly higher than into erythrocytes containing 3β-hydroxy steroids. The increased osmotic fragility of the depleted cells may be the consequence of increased cell volume in the isotonic medium as a result of increased ion content of the cells due to enhanced permeability. Both 25-hydroxycholesterol and 25-hydroxyvitamin D_3 were shown to enhance the permeability of phospholipid liposomes to Ca^{2+}, Na^+, and other cations.[67] The lack of specificity of the permeability change towards cations with these sterols and their low ionophoretic properties suggest that a general perturbance of membrane properties occurs rather than there being a specific sterol-cation interaction. Physical-chemical investigations of artificial bilayer lipid membranes suggest that incorporation of cholesterol oxides into the membrane affects the potential energy barrier to inorganic ion conduction through the membrane, as well as membrane fluidity and molecular packing characteristics.[73] Factors influencing ion conductance in the so called ''black lipid membrane'' model (containing cholesterol oxides) include a negative surface charge[74] and an increase in elasticity caused by Ca^{2+} possibly absorbed to the cholesterol oxides in the membrane by a chelation effect.[75]

It is well established that an essential function of cholesterol is to contribute to the barrier properties of phospholipids. Cholesterol has been shown to reduce the permeability of biomembranes due to its condensing effect. The membrane properties of the cholesterol autoxidation products including 7-ketocholesterol, 7β-hydroxycholesterol, 7α-hydroxycholesterol, and 25-hydroxycholesterol were examined by Theunissen et al.[24] using liposome preparations. 7-Ketocholesterol, 7β-hydroxycholesterol, and 7α-hydroxycholesterol have cholesterol-like condensing effects in this order, although to a lesser extent than that observed for cholesterol. In liposomes, these cholesterol oxides also reduced glucose permeability and in the same relative order as their condensing effect. On the other hand, 25-hydroxycholesterol showed no condensing effects, but acted as a spacer molecule in monomolecular layers, and glucose permeability in liposomes is enormously increased. The large increase in permeability to glucose at low 25-hydroxycholesterol concentrations is in agreement with that found for Ca^{2+}. As a result of the perpendicular orientation of 25-hydroxycholesterol, the hydroxyl group at C25 is present in the hydrophobic core of the bilayer. The perpendicular shuttling of 25-hydroxycholesterol between the inner and outer monolayer of the membrane is believed to cause a local disordering of the membrane lipids, leading to the observed high permeability.

E. EFFECTS ON HORSERADISH PEROXIDASE UPTAKE BY NON-RECEPTOR-MEDIATED ENDOCYTOSIS

A variety of cells directly interiorize soluble protein (such as horseradish peroxidase) by invagination of the plasma membrane without previous binding of the proteins to cell surface receptors. Mouse fibroblast L-cells were incubated for various periods of time with cholesterol oxides and were then reincubated for 1 to 3 hours with horseradish peroxidase dissolved in fresh culture medium. The cells were subsequently lysed by incubation with

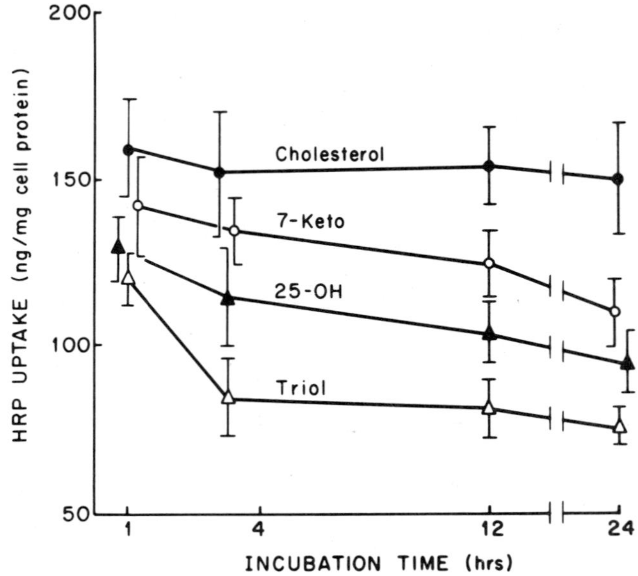

FIGURE 2. Effects of preincubation of cultured human umbilical endothelial cells for 1, 4, 12, or 24 h with cholesterol oxides on the subsequent uptake of horseradish peroxidase. (Reprinted from Peng, S. K., Hu, B., Hsiao, W., and Morin, R. J., *Artery*, 17, 84, 1990. With permission.)

an aqueous solution of 0.05% Triton X-100 for 2 hours at room temperature. The amount of peroxidase in the lysate indicated that there were no cell surface receptors for the probe (peroxidase), and that the rate of endocytosis increased linearly with temperature. 25-Hydroxycholesterol and 20α-hydroxycholesterol depressed the rate of endocytosis by 54% and 42% respectively, whereas 7-ketocholesterol and 6-ketocholestanol were less active.[76] The relative potencies of these sterols as suppressors of endocytosis were similar to their reported potencies as inhibitors of sterol synthesis. Purified cholesterol at a higher dose (100 μg/ml) had no effect on endocytosis. The rate of uptake of peroxidase in the control cultures averaged 167 ng per mg cellular protein per hour. In other studies, rabbit vascular endothelial cells were incubated with cholestane-3β,5α,6β-triol, 7-ketocholesterol, 25-hydroxycholesterol, and cholesterol 5α,6α-epoxide for various periods of time.[77] Incubation with cholestanetriol at 40 μg/ml decreased the rate of endocytosis in 15 minutes. After one hour of incubation, the effect of cholestanetriol leveled off at about 35% of the control rate. 7-Ketocholesterol had a slower and less dramatic effect, leveling off at 55% of the control rate after 24 hours of incubation. 25-Hydroxycholesterol and cholesterol 5α,6α-epoxide had less immediate effects on edocytosis and purified cholesterol at the same concentration had no effect on the rate of endocytosis. Human umbilical vein endothelial cells were also studied and incubated for intervals up to 24 hours with varying concentrations of cholesterol, 7-ketocholesterol, 25-hydroxycholesterol, cholestane-3β,5α,6β-triol, or cholesterol 5α,6α-epoxide.[78] Endocytosis, as measured by uptake of horseradish peroxidase, was inhibited in a dose- and time-dependent manner by cholestane-3β,5α,6β-triol and 25-hydroxycholesterol. Both sterols showed significant inhibition at a concentration of 5 μg/ml after 24 hours of incubation and 7-ketocholesterol produced inhibition at a concentration of 20 μg/ml. Cholestanetriol was again shown to be the most potent inhibitor of horseradish peroxidase uptake. After 24 hours of incubation, cholestanetriol at a concentration of 20 μg/ml produced 52.4% inhibition, followed by 25-hydroxycholesterol with 40.5% inhibition and 7-ketocholesterol with 31.3% inhibition (Figure 2). The rate of uptake of peroxidase in the control cultured endothelial cells averged 160 ng per mg protein per hour.

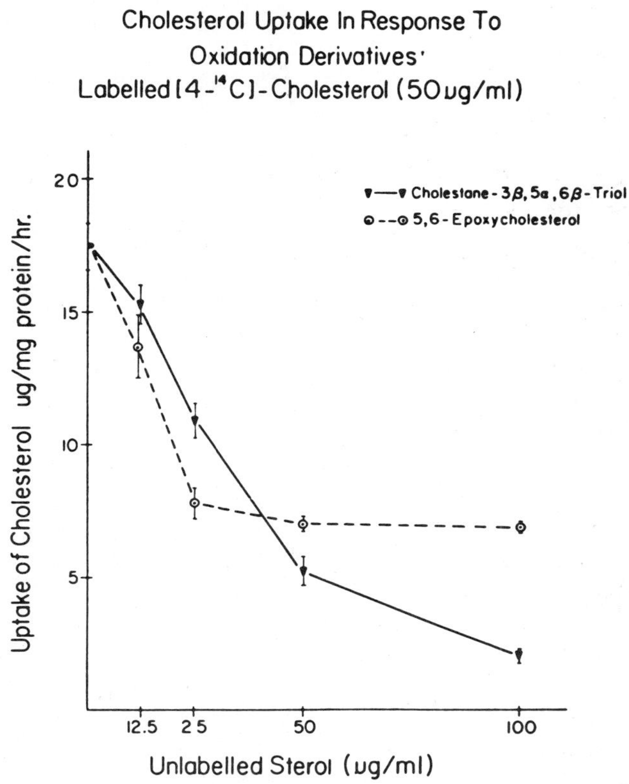

FIGURE 3. Uptake of cholesterol in the presence of cholestane-3β,5α,6β-triol and cholesterol 5α,6α-epoxide. Standard uptake assay conditions with 50 μg/ml of labeled cholesterol and different concentrations of cholesterol oxides from 0 to 100 μg/ml. (Reprinted from Peng, S. K., Morin, R. J., Tham, P., and Taylor, C. B., *Artery,* 13, 144, 1985. With permission.)

Neither purified cholesterol nor cholesterol epoxide showed significant inhibitory effects. Similar results were also obtained in cultured rabbit aortic smooth muscle cells.[78,79] The rate of uptake of peroxidase in the control smooth muscle cells was also very similar to that of endothelial cells, with an average uptake of 159 ng per mg protein per hour. Cholestane-3β,5α,6β-triol showed earlier inhibitory effects on horseradish peroxidase uptake, with 34.4% inhibition at 5 μg/ml after 4 hours of incubation and 58.5% inhibition after 24 hours of incubation. 25-Hydroxycholesterol showed a significant inhibition of 45.3% at 5 μg/ml after 24 hours of incubation, but no significant effect at 4 hours. Purified cholesterol had no inhibitory effect at the same concentration.

F. EFFECTS ON CHOLESTEROL UPTAKE AND RECEPTOR-MEDIATED ENDOCYTOSIS

Cultured rabbit aortic smooth muscle cells were used in studying cholesterol uptake.[80] Labeled [4-^{14}C]-cholesterol was added to cells in culture media containing various concentrations (0 to 50 μg/ml) of cholesterol oxides and incubated for one hour at 37°C. For removal of nonspecific surface-bound sterol from the cells, the cultured cells were incubated at 37°C for 45 min in media containing pronase. The nonspecific surface-bound cholesterol amounted to 35% of total cholesterol uptake. The most potent inhibitor of cholesterol uptake was cholestane-3β,5α,6β-triol. When cholestanetriol was present at 50 μg/ml of medium, uptake of cholesterol by cultured cells decreased more than 70% (Figure 3). The next most potent inhibitor of cholesterol uptake was cholesterol 5α,6α-epoxide, which reduced uptake of cholesterol by 60%. 25-Hydroxycholesterol also inhibited uptake of cholesterol up to 60%. 7-Ketocholesterol and 7α- and 7β-hydroxycholesterol inhibited cholesterol uptake to a similar extent of between 30 to 50%. Uptake of cholesterol in mammalian cells was shown to be mediated by cell surface low-density lipoprotein (LDL) receptors.[81] Brown and Goldstein showed a diminished uptake of LDL in normal human fibroblasts that had been preincubated with 25-hydroxycholesterol.[82] Lorenzo et al.[83] also demonstrated that 26-hydroxycholesterol inhibited the specific binding, uptake, and degradation of ^{125}I-LDL in human fibroblasts. Cultured rabbit aortic smooth muscle cells showed similar responses.[84] The cultured cells were preincubated in 5% lipoprotein-deficient medium for 24 h and then incubated with 1 to 5 μg/ml of either pure cholesterol, 25-hydroxycholesterol, 7-ketocholesterol, cholestane-3β,5α,6β-triol, or cholesterol-5α,6α-epoxide for 12 to 24 h. The uptake of ^{125}I-LDL after 25-hydroxycholesterol or cholesterol 5α,6α-epoxide at 1 μg/ml for 12 h of incubation was suppressed to 142 ± 65 and 169 ± 70 respectively, and after cholestane-3β,5α,6β-triol or 7-ketocholesterol at 5 μg/ml for 12 h was 158 ± 40 and 157 ± 29 respectively, when compared with controls at 219 ± 89 cell bound ^{125}I-LDL μg/mg protein.

Cholesterol oxides also affect the uptake of corticosterone.[85] Isolated rat liver cells were preincubated for 15 min in the absence or presence of 60 μ*M* 25-hydroxycholesterol. The phase transition was shifted to a higher temperature (32°C), at the same time the uptake of corticosterone was diminished. The mechanism might involve inhibition of an active uptake mechanism caused by a change in the plasma membrane.

G. EFFECTS ON MEMBRANE-ASSOCIATED IMMUNOLOGICAL FUNCTIONS OF LYMPHOCYTES

Lymphocyte E (erythrocyte)-rosette formation with sheep red cells by means of membrane-associated receptors has been used for a surface membrane marker of T-lymphocyte. Various cholesterol oxides when added to cultures of human lymphocytes in lipoprotein-depleted medium at a concentration of 25 μ*M* inhibited E-rosette formation with sheep red cells.[86] 20α-Hydroxycholesterol, 7α-hydroxycholesterol, and cholestan-3β,5α-diol-6-one are more potent than 25-hydroxycholesterol. The inhibitory effect of cholestan-3β,5α-diol-6-one on E-rosette formation appeared after 15 min of exposure; with the other three cholesterol oxides, an exposure time of 18 h was necessary. The inhibitory effect of cholestan-3β,5α-diol-6-one on E-rosette formation could be abolished by addition of free cholesterol, LDL, and HDL (high-density lipoprotein) to the medium but not by the addition of mevalonic acid. These observations suggest that the capacity of cholesterol oxides to inhibit E-rosette formation is independent of their inhibitory effect on cholesterol biosynthesis, even if lymphocyte cholesterol biosynthesis was immediately and totally suppressed, a cellular deficiency of cholesterol would take many hours to develop. It is possible that cholesterol oxides inhibit E-rosette formation as a consequence of their insertion into lymphocyte membrane as cholesterol analogs.

Another effect of cholesterol oxides on the immune system is suppression of the plaque-forming cell response to a sepharose 4B to which ε-dinitrophenyl lysine has been covalently attached acting as an *in vitro* T cell-independent antigen. When mouse spleen cells were incubated with liposome containing air-aged impure cholesterol, the plaque-forming cells were totally inhibited.[87] After purification, cholesterol no longer had such immunosuppressant effect. The major inhibitory components in the impure cholesterol were identified as 25-hydroperoxide of cholesterol and 25-hydroxycholesterol. 7α-Hydroxycholesterol, 7β-hydroxycholesterol, and 7-ketocholesterol also showed some immunosuppressant effect. Since this effect was not mitigated by added mevalonate, the immunosuppression appears not to involve sterol biosynthesis.

H. EFFECTS ON CHEMOTAXIS OF POLYMORPHONUCLEAR LEUKOCYTES

The migration of human polymorphonuclear leukocytes toward the chemotactic material was inhibited when these cells were incubated with cholesterol oxides at concentration as low as 6.25 μM in lipoprotein-depleted media for one hour, no significant cell loss occurred, and only a marginal decrease in viability was noted (<5%).[40] In contrast, the capacity of leukocytes to migrate at random around the perimeter of the test well was unimpaired. Cholestan-3β,5α-diol-6-one is the most potent inhibitor of chemotaxis. The cells treated with this compound were found to be almost totally incapable of migration toward the chemotactic material (89%). 6-Ketocholestanol was also capable of producing virtually complete abolition of leukocyte chemotaxis (81%) whereas cholestane-3β,5α,6β-triol, 20α-hydroxycholesterol, and 7β-hydroxycholesterol showed partial inhibitory activity (56%, 55%, and 31% respectively). Cholesterol itself showed no effect. If the preincubation of polymorphonuclear leukocytes was carried out in lipoprotein-containing medium, the cells were only partially protected against the development of the chemotactic defect. The capacity of cholestane-3β,5α,6β-triol to inhibit chemotaxis was also only partially reversed by addition of free cholesterol. As mature human polymorphonuclear leukocytes do not synthesize cholesterol,[88] the inhibition of chemotaxis by cholesterol oxides cannot be attributed entirely to their inhibitory effect on sterol biosynthesis. The rapidity of onset of inhibition of chemotaxis and the lack of correspondence between the effect and the inhibition of HMG-CoA reductase activity suggest a direct membrane effect by cholesterol oxides.

V. SUMMARY

Cholesterol is a major constituent of the membranes of mammalian cells and an important modulator of membrane functions. It is well established that its interaction with phospholipids is very crucial for maintenance of normal membrane structure and function. The structure and dynamics of the lipid core of the cell membrane is a function influenced by the presence of cholesterol. Exposure of cholesterol oxides to cultured cells may result in an alteration of the molar ratio of sterol to phospholipid either by direct insertion of cholesterol oxides into cell membranes or indirectly by inhibition of cholesterol biosynthesis and a reduction in the cholesterol available for membrane synthesis. Using phospholipid liposomes as synthetic model membranes, cholesterol oxides have been shown to reduce the temperature and enthalpy for the gel-to-liquid-crystalline and bilayer-to-hexagonal phase transitions. The effects depend on the specific cholesterol oxide and its amount as well as the specific phospholipid composition. This results in decreased membrane fluidity, with attendant altered membrane permeability and decreased stability. Displacement of membrane cholesterol by cholesterol oxides could affect the surface shape of several cell types, including echinocyte formation of red blood cells with attendant changes in osmotic fragility, and a loss of cell surface microvilli of lymphocytes and mouse L cells.

Proteins that have important cellular functions, such as transporters, receptors, and enzymes, are embedded in the cholesterol-phospholipid bilayer of cell membranes. Cholesterol oxides which could alter membrane fluidity may, therefore, affect the positioning of membrane proteins and alter membrane-associated functions. There is a decrease in the rate of carrier-mediated hexose transport in cultured aortic smooth muscle cells by cholestane-3β,5α,6β-triol after only 15 min of incubation, and a decrease in the membrane bound enzyme 5′-nucleotidase activity after 24 to 48 h of incubation with either cholestane-3β,5α,6β-triol or 25-hydroxycholesterol. The effects on Na^+,K^+-ATPase vary depending on the types of cells and the duration of incubation. There is an increase in the rate of $^{86}Rb^+$ transport by Na^+,K^+-activated ATPase. Cholesterol oxides could perturbate Ca^{2+} homeostasis largely via their effects on calcium channels. Most of the cholesterol oxides increase Ca^{2+} influx to cultured cells and Ca^{2+} uptake into phospholipid liposomes, which may partly contribute to their cytotoxicity and atherogenicity. The permeability to other cations and glycerol is also increased by cholesterol oxides. The effect of cholesterol oxides on permeability appears to be related to their condensing effect, since the essential property of cholesterol is to contribute to the barrier properties of phospholipids through its condensing effect. Incorporation of cholesterol oxides into the membrane could reduce their condensing effect, thus increasing membrane permeability. The cells exposed to cholesterol oxides also show reduced ability to take up soluble horseradish peroxidase by pinocytosis, indicating that vesicle formation and internalization of the plasma membrane is dependent on the maintenance of a suitable level of rigidity in the lipid core of the membrane. Receptor-mediated endocytosis, such as uptake of cholesterol and corticosterone, may also be affected by cholesterol oxide-induced membrane changes. There are at least two possible mechanisms whereby the cholesterol oxides could influence the cholesterol content in the cell membrane, resulting in membrane dysfunction. Many cholesterol oxides such as 25-hydroxycholesterol or 20α-hydroxycholesterol, are potent inhibitors of cholesterol biosynthesis. Consequently, the cellular cholesterol level is decreased and the cholesterol available for membrane synthesis is depleted. The cellular deficiency of cholesterol take many hours to develop, but some effects of cholesterol oxides on membrane morphology and functions develop immediately, such as echinocyte formation of red blood cells, carrier-mediated hexose transport, horseradish peroxidase endocytosis, lymphocyte E-rosette formation, and leukocyte chemotactic movement. Furthermore, mature red blood cells and polymorphonuclear leukocytes do not synthesize cholesterol. The inhibitory effects of cholesterol oxides on cholesterol biosynthesis alone, therefore, cannot be entirely attributed to their membrane effects. Another explanation is that cholesterol oxides may be incorporated into cell membranes, thereby altering membrane lipid composition and membrane fluidity. Cholestane-3β,5α,6β-triol and cholestan-3β,5α-diol-6-one appear to have the most prominent direct membrane effects and 7-oxygenated cholesterol oxides seem to exert their membrane effects via both mechanisms.

REFERENCES

1. **Singer, S. J. and Nicolson, G. L.,** The fluid mosaic model of the structure of cell membranes, *Science,* 175, 720, 1972.
2. **Singer, S. J.,** The molecular organization of membranes, *Annu. Rev. Biochem.,* 43, 805, 1974.
3. **Malhotra, S. K.,** Molecular structure of biological membranes: functional characterization, *Subcell. Biochem.,* 5, 221, 1978.
4. **Frye, L. D. and Edidin, M.,** The rapid intermixing of cell surface antigens after formation of mouse-human heterokaryons, *J. Cell. Sci.,* 7, 319, 1970.
5. **Quinn, P. J. and Chapman, D.,** The dynamics of membrane structure, *Crit. Rev. Biochem.,* 8, 1, 1980.
6. **Turner, J. C.,** Absence of lecithin from the stromata of the cells of certain animals (ruminants) and its relation to venom hydrolysis, *J. Exp. Med.,* 105, 189, 1957.
7. **Maddy, A. H.,** A fluorescent label for the outer components of the plasma membrane, *Biochim. Biophys. Acta,* 88, 390, 1964.
8. **Brulet, P. and McConnell, H. M.,** Magnetic resonance spectrum of membranes, *Proc. Natl. Acad. Sci. U.S.A.,* 72, 1451, 1975.
9. **Scandella, C. J., Devaux, P., and McConnell, H. M.,** Rapid lateral diffusion of phospholipids in rabbit sarcoplasmic reticulum, *Proc. Natl. Acad. Sci. U.S.A.,* 69, 2056, 1972.
10. **Greenhut, S. F. and Roseman, M. A.,** Cytochrome b_5 induced flip-flop of phospholipids in sonicated vesicles, *Biochemistry,* 24, 152, 1985.
11. **Noondam, P. C., Van Echteld, C. J. A., De Krujff, B., and De Gier, J.,** Rapid transbilayer movement of phosphatidylcholine in unsaturated phosphatidylethanolamine containing model membranes, *Biochim. Biophys. Acta,* 646, 483, 1981.
12. **Chen, H. W., Hans-Jörg, H., and Kandutsch, A. A.,** Relationship between sterol synthesis and DNA synthesis in phytohemagglutinin-stimulated mouse lymphocytes, *Proc. Natl. Acad. Sci. U.S.A.,* 72, 1950, 1975.

13. **Murphy, J. R.,** Erythrocyte metabolism. VI. Cell shape and the location of cholesterol in the erythrocyte membrane, *J. Lab. Clin. Med.,* 65, 756, 1965.
14. **Montesano, R.,** Inhomogenous distribution of filipin-sterol complexes in smooth muscle cell plasma membrane, *Nature,* 280, 328, 1970.
15. **Shinitzky, M. and Henkart, P.,** Fluidity of cell membranes-current concepts and trends, *Int. Rev. Cytol.,* 60, 121, 1979.
16. **Phillips, M. C.,** The physical state of phospholipids and cholesterol in monolayers, bilayers and mebranes, *Prog. Surface Membrane Sci.,* 5, 139, 1972.
17. **Fionini, R. M., Valentino, M., Glaser, M., Gratton, E., and Ceratola, G.,** Fluorescence lifetime distributions of 1,6-diphenyl-hexatriene reveal the effect of cholesterol on the microheterogeneity of erythrocyte membrane, *Biochim. Biophys. Acta,* 939, 485, 1988.
18. **Cooper, R. A.,** Influence of increased membrane cholesterol on membrane fluidity and cell function in human red blood cells, *J. Supramol. Struct.,* 8, 413, 1978.
19. **Demel, R. A., Bruckdorfer, K. R., and Van Deenen, L. L. M.,** Structural requirements of sterols for the interaction with lecithin at the air water interface, *Biochim. Biophys. Acta,* 255, 311, 1972.
20. **Demel, R. A. and De Kruyff, B.,** The function of sterols in membranes, *Biochim. Biophys. Acta,* 457, 109, 1976.
21. **Huang, C.-H.,** A structural model for the cholesterol phosphatidylcholine complexes in bilayer membranes, *Lipids,* 12, 348, 1979.
22. **Rooney, M. W., Yachnin, S., Kucuk, O., Lis, L. J., and Kauffman, J. W.,** Oxygenated cholesterols synergistically immobilize acyl chains and enhance protein helical structure in human erythrocyte membranes, *Biochim. Biophys. Acta,* 820, 33, 1985.
23. **Egli, U. H., Streuli, R. A., and Dubler, E.,** Influence of oxygenated sterol compounds of phase transitions in model membranes. A study by differential scanning calorimetry, *Biochemistry,* 23, 148, 1984.
24. **Theunissen, J. J. H., Jackson, R. L., Kempen, H. J. M., and Demel, R. A.,** Membrane properties of oxysterols. Interfacial orientation, influence on membrane permeability and redistribution between membranes, *Biochim. Biophys. Acta,* 860, 66, 1986.
25. **Hagiwara, H., Nagasaki, T., Inada, Y., Saito, Y., Yasuda, T., Kojima, H., Morisaki, M., and Ikekawa, N.,** The interaction of cholesterol analogues with phospholipid — the effect of length of side-chain and configuration of a hydroxyl group introduced in the side-chain, *Biochem. Int.,* 5, 329, 1982.
26. **Gallay, J., De Krujff, B., and Demel, R. A.,** Sterol-phospholipid interactions in model membranes. Effect of polar group substitutions in the cholesterol side-chain at C_{20} and C_{22}, *Biochim. Biophys. Acta,* 769, 96, 1984.
27. **Vincent, M. and Gallay, J.,** Steroid-lipid interactions in sonicated dipalmitoyl phosphatidyl choline vesicles: a steady-state and time-resolved fluorescence anisotrophy study with all trans-1,6-diphenyl-1,3,5-hexatriene as probe, *Biochem. Biophys. Res. Commun.,* 113, 799, 1983.
28. **Benga, G., Hodarnau, A., Ionescu, M., Pop, V. I., Frangopol, P. T., Strujan, V., Holmes, R. P., and Kummerow, F. A.,** A comparison of the effects of cholesterol and 25-hydroxycholesterol on egg yolk lecithin liposomes: spin label studies, *Ann. N.Y. Acad. Sci.,* 414, 140, 1983.
29. **Epand, R. M. and Bottega, R.,** Modulation of the phase transition behavior of phosphatidylethanolamine by cholesterol and oxysterols, *Biochemistry,* 26, 1820, 1987.
30. **Rooney, M., Tamura-Lis, W., Lis, L. J., Yachnin, S., Kucuk, O., and Kauffman, J. W.,** The influence of oxygenated sterol compounds in dipalmitoylphosphatidylcholine bilayer structure and packing, *Chem. Phys. Lipids,* 41, 81, 1986.
31. **Hsu, R. C., Kanofsky, J. R., and Yachnin, S.,** The formation of echinocytes by insertion of oxygenated sterol compounds in red cell membranes, *Blood,* 56, 109, 1980.
32. **Streuli, R. A., Chung, J., Scanu, A. M., and Yachnin, S.,** Serum lipoproteins modulate oxygenated sterol insertion into human red cell membranes, *Science,* 212, 1294, 1981.
33. **Plesser, Y. M., Doljanski, F., and Polliack, A.,** Alteration in lymphocyte surface morphology and membrane fluidity induced by cholesterol depletion, *Cell Mol. Biol.,* 25, 203, 1979.
34. **Lin, P.-S. and Chen, H. W.,** Diminution of L cell microvilli following exposure to 25-hydroxycholesterol, *Cell Biol. Int. Rep.,* 3, 51, 1979.
35. **Luu, B.,** Use of in vitro cell cultures to study cytotoxic properties of natural products, in *Advances in Medicinal Phytochemistry,* Barton, D. and Ollis, W. D., Eds., John Libbey, London, 1986, 97.
36. **Saito, Y., Shimada, H., Imada, T., Kikuchi, T., Ikekawa, N., and Inada, Y.,** Lysis of platelets and erythrocytes by the incorporation of a unique oxygenated sterol: 22R-hydroxycholesterol, *J. Membrane Biol.,* 83, 187, 1985.
37. **Richert, L., Castagna, M., Beck, J., Rong, S., Luu, B., and Ourisson, G.,** Growth-rate-related and hydroxysterol-induced changes in membrane fluidity of cultured hepatoma cells: correlation with 3-hydroxy-3-methyl glutaryl CoA reductase activity, *Biochem. Biophys. Res. Commun.,* 120, 192, 1984.
38. **Peng, S. K., Tham, P., Taylor, C. B., and Mikkelson, B.,** Cytotoxicity of cholesterol oxidation derivatives on cultured aortic smooth muscle cells and their effect on cholesterol biosynthesis, *Am. J. Clin. Nutr.,* 32, 1033, 1979.

39. **Cornell, R. B. and Horwitz, A. F.,** Apparent coordination of the biosynthesis of lipid in cultured cells: its relationship to the regulation of the membrane sterol: phospholipid ratio and cell cycling, *J. Cell Biol.,* 86, 810, 1980.
40. **Yachnin, S., Streuli, R. A., Gordon, L. I., and Hsu, R. C.,** Alteration of peripheral blood cell membrane function and morphology by oxygenated sterols; a membrane insertion hypothesis, *Curr. Top. Hematol.,* 2, 245, 1979.
41. **Santillan, G. G., Zak, I., and Bing, R. J.,** Incorporation of 7-ketocholesterol by plasma membranes from porcine aortic endothelial cells, *Artery,* 11, 119, 1982.
42. **Wharton, S. A. and Green, C.,** Effect of sterol structure on the transfer of sterols and phospholipids from liposomes to erythrocytes in vitro, *Biochim. Biophys. Acta,* 711, 398, 1982.
43. **Scandella, C. J., Hayward, J. A., and Lee, N.,** Cholesterol levels and plasma membrane fluidity in 3T3 and SV101-3T3 cells, *J. Supramol. Struct.,* 11, 477, 1979.
44. **Peng, S. K., Hill, J. C., Morin, R. J., and Taylor, C. B.,** Influence of cholesterol oxidation derivatives on membrane bound enzymes in cultured aortic smooth muscle cells, *Proc. Soc. Exp. Biol. Med.,* 180, 126, 1985.
45. **Kandutsch, A. W., Chen, H. W., and Heiniger, H.-J.,** Biological activity of some oxygenated sterols, *Science,* 201, 498, 1978.
46. **Hill, J. C., Peng, S. K., Morin, R. J., and Taylor, C. B.,** Effects of cholesterol autoxidation derivatives on hexose transport in cultured aortic smooth muscle cells, *Exp. Mol. Pathol.,* 41, 249, 1984.
47. **Howard, B. V., Mott, D. M., Fields, R. M., and Bennett, P. H.,** Insulin stimulation of glucose entry in cultured human fibroblasts, *J. Cell Physiol.,* 101, 129, 1979.
48. **DePierre, J. W. and Karnovsky, M. L.,** Plasma membranes of mammalian cells: a review of methods for their characterization and isolation, *J. Cell Biol.,* 56, 275, 1973.
49. **Arche, J. R. S. and Newsholme, E. A.,** Activities and some properties of 5′-nucleotidase, adenosine kinase and adenosine deaminase in tissues from vertebrates and invertebrates in relation to the control of the concentration and the physiological role of adenosine, *Biochem. J.,* 174, 965, 1978.
50. **Bernan, J. J., Tong, C., and Williams, G. M.,** 5′-Nucleotidase activities in cultured rat liver epithelial and fibroblast cells, *J. Histochem. Cytochem.,* 28, 174, 1980.
51. **Robinson, J. D. and Flashner, M. S.,** The Na^+K^+ activated ATPase: enzymatic and transport properties, *Biochim. Biophys. Acta,* 549, 145, 1979.
52. **Peng, S. K. and Morin, R. J.,** Effects on membrane function by cholesterol oxidation derivatives in cultured aortic smooth muscle cells, *Artery,* 14, 85, 1987.
53. **Peng, S. K., Hill, J. C., Morin, R. J., and Safarik, J.,** Electron microscopic cytochemistry of membrane-bound enzymes of cultured aortic cells treated with cholesterol oxidation products, in *Proc. 42nd Annu. Meet. Electron Microscopy Society of America,* Bailey, G. W., Ed., San Francisco Press, San Francisco, 1984, 714.
54. **Peng, S. K., Riley, N. E., and Morin, R. J.,** Effects of cholesterol oxides on cholesterol metabolism and membrane function, 101st Assoc. Official Anal. Chemists Meet., San Francisco, CA, Sept. 14-17, 1987, Abstracts, p. 108.
55. **Chen, H. W., Heiniger, H. J., and Kandutsch, A. A.,** Alteration of $^{86}Rb^+$ influx and efflux following depletion of membrane sterol in L-cells, *J. Biol. Chem.,* 258, 3180, 1978.
56. **Seiler, D. and Fiehn, W.,** Effect of cholesterol oxidation on Na^+K^+-ATPase activity of erythrocyte membranes, *Experientia,* 32, 849, 1975.
57. **Proverbio, F., Proverbio, T., and Rawlins, F.,** Effect of the partial replacement of membrane cholesterol by other sterols in the Na^+K^+-ATPase activity of microsomal fractions from rat brain cortex, *Acta Cient. Venez.,* 31, 219, 1980.
58. **Tugendhaft Mills, J., Mehler, M. F., and Adamany, A. M.,** Effect of 25-hydroxycholesterol on disposition and turnover of membrane components in cultured aortic smooth muscle cells, *Fed. Proc. Fed. Am. Soc. Exp. Biol.,* 39, 3009, 1980.
59. **Borochov, H. and Shinitzky, M.,** Vertical displacement of membrane proteins mediated by changes in microviscosity, *Proc. Natl. Acad. Sci. U.S.A.,* 73, 4526, 1976.
60. **Schanne, F., Kane, A. B., Young, E. E., and Farber, J. L.,** Calcium dependence of toxic cell death: a final common pathway, *Science,* 206, 700, 1979.
61. **Sevanian, A.,** Cytotoxic and genotoxic effects of cholesterol oxidation products, 101st Assoc. Official Anal. Chemists Meet., San Francisco, CA, Sept. 14-17, 1987, 107.
62. **Sevanian, A., Peterson, H., and Coates, T.,** Cytotoxicity of cholesterol oxidation products in bovine and rabbit aortic endothelial cells, Eighth Int. Symp. Atherosclerosis, Venice, Italy, Oct. 7-8, 1988, 103.
63. **Neyses, L., Stimpel, M., Locher, R., Streuli, R., Kueffer, B., and Vetter, W.,** Cholesterol and its oxidized derivatives modulate the calcium channel in human red blood cells, *J. Hypertension,* 2 (Suppl. 3), 489, 1984.
64. **Holmes, R. P., Kou, I.-L. and Yoss, N. L.,** The effect of 26-hydroxycholesterol on the permeability of membranes to Ca^{2+}, *Fed. Proc. Fed. Am. Soc. Exp. Biol.,* 45, 310, 1986.

65. **Boissonneault, G. A. and Heiniger, H.-J.,** 25-Hydroxycholesterol-induced elevation in ^{45}Ca uptake: correlation with depressed DNA synthesis, *J. Cell Physiol.,* 120, 151, 1984.
66. **Neyses, L., Locher, R., Stimpel, M., Streuli, R., and Vetter, W.,** Stereospecific modulation of the calcium channel in human erythrocytes by cholesterol and its oxidized derivatives, *Biochem. J.,* 227, 105, 1985.
67. **Holmes, R. P. and Yoss, N. L.,** 25-Hydroxysterols increase the permeability of liposomes to Ca^{2+} and other cations, *Biochim. Biophys. Acta,* 770, 15, 1984.
68. **Tipton, C. L., Leung, P. C., Johnson, J. S., Brooks, R. J., and Beitz, D. C.,** Cholesterol hydroperoxidase inhibit calmodulin and suppress atherogenesis in rabbits, *Biochem. Biophys. Res. Commun.,* 146, 1166, 1987.
69. **Beg, Z. H., Stonik, J. A., and Brewer, H. B.,** Modulation of the enzymic activity of 3-hydroxy-3-methylglutaryl coenzyme A reductase by multiple kinase systems involving reversible phosphorlyation: a review, *Metabolism,* 36, 900, 1987.
70. **Streuli, R. A., Kanofsky, J. R., Gunn, R. B., and Yachnin, S.,** Diminished osmotic fragility of human erythrocytes following the membrane insertion of oxygenated sterol compounds, *Blood,* 58, 317, 1981.
71. **Streuli, R. A., Kanofsky, J. R., Gunn, R. B., and Yachnin, S.,** Correction of red blood cell (RBC) osmotic fragility (OF) in hereditary spherocytosis (HS) by oxygenated sterol compounds (OSC), *Clin. Res.,* 28, 549A, 1980.
72. **Bruckdorfer, K. R., Demel, R. A., De Gier, J., and Van Deenen, L. L. M.,** The effect of partial replacements of membrane cholesterol by other sterols on the osmotic fragility and glycerol permeability of erythrocytes, *Biochim. Biophys. Acta,* 183, 334, 1969.
73. **Krull, U. J., Thompson, M., Vandenberg, E. T., and Wong, H. E.,** Langmuir-Blodgett film characteristics and phospholipid membrane ion conduction. I. Modification by cholesterol and oxidized derivatives, *Anal. Chim. Acta,* 174, 83, 1985.
74. **Berczi, A. and Szundi, I.,** Ion permeability and surface charge properties of oxidized cholesterol model membranes, *Stud. Biophys.,* 77, 101, 1979.
75. **Hianik, T., Miklovichova, J., Bajchi, A., Chorvat, D., and Schajter, V.,** Calcium and pH-induced changes in oxidized cholesterol bilayers, *Stud. Biophys.,* 99, 9, 1984.
76. **Heiniger, H. J., Kandutsch, A. A., and Chen, H. W.,** Depletion of L-cell sterol depress endocytosis, *Nature,* 263, 515, 1976.
77. **Sentovich, S., Morin, R. J., and Peng, S. K.,** Effect of cholesterol oxidation derivatives on endocytosis of rabbit vascular endothelium, *Fed. Proc. Fed. Am. Soc. Exp. Biol.,* 44, A1263, 1985.
78. **Peng, S. K., Hu, B., Hsiao, W., and Morin, R. J.,** Influence of cholesterol oxides on endocytosis of cultured endothelial and smooth muscle cells, *Artery,* 17, 84, 1990.
79. **Peng, S. K., Riley, N. E., Gao, D. W., and Peng, S. K.,** Effects of cholesterol oxides on enzymes of cholesterol ester metabolism in rabbit aortic smooth muscle cells, *FASEB J.,* 2, A1216, 1988.
80. **Peng, S. K., Morin, R. J., Tham, P., and Taylor, C. B.,** The effects of oxygenated derivatives of cholesterol on cholesterol uptake by aortic smooth muscle cells, *Artery,* 13, 144, 1985.
81. **Goldstein, J. L. and Brown, M. S.,** The low density lipoprotein pathway and its relation to atherosclerosis, *Annu. Rev. Biochem.,* 46, 897, 1977.
82. **Brown, M. S. and Goldstein, J. L.,** Regulation of the activity of the low density lipoprotein receptor in human fibroblasts, *Cell,* 6, 307, 1975.
83. **Lorenzo, J. L., Allorio, M., Bernini, F., Corsini, A., and Fumagalli, R.,** Regulation of low density lipoprotein metabolism by 25-hydroxycholesterol in human fibroblasts, *Fed. Eur. Biochem. Soc.,* 218, 77, 1987.
84. **Hu, B., Peng, S. K., Chen, H., Peng, A., and Morin, R. J.,** Influence of cholesterol oxides on LDL receptor functions in culture, *FASEB J.,* 5, A1252, 1991.
85. **Huijmans, J. G. M., Degenhart, H. J., and Kortleve, D. J.,** Effect of 25-hydroxycholesterol on the uptake of corticosterone in isolated rat liver cells, *Horm. Metab. Res.,* 20, 28, 1988.
86. **Streuli, R. A., Chung, J., Scanu, A. M., and Yachnin, S.,** Inhibition of human lymphocyte E-rosette formation by oxygenated sterols, *J. Immunol.,* 123, 2897, 1979.
87. **Humphries, G. M. K. and McConnell, H. M.,** Potent immunosuppression by oxidized cholesterol, *J. Immunol.,* 122, 121, 1979.
88. **Fogelman, A. M., Seager, J., Edwards, P. A., Hokom, M., and Popjak, G.,** Cholesterol biosynthesis in human lymphocytes, monocytes, and granulocytes, *Biochem. Biophys. Res. Commun.,* 76, 167, 1977.

Chapter 8

CYTOTOXICITY OF CHOLESTEROL OXIDES

Shi-Kaung Peng, Alex Sevanian, and Robert J. Morin

TABLE OF CONTENTS

I. INTRODUCTION

Many important biological activities are attributed to the presence of cholesterol in the cells and cell membranes. Displacement of cholesterol by cholesterol oxides at these sites may result in adverse effects on cell functions, and can occur under various conditions. There are several possible sources of cholesterol oxide formation. Cholesterol in the diet can readily autoxidize in the presence of air at room temperature[1] and the formation of cholesterol oxides can be significantly accelerated by high temperatures during broiling or frying. These compounds can be absorbed and transported in plasma lipoproteins.[2] Cholesterol oxides can also be endogenously produced *in vivo,* either via free radical-induced chain reactions, such as lipid peroxidation[3-5] or by enzyme-catalyzed reactions, notably through microsomal and mitochondrial cytochrome P-450 systems, which metabolize cholesterol to a variety of steroid hormones and bile acids.[6] The effects of cholesterol oxides on cholesterol metabolism, and on cell membrane morphology and functions as discussed in previous chapters, could eventually influence cell and tissue viability, resulting in disease processes such as atherosclerosis and cancer.

Cytotoxic effects of cholesterol oxides were observed as early as 1911, when necrotizing effects on frog heart preparations were noted to be induced by oxidative derivatives of cholesterol.[7] Later Biswas et al.[8] in 1964 accumulated a large number of chemical compounds related to cholesterol for testing their toxicity, but since many were available only in milligram amounts, they excluded whole-animal experiments and used organ cultures of chicken heart explants for their first large-scale screening. Of 103 compounds tested, 36 steroids were cytotoxic when 63 μg of the materials were used per culture, and 15 of these produced total inhibition of cell growth. In cases of partial inhibition, the cells frequently showed cytological changes such as swelling, increased vacuolation, predominance of round cells instead of the usual spindle-shaped cells, and increase of intracellular fat. The severity of toxic effects was related to the amount of steroid present in the culture medium. Many of them were oxidative derivatives of cholesterol. At 63 μg per culture, cholestane-3β,5α,6β-triol and 7β-hydroxycholesterol were highly toxic. 7α-Hydroxycholesterol, 24-, 25-, and 26-hydroxycholesterol, cholest-5-en-3-one, and cholest-7-en-3β-ol were moderately toxic and cholesterol was nontoxic. At lower levels, 25-hydroxycholesterol and 26-hydroxycholesterol were still toxic at 6 μg per culture, and 7β-hydroxycholesterol was slightly toxic at 6 μg. Cholestane-3β,5α,6β-triol inhibited growth entirely at 13 μg, cholest-5-en-3-one was moderately toxic at 13 μg, and 7α-hydroxycholesterol was slightly toxic at 13 μg (Table 1). McDougall et al.[9] chose 9 of these compounds for further study because of their close association with cholesterol. These were cholestanol, cholestane-3β,5α,6β-triol, cholest-7-en-3β-ol (lathosterol), 7α-hydroxycholesterol, 7β-hydroxycholesterol, 25-hydroxycholesterol, 26-hydroxycholesterol, and cholest-5-en-3-one, along with cholesterol. This time they used organ cultures of rabbit aorta and demonstrated that compounds always associated with cholesterol in tissues, i.e., cholestanol, cholestane-3β,5α,6β-triol, lathosterol, and 26-hydroxycholesterol had the most marked effects on the aortic tissue. 7β-Hydroxycholesterol, 25-hydroxycholesterol, cholest-5-en-3-one, and cholesterol itself caused a lesser degree of damage to the aortic tissue and 7α-hydroxycholesterol was nontoxic. The changes in aortic tissue included lipid infiltration, pyknosis of nuclei, disorganization of elastic tissue, alteration of the ground substance, and changes in the acid phosphatase content of the tissue.

More recent studies on cytotoxicity of cholesterol oxides have been largely done on cultured mammalian cells. The cytotoxic effects of cholesterol oxides appear to vary depending upon the type of cell line used, the status of the cell cultures (whether in the proliferative or stationary phase), the composition of the incubation medium, the dose and solubility of the compound tested, the duration of exposure time, etc. Furthermore, whether these *in vitro* cytotoxic effects have any relationship to physiological function *in vivo* remains

TABLE 1
Toxicity of Some Steroids Associated with Cholesterol in Organ Explant Cultures

	Chick heart explants	Rabbit aorta explants
Cholesterol	Nontoxic at 63 μg	Moderately toxic
5α-Cholest-7-en-3β-ol	Nontoxic at 13 μg	Highly toxic
Cholestanol	Nontoxic at 13 μg	Highly toxic
Cholestane-3β,5α,6β-triol	No growth at 13 μg	Highly toxic
7α-Hydroxycholesterol	Slightly toxic at 13 μg	Nontoxic
7β-Hydroxycholesterol	Slightly toxic at 6 μg	Moderately toxic
25-Hydroxycholesterol	Toxic at 6 μg	Moderately toxic
26-Hydroxycholesterol	Toxic at 6 μg	Highly toxic
Cholest-5-en-3-one	Toxic at 13 μg	Moderately toxic

to be established. A few studies on *in vivo* toxic effects of cholesterol oxides have also been performed, and these will be reviewed separately.

II. *IN VITRO* CYTOTOXICITY

A. VASCULAR SMOOTH MUSCLE AND ENDOTHELIAL CELL CULTURES

Cytotoxicity to arterial smooth muscle cells and endothelial cells appear to have significant implications since these two cell types play major roles in the development of atherosclerotic lesions. Peng et al.[1] in 1978 studied the effects of autoxidation products of cholesterol which were identified and separated from U.S.P.-grade cholesterol which had been stored at room temperature in the presence of air for 5 years, along with the 12 most common cholesterol oxides which were commercially available.[10] U.S.P. Grade cholesterol was dissolved in methanol and recrystallized. After repeating the same procedure three times, the methanol mother liquor was retained and evaporated under nitrogen until a thick syrup formed, then the syrup was dried under vacuum. The autoxidation products of cholesterol which were concentrated in the methanol mother liquor were separated and identified by employing thin-layer chromatography and gas-liquid chromatography as shown in Table 2. These autoxidation products, along with 12 commercially available cholesterol oxides, which were further purified by thin-layer chromatography, were tested for cytotoxicity using cultured rabbit aortic smooth muscle cells. All cholesterol oxides were first dissolved in a small amount of ethanol and then added to the culture media at a level not exceeding 0.8% ethanol. The medium containing 10% fetal calf serum served as a solubilizing vehicle. Twenty-four hours after adding the test compounds, the viability of the cultured cells, including dying and dead cells, was measured. Using these bioassay tests, bands I and III separated from the cholesterol autoxidation mixture by thin-layer chromatography were the most toxic (Table 3). Band I contained cholestane-3β,5α,6β-triol and band III contained 25-hydroxycholesterol. These findings were further confirmed by the results of testing 12 purified cholesterol oxides, which showed that 25-hydroxycholesterol and cholestane-3β,5α,6β-triol were the most toxic, with cytotoxicity at 10 μg/ml concentration in the culture medium after 24 h of incubation as shown in Table 4. 20α-Hydroxycholesterol, cholestan-3β,5α-diol-6-one and cholestan-3β-ol-6-one were the next most toxic at 20 μg/ml, and 7α- and 7β-hydroxycho-lesterol, 7-ketocholestanol, and 4-cholesten-3-one were only mildly toxic at 50 μg/ml. Cholesta-3,5,-dien-7-one, 5,6-epoxycholesterol, cholestan-3,6-dione, and purified cholesterol were innocuous at the concentration of 100 μg/ml in culture medium. The minimal concentration of the most toxic compounds required to induce aortic smooth muscle cell death within 24 h is 5 to 10 μg/ml; this is approximately one two-hundredth of the usual human plasma cholesterol concentration. In other words, if only 0.5% of plasma cholesterol

TABLE 2
Major Autoxidation Products of the Concentrate from USP-Grade Cholesterol

TLC spots R_F[a]color[b]	Identified sterols	GLC quantitation, %	TLC fractionation
1.00	Solvent front		VI (0.80—1.00)
0.86 Beige	Cholesta-3,5-dien-7-one		
0.75 Magenta	Cholesterol	38.0	V (0.70—0.80)
0.65 Violet	Unidentified	3.8	IV (0.55—0.70)
0.61 Blue	5 & 7-Hydroperoxide of cholesterol		
0.50 Magenta to gray blue	25-Hydroxycholesterol	13.4	III (0.40—0.55)
0.44[c]	7-Ketocholesterol	12.9	II (0.20—0.40)
0.33 Blue	7β-Hydroxycholesterol	6.7	
0.27 Blue	7α-hydroxycholesterol		
0.08 Yellow	Cholestane-3β,5α,6β-triol	5.1	I (0.00—0.20)
0.00 Brown	Origin compounds		

[a] Mobility relative to solvent front.
[b] Color development with 50% sulfuric acid.
[c] Detected by ultraviolet light absorption.

Reprinted from Peng, S. K., Taylor, C. B., Tham, P., Werthessen, N. T., and Mikkelson, B., *Arch. Pathol. Lab. Med.*, 102, 57, 1978. Copyright 1978, American Medical Association. With permission.

is in the form of toxic oxidation derivatives which might be ingested in food containing cholesterol or endogenously produced by free radical-induced reactions in lipoproteins, the aortic smooth muscle cells could be injured within 24 h.

Using pig vascular smooth muscle cell cultures, Baranowski et al.[11] showed that cholestane-3β,5α,6β-triol was the most cytotoxic among the cholesterol oxides when added to primary cultures on the second day; cells were killed within 24 h after such addition. Cholesterol 5,6-epoxide and 25-hydroxycholesterol were less toxic, causing cellular damage and death by the third day after addition. Intracellular refractile lipid droplets appeared as early evidence of toxicity. Air-oxidized cholesterol produced only minor changes, while addition of purified cholesterol resulted in normal growth, similar to that seen in the controls. 25-Hydroxycholesterol was also examined in sparse and confluent cultures of rabbit aortic smooth muscle cells and human umbilical vein endothelial cells.[12,13] Sparse smooth muscle cell cultures have been postulated to approximate the conditions found as smooth muscle cells begin to migrate from the media into the intima; confluent smooth muscle cell cultures resemble the aortic media in that they are multilayered and relatively nonproliferative. Similarly, sparse endothelial cell cultures could be an approximate model for areas of intimal injury or denudation, while confluent endothelial cell cultures approximate the intact intima. 25-Hydroxycholesterol inhibited confluent smooth muscle cell growth at 10^{-4} *M*, inhibited sparse smooth muscle cell growth at much lower concentrations in a dose-related manner at 10^{-4}, 10^{-5}, and 10^{-6} *M*, inhibited confluent human umbilical vein endothelial cell growth at 10^{-4} *M*, and inhibited sparse human umbilical vein endothelial cell growth also at 10^{-4} *M* and at the lower concentrations of 10^{-5} and 10^{-6} *M* in an ethanol vehicle. In a fetal bovine serum vehicle 25-hydroxycholesterol inhibited cell growth at only 10^{-4} and 10^{-5} *M*. Thus, rabbit aortic smooth muscle cell growth was more sensitive to inhibition by 25-hydroxycholesterol than human umbilical vein endothelial cell growth and both cultured cells were considerably more sensitive to inhibition by 25-hydroxycholesterol in sparse cell cultures than in confluent cell cultures. Cholesterol did not inhibit cell growth under any experimental condition.

In cultured human aortic smooth muscle cell monolayers, cholesterol oxides, including 25-hydroxycholesterol, 7-ketocholesterol, or cholesterol 5,6-epoxide, markedly inhibited

TABLE 3
Smooth Muscle Cell Toxicity Among the Fractions of the Concentrate[a]

Fraction no.	Dying cells, %	Dead cells, %	Total cells counted
Concentrate of 1.0 mg Used for TLC Fractionation[b]			
I	8.2[c]	5.8[c]	1,212
II	6.2	3.8	795
III	34.6	36.0	696
IV	6.3	2.1	1,662
V	3.8	1.6	950
VI	3.3	1.5	1,105
Concentrate of 1.75 mg Used for TLC Fractionation			
I	19.9	23.4	898
II	7.1	4.9	874
III	31.8	51.7	1,015
IV	2.2	1.9	994
V	3.8	3.5	796
VI	4.1	2.0	936
Concentrate of 2.50 mg Used for TLC Fractionation			
I	18.5	38.4	2,369
II	7.7	6.0	681
III	38.1	54.0	2,219
IV	15.6	18.1	1,873
V	3.2	2.3	1,688
VI	2.0	1.7	1,227

[a] Cell counts done by three observers as unknowns; identity of slides revealed after counting.
[b] The concentration of each fraction can be estimated from Table 2.
[c] Percent of dying and dead cells represents an average of the counts of three observers.

Reprinted from Peng, S. K., Taylor, C. B., Tham, P., Werthessen, M. T., and Mikkelson, B., *Arch. Pathol. Lab. Med.*, 102, 57, 1978. Copyright 1978, American Medical Association. With permission.

growth at 5 μg/ml concentration, with no effects on the viability of the cultures. Higher concentrations (more than 10 μg/ml) were toxic and cell viability declined.[14] The effects on growth were concentration dependent, and the most potent inhibitor of cellular growth, as measured by decreased incorporation of thymidine into DNA and suppression of hydroxymethylglutaryl coenzyme A (HMG-CoA) reductase activity, was 25-hydroxycholesterol, followed in order of decreasing effect by cholesterol 5,6-epoxide and 7-ketocholesterol. The inhibitory effect of 25-hydroxycholesterol on cellular growth was not reversible when the cultures were grown in medium with normal fetal calf serum. In media with delipidated serum, however, addition of purified cholesterol partially prevented growth inhibition induced by 25-hydroxycholesterol. Purified cholesterol, independently or in combination with tocopherol, had no toxic effects on cellular growth. Additionally, when human aortic smooth muscle cells were incubated with different concentrations of 25-hydroxycholesterol, cholesterol 5,6-epoxide, or 7-ketocholesterol, an increase in acid phosphatase activity was observed in the incubation medium. The most toxic cholesterol oxide, 25-hydroxycholesterol,

TABLE 4
The Cytotoxic Effect of Oxidation Derivatives of Cholesterol on Cultured Rabbit Aortic Smooth Muscle Cells

	Concentration in culture medium, μg/ml			
Compounds	**10**	**20**	**50**	**100**
Grade of cytoxicity[a]				
25-OH cholesterol	1	2	4	4
20α-OH cholesterol	0	1	2	4
4-Cholesten-3-one	0	0	1	3
7α-OH cholesterol	0	0	1	2
7β-OH cholesterol	0	0	1	3
7-Ketocholesterol	0	0	1	3
3.5-Cholestadien-7-one	0	0	0	0
5,6-Epoxycholesterol	0	0	0	0
Cholestane-3β,5α,6β-triol	1	2	4	4
Cholestan-3β,5α-diol-6-one	0	1	2	3
Cholestan-3β-ol-6-one	0	1	2	4
Cholestan-3,6-dione	0	0	0	0
Purified cholesterol	0	0	0	0

[a] Degrees of cytotoxicity were graded as percent of dying and dead cells: (0) less than 5%; (1) 5—25%; (2) 25—50%; (3) 50—75%; (4) 75—100%. Results based upon average of 5 sets of studies. Degrees of cytotoxicity were very consistent.

Reprinted from Peng, S. K., Tham, P., Taylor, C. B., and Mikkelson, B., *Am. J. Clin. Nutr.*, 32, 1033, 1979. With permission.

resulted in the greatest release of acid phosphatase, and the least toxic, 7-ketocholesterol, showed statistically significant release only at the highest level (10 μg/ml). The effect was concentration dependent and was inversely related to cellular growth. It has been demonstrated both histochemically and biochemically that acid phosphatase activity is significantly increased in human and experimental atherosclerotic lesions and in diet-induced aortic atherosclerotic lesions but not in unaffected regions of the same aorta.[15,16] It has also been postulated that cholesterol oxides may cause an increased stimulation of cholesterol esters within the lysosomes, which then activates this lysosomal enzyme.[17]

In cultured rabbit aortic endothelial cells, the cytotoxicity of cholestane-3β,5α,6β-triol was compared with that of purified cholesterol.[18] Cells incubated with 10 μg/ml of cholestanetriol for 12 or 24 h showed an increased rate of cell death and decreased [^{3}H]-thymidine incorporation into DNA as compared to either the control or purified cholesterol. These effects were more marked at 24 h than at 12 h. Purified cholesterol had no significant effects at either time interval. The uptake and cytotoxicity of the isomeric cholesterol 5,6-epoxides and cholestane-3β,5α,6β-triol were examined by Sevanian et al.[19] in cultured rabbit aortic endothelial cells. The isomeric cholesterol epoxides were incorporated with equal facility and in a dose-dependent manner. The pattern of uptake, which was markedly influenced by media serum concentration and by temperature, suggested that these compounds were partly incorporated through association with serum lipoproteins. Following incorporation, both epoxide isomers were rapidly converted to cholestanetriol which readily exited the cells. Cholestanetriol was further metabolized to an ester-type product, representing up to 10% of the added cholesterol epoxides, by 24 hours of incubation. The order of cytotoxic potency of these cholesterol oxides was cholestane-3β,5α,6β-triol > cholesterol 5β,6β-epoxide > cholesterol 5α,6α-epoxide with LD_{50} concentrations ranging from 23 to 150 μM in confluent cells. It appeared that these cytotoxic cholesterol oxides were incorporated by endothelial cells through a combination of receptor-mediated and nonspecific or passive mechanisms, however, the efficacy of uptake and resulting toxicity was substantially influenced by serum lipoproteins. In bovine endothelial cell cultures, cholesterol oxides produced a linear increase in toxicity with increasing dose.[20] Cholestane-3β,5α,6β-triol was the most potent cytotoxic compound among the cholesterol oxides with LD_{50} concentration of 30 μ*M*. Somewhat lower cytotoxic potencies were observed for cholesterol β-epoxide, 7-ketocholesterol, 3,5-cholestadiene-7-one, cholesterol α-epoxide, and 7α-hydroxycholesterol, with LD_{50} concentrations of 40, 42, 45, 58, and 72 μ*M*, respectively. In cultured human umbilical endothelial cells, 25-hydroxycholesterol inhibited cell growth at 10^{-4} *M* in confluent cell cultures and at 10^{-6} *M* in sparse cell cultures as previously mentioned.[12,13]

Cytotoxicity to bovine aortic endothelial cells was also tested by enzymatically treating low-density lipoproteins (LDL) and high-density lipoproteins (HDL) with cholesterol oxidase, which catalyzes the oxidation of cholesterol to cholest-4-en-3-one. Untreated LDL and HDL were used as controls. Only the cells incubated with cholesterol oxidase-treated LDL showed significant cytotoxicity after 48 h.[21] Cells incubated with untreated LDL or HDL or cholesterol oxidase-treated HDL all showed no apparent effects. Interestingly, it has also been demonstrated that oxidized LDL formed by free radical-induced lipid peroxidation was injurious or toxic to cultured endothelial cells and arterial smooth muscle cells.[22,23] The mechanism by which the oxidized lipid fraction of LDL might damage cultured cells and the identity of the toxic agent remain subjects of speculation. It is very possible that these toxic components are cholesterol oxides.[24,25]

B. FIBROBLAST CULTURES

Reduction or cessation of cell growth of cultured mouse fibroblast L cells was reported after prolonged incubation with cholesterol oxides.[26] The most potent inhibitor of cell culture growth was 25-hydroxycholesterol, followed in order of decreasing activity by 20α-hydroxycholesterol, 7-ketocholesterol, 7β-hydroxycholesterol, and 7α-hydroxycholesterol. The first three sterols inhibited growth of cultures by more than 90% at 1, 2.5, and 10 μg/ml, respectively; 7α-hydroxycholesterol at 50 μg/ml inhibited growth by 50%. L cells incubated with these cholesterol oxides (which specifically inhibit sterol biosynthesis at the level of HMG-CoA reductase) became severely deficient in desmosterol. The growth of the cultures was then greatly depressed unless an appropriate sterol or sterol precursor was supplied exogenously. Higley and Taylor[27] measured cellular protein as an indicator of the cytotoxicity of cholesterol oxides. When monolayers of mouse fibroblast L cells were incubated with cholesterol oxides, the site of oxidation of the cholesterol molecule was not related to the degree of cytotoxicity. Of the 12 cholesterol oxides tested, cholestane-3β,5α,6β-triol and cholestan-3,5-dien-7-one exhibited the greatest cytotoxicity, inducing approximately 45% reduction in cellular protein compared to the control at concentrations of 10 μg/ml after 1 d of incubation with the compounds. 25-Hydroxycholesterol and 7-ketocholesterol were the most cytotoxic after 5 d of exposure, with concentrations of 10 μg/ml reducing the cellular protein content to approximately 10% of the control. Lipid accumulation (steatosis) in the mouse fibroblasts, as measured by Oil Red O-stainable droplets formed in response to cholesterol oxides was also investigated. Grade 3 steatosis, with more than 75% of the L cell cytoplasm occupied by lipid droplets after 1 d of exposure to cholesterol oxides was generally proportionate to the cytotoxic effects of the same cholesterol oxides after 5 d of exposure. Some of the steatotic cholesterol oxides, however, such as the α and β isomers of cholesterol epoxide, were highly active with respect to steatosis, increasing the grade 3 value to 209 and 390% of control, respectively, but were not cytotoxic. These findings appear to coincide with the report by Brown et al.,[28] who observed an increase in the cholesterol ester content of cultured human fibroblasts following exposure to 25-hydroxycholesterol, 6-ketocholestanol, and 7-ketocholesterol resulting from an increased activity of the membrane-bound fatty acyl-CoA-cholesteryl acyltransferase (ACAT). Baranowski et al.[11]

used cultured mouse embryo fibroblasts to test cholesterol oxides and showed similar results. Fibroblast growth was markedly inhibited by cholestane-3β,5α,6β-triol, but less so by 25-hydroxycholesterol and cholesterol 5,6-epoxide. After addition of sterols to the primary cultures, cytotoxicity occurred earlier with cholestanetriol than with the other two cholesterol oxides. Addition of cholesterol oxides to the primary cultures and then to the subcultures again showed cholestanetriol to be the most cytotoxic of these three cholesterol oxides. Changes in total DNA in culture wells after addition of cholesterol oxides confirmed the macroscopic and microscopic observation on the cultured cells.

Sevanian and Peterson[29] and Peterson et al.[30] used cultured Chinese hamster lung fibroblast V79 cells to study the cytotoxicity and mutagenic properties of cholesterol oxides. Four cholesterol oxidation products that occur most consistently in membrane preparations subjected to free radical-induced lipid peroxidation, including cholesterol 5α,6α- and 5β,6β-epoxides, their common hydration product cholestane-3β,5α,6β-triol, and 7-ketocholesterol, were tested in the cell cultures. The relative cytotoxic potency in order of decreasing effect was as follows: cholestanetriol, toxic at 12 μ*M;* cholesterol 5β,6β-epoxide, toxic at 20 μ*M;* 7-ketocholesterol, toxic at 25 μ*M;* and 5α,6α-epoxide, toxic at 50 μ*M*. The inhibition of DNA synthesis, as measured by [^{3}H]-thymidine incorporation into DNA, was well correlated with the degree of cytotoxicity for each compound tested. For mutations per 10^5 survivors, only cholesterol epoxides were weakly mutagenic, with the β isomer apparently more potent than the α isomer. Cholesterol β-epoxide induced colonies of cells resistant to 6-thioguanine at 2 to 3 times the frequencies observed with the α isomer, but neither compound produced ouabain-resistant colonies.[30] The epoxides were assimilated by cells to an equal extent, however, β-epoxide was metabolized to cholestane-3β,5α,6β-triol twice as rapidly as the α isomer. These interesting findings suggest that the action of cholesterol epoxide hydrase, which converts the epoxides to cholestanetriol, is important in the mutagenicity of cholesterol epoxide, and inhibition of this enzyme or prevention of cholestanetriol accumulation should reduce the cytotoxicity of the epoxide and enhance its mutagenic effect. The weak mutagenic activity of cholesterol epoxides is presumably based on their electrophilic properties.[31] Both epoxides reacted with 4-(4′-nitrobenzyl)-pyridine to a similar extent, and with identical nucleophilic selectivity at pH 7.4, but their alkylating activity was estimated on this basis to be two orders of magnitude less than methyl methanesulfonate. Binding experiments with the DNA or cultured V79 cells or with calf thymus DNA indicated that interaction was noncovalent and DNA binding did not correlate with the potency of the epoxides to induce the 6-thioguanine-resistant phenotype.[29]

C. MONOCYTE/MACROPHAGES

It is known that the arterial cellular response to injury involves not only vascular endothelial and smooth muscle cells, but also macrophages. In fact, cholesteryl ester accumulation in arterial wall macrophages (foam cells) is a prominent feature of atherosclerotic lesions. It has been shown that oxidatively modified LDL, which may contain exogenous or endogenously produced cholesterol oxides, can be internalized by resident monocyte/macrophages in large quantities via a high-affinity scavenger cell receptor which is not subject to down-regulation by the intracellular content of cholesterol.[32] This may in part account for the accumulation of cholesteryl esters in macrophage-derived foam cells and the eventual destruction of the cell by this accumulation. It has also been shown that uptake of LDL by J774 macrophages is not down-regulated by 25-hydroxycholesterol when compared to fibroblasts.[33] LDL oxidized enzymatically by cholesterol oxidase, which catalyzes the oxidation of cholesterol to cholest-4-en-3-one, proved toxic for cultured human macrophage-like histiocytes (U937), whereas cholesterol oxidase treated HDL was not toxic.[21] Neither enzyme nor lipoprotein alone was cytotoxic.

Mouse peritoneal macrophages were also tested for cytocoxicity induced by cholesterol oxides. These cells were most sensitive to the cholestane-3β,5α,6β-triol at 1 to 2 d after addition to the medium, as judged by morphological appearances and by uptake of trypan blue after 4 d in culture.[11] Cholesterol 5,6-epoxide caused cell damage by the second day, while 25-hydroxycholesterol was less toxic. Cholesterol showed no apparent effect.

D. LYMPHOCYTES

Yachnin et al.[34] had demonstrated that certain cholesterol oxides could, to varying degrees, kill lymphocytes, reduce their number in tissue culture (lysis), inhibit E-rosette formation by T-lymphocytes in a time- and concentration-dependent manner, abolish the chemotactic migration of polymorphonuclear leukocytes, and produce echinocytic changes in red cell morphology. The effects of cholesterol oxides on lymphocyte E-rosette formation, leukocyte chemotaxis, and red cell echinocyte formation have been described in a previous chapter. Cholestan-3β,5α-diol-6-one has been shown to be the most potent inducer of lymphocytic cytotoxicity and lymphocytolysis, followed by 7α-hydroxycholesterol and 7β-hydroxycholesterol. 25-Hydroxycholesterol and 20α-hydroxycholesterol also showed significant lymphocyte cytotoxicity but no evidence of lymphocytolysis.

Incubation of peripheral blood or isolated lymphocytes of C57L/J mice with phytohemagglutinin stimulated the incorporation of thymidine into DNA of lymphocytes as they transformed into large lymphoblasts. DNA synthesis began after about 24 h of incubation and reached a peak at 48 h. The *de novo* synthesis of sterols from acetate was stimulated much earlier, at 4 h of incubation, and the rate reached a maximum at 24 h, approximately at the time DNA synthesis began. Phytohemagglutinin stimulation of sterol synthesis could be abolished by the addition of certain cholesterol oxides such as 25-hydroxycholesterol and 20α-hydroxycholesterol.[35] This treatment also abolished DNA synthesis and blastogenesis of lymphocytes. The inhibitory effect on DNA synthesis is closely related to sterol structure. Contrary to the observations with fibroblast cultures, the 7-oxygenated cholesterol oxides showed no effect in human lymphocytes. Only 25-hydroxylated cholesterol oxides inhibited phytohemagglutinin-stimulated ^{3}H-thymidine uptake of peripheral blood lymphocytes, which could be mitigated by the addition of cholesterol or mevalonic acid, suggesting a relationship to inhibitory effects of these cholesterol oxides on sterol biosynthesis. None of the sterols tested showed cytotoxicity toward normal phytoagglutinin-stimulated lymphocytes at the cconcentration of 1 μg/ml sterols. The inhibitory effects of sterols on DNA synthesis were strictly related to the position and the configuration of hydroxyl group on the side chain. The inhibitory power increased as the distance of the hydroxyl group from the sterol nucleus increased. The S compounds of side-chain cholesterol oxides were more efficient than R epimers. 7-Ketocholesterol was shown to block the blastogenesis of mouse spleen lymphocytes in response to concanavalin A. This effect could be reversed by liposomes containing cholesterol, and cholesterol-rich liposomes enhanced lymphocyte blastogenesis even in the presence of 7-ketocholesterol. The inhibitory effect of cholesterol oxides on DNA synthesis in concanavalin-A-stimulated human peripheral blood lymphocytes is closely related to their inhibitory effect on sterol biosynthesis.[36] 25-Hydroxycholesterol and 20α-hydroxycholesterol are the most potent inhibitors of both. Inhibition of sterol biosynthesis by 25-hydroxycholesterol also inhibited splenic lymphocyte differentiation into cytotoxic effector cells as well as cellular proliferation. The inhibition of differentiation and proliferation correlated well with the dose of 25-hydroxycholesterol administered and could be counteracted by addition of cholesterol-containing liposomes. Both 7-ketocholesterol and 25-hydroxycholesterol markedly decreased mitogen-induced human T lymphocyte activation after 72 or 96 h of incubation, which could be prevented by supplementation of the culture with mevalonate. Cholesterol oxides did not inhibit the early event in lymphocyte activation, however, but rather altered events occurring afterward. It was suggested that the decrease in lymphocyte responses observed later in culture is a secondary phenomenon resulting from inhibition of the capacity of the cells to synthesize adequate membrane for enlargement and subsequent division.

The cytotoxic effects of 7β-hydroxycholesterol on lymphoma YAC-1, EL-4, and RDM-4 cells in culture, on lymphocytes entering blastic transformation, and on quiescent murine spleen lymphocytes have been investigated.[37] The early events of blastogenesis as well as lymphoma YAC-1, EL-4, and RDM-4 cells were shown to be very sensitive to this sterol at 3-μ*M* concentration, whereas constituted lymphoblasts and normal lymphocytes remained insensitive at 50 times higher concentrations. The cytotoxic effects of this compound were less in medium supplemented with fetal calf serum than in lipoprotein-deficient serum (12-μ*M* vs. 3-μ*M* concentration of 7β-hydroxycholesterol). This selective killing of cancerous cells by 7β-hydroxycholesterol may have an antitumor therapeutic potential. When compared with other anticancerous drugs, such as mitomycin-C, adriamycin, methotrexate, there is a major difference in that these drugs have not been observed to be sensitive to the presence of supplemental serum in the growth medium. 7β-Hydroxycholesterol, when added to lymphocytes simultaneously with concavalin-A or phytohemagglutinin inhibited blastogenesis. Since this sterol is not effective on already constituted blasts and exerted its toxicity only on lymphocytes entering blastic transformation, it seems that its effects are due to a selective impairment of the early events in blastogenesis, prior to maximum [^{3}H]-thymidine incorporation into DNA and probably even prior to maximum cholesterol biosynthesis, which occurs 24 h before DNA synthesis. Mitogenic stimulation of lymphocytes was also significantly inhibited by addition of human serum LDL that had undergone autoxidation, while no significant effect was seen with nonoxidized LDL.[38] The inhibition was effective for cells stimulated either by phytohemagglutinin or enzymatically by neuraminidase-galactose oxidase treatment. Inhibition was markedly attenuated, however, when oxidized LDL was added to cells that had already been triggered 24 h earlier. Lipid extracts from oxidized LDL were similarly inhibitory, indicating that the effect was mediated by an oxidized lipid fraction. In another study, enzymatically oxidized LDL treated with cholesterol oxidase which produced cholest-4-en-3-one also proved toxic to B-lymphoblastoid cells (Daudi).[21] Cholesterol-oxidase-treated HDL showed only questionable toxicity and untreated LDL or HDL showed no effects.

E. ERYTHROCYTES AND PLATELETS

One of the striking effects of cholesterol oxides on erythrocytes is the echinocyte formation which has been discussed in the previous chapter. The erythrocytes, however, remain intact. Some side-chain cholesterol oxides, particularly 22R-hydroxycholesterol at 40 μ*M*, could produce lysis of not only erythrocytes but also platelets in a dose-dependent manner.[39] The lysis of erythrocytes and platelets was evidenced by the release of intracellular proteins, hemoglobin and lactate dehydrogenase respectively. Elevated temperature was required for the lysis, probably to redistribute the sterol in the lipid bilayer of the plasma membranes. When the sterol was incorporated at low temperature, the temperature had to be increased to readily lyse cells. This lytic effect was, surprisingly enough, restricted to the R isomer; the S isomer was only marginally effective. The difference between these two isomers could not be attributed to their amounts incorporated in cells, since both sterols were incorporated to comparable extents. Other cholesterol oxides with hydroxyl group at different positions of the side chain of cholesterol were much less lytic (only one-fourth as effective as 22R-hydroxycholesterol) regardless of the hydroxyl group introduced. Although cholestane-3β,5α,6β-triol, cholestan-3β,5α-diol-6-one, 7-ketocholesterol, 6-ketocholestanol, and 6β-hydroxycholestanol were shown to be capable of producing echinocytic effects on erythrocytes, they had very little lytic effect on erythrocytes. Pure cholesterol was totally ineffective. A possible mechanism for this interesting phenomenon is that 22R-hydroxycholesterol has very weak interactions with membrane phospholipid and does not form and maintain a stable and functional membrane. This is probably due to the "kink" in the side chain resulting from the introduction of a hydroxyl group with R configuration specifically at the C22 position.

F. BONE MARROW GRANULOCYTIC PROGENITOR CELLS

Various cholesterol oxides were found to inhibit human bone marrow granulocyte (CFU-C) colony formation, and this inhibition was concentration dependent.[40] The compound showing the greatest inhibitory potency was 25-hydroxycholesterol. The 50% inhibitory concentration for CFU-C growth by 25-hydroxycholesterol was approximately 10^{-7} *M*, the same as that for similar inhibition of cholesterol synthesis in marrow mononuclear cells. Such inhibition leads to a change in the properties of cell membranes and could result in impaired cell division, inhibition of DNA synthesis, and subsequent inhibition of colony growth. This compound, as well as 6-ketocholestanol, 7-ketocholesterol, and 20α-hydroxycholesterol, also demonstrated marked inhibition of CFU-C proliferation. The latter effect, which was not a result of direct cytotoxicity of the compounds, was reversible by cholesterol, but not by mevalonic acid. The suppression of CFU-C proliferation cannot be sufficiently explained by the inhibition of cholesterol synthesis. Another mechanism could be their insertion into the membrane lipid bilayer as cholesterol analogs. This possibility is enhanced by the demonstration that both 25-hydroxycholesterol and 20α-hydroxycholesterol may replace a large proportion of mammalian cell sterol following prolonged *in vitro* incubation.[41]

G. HEPATOMA CELL CULTURES

In rat Morris hepatoma HTC cell cultures containing 10% whole calf serum, 80 μ*M* of 7β-hydroxycholesterol caused the lysis of all cells within 3 d, while 160 μ*M* had the same effect within 24 h.[42] If lipoprotein-deprived calf serum was used, cell mortality occurred at far lower doses; with 8 μ*M* of 7β-hydroxycholesterol, all cells were lysed in 3 d, and with 16 μ*M*, all were lysed in 24 h. In other words, 7β-hydroxycholesterol is 10 times more toxic for HTC cells cultured in a lipoprotein-deprived medium than in calf serum. The effect is proportional to the dose and exposure time. 7β-Hydroxycholesterol inhibited DNA synthesis before cell lysis occurred, and to the same degree whether the cells were cultured with serum or in lipoprotein-deficient medium. Thus the absence of serum lipoproteins enhanced the lytic effect but did not increase their intracellular action on DNA synthesis. 7β-Hydroxycholesterol did not affect protein synthesis until the 80-μ*M* concentration. It has also been shown that 40 μ*M* of 22R-hydroxydesmosterol caused total lysis of all cells within 3 d, and at 80 μ*M* of the same compound caused lysis of all cells within 24 h.[43] The cytolytic action of 22R-hydroxydesmosterol on HTC cells cultured in a medium containing 10% newborn-calf serum could be reversed within certain concentration limits by adding cholesterol to the cultured medium. In contrast, under the same conditions, the cytotoxicity of 7β-hydroxycholesterol could not be reversed, whatever the concentration of cholesterol added. In a lipoprotein-poor and in a chemically defined medium, however, the cytolytic action of both hydroxysterols could be reversed by adding cholesterol, but growth inhibition could not be suppressed. This demonstrates the importance of serum lipoproteins for the cytotoxicity of cholesterol oxides and for the ameliorating effect of cholesterol. The growth of HTC cells treated with 80 μ*M* of 7β-hydroxycholesterol could be restored if 7β-hydroxycholesterol was removed from the medium within 24 h. The mechanisms of action of 7β-hydroxycholesterol and 22R-hydroxycholesterol on HTC hepatoma cells seem to be different. In addition to the inhibition of cholesterol biosynthesis by both hydroxysterols, 7β-hydroxycholesterol might be able to insert into the cell membranes and serum lipoproteins may facilitate this insertion.[44]

H. GLIOMA CELL CULTURES

Prolonged incubation of proliferating C-6 rat glioma cells with 25-hydroxycholesterol resulted in a loss of morphological integrity and an 80% decline in cell viability, determined by trypan blue dye exclusion.[45] In contrast, C-6 cells that were induced to enter a quiescent state by removal of serum from the medium remained viable and morphologically differentiated in the presence of 25-hydroxycholesterol. The selective killing of proliferating C-6 glioma cells by 25-hydroxycholesterol was correlated with a 45 to 50% decline in the sterol/phospholipid molar ratio, whereas the sterol/phospholipid ratio in the quiescent cells was not affected by 25-hydroxycholesterol. The results suggest that inhibitors of sterol synthesis may have potential as agents that might selectively decrease the growth and viability of glioma cells without detriment to the normal nondividing neural cells.

I. HEART CELL CULTURES

In addition to toxic effects of cholesterol oxides shown in preparations from frog[7] and chicken hearts,[8] cultured heart cells were also employed for the studies on the effects of cholesterol oxides.[46] Primary cultures derived from neonatal rat heart yielded both fibroblasts and myocardial cells. Treatment with 7β-hydroxycholesterol of either mixed primary culture (4 days in culture) or fibroblast secondary cultures (3 days in culture), to a final concentration of 18 μM, had drastic effects on fibroblasts. The fibroblasts no longer adhered to the solid substratum and were all removed by simply washing the cell cultures. When enriched myocardial cell cultures were treated similarly, detachment of fibroblasts was not observed, probably due to the thick syncytia entrapping the fibroblastic areas. Myocardial cells behave differently on treatment with 7β-hydroxycholesterol. They continued to adhere to the support, but their beatings progressively became arrhythmic and asynchronous, and some of the syncytia even stopped beating completely, which however, could be restored by replacing the 7β-hydroxycholesterol-containing medium with fresh culture medium. ^{14}C-labeled 7β-hydroxycholesterol was used to follow its metabolic pathway in myocardial and fibroblast cultures. A major ^{14}C-labeled metabolite could be extracted from fibroblast cultures after 12 h of treatment, and its concentration remained constant for 48 h. In contrast, in enriched myocardial cell cultures, only a small amount of ^{14}C-labeled metabolite could be detected. Using gas liquid chromatography-mass spectrometry, this metabolite appeared to be a steroidal derivative. These observations suggested that 7β-hydroxycholesterol was differentially metabolized by heart fibroblast-like and myocardial cell cultures and the cytotoxic effect of 7β-hydroxycholesterol on the fibroblasts was not due to 7β-hydroxycholesterol but to a metabolite. On the other hand, myocardial cells were unable to metabolize 7β-hydroxycholesterol, it might be assumed that 7β-hydroxycholesterol itself affected the cardiac beating rhythm.

J. Skin Cell Cultures

Cultured human keratinocytes and dermal fibroblasts from skin tissue were incubated with 1.25 μM concentration of 25-hydroxycholesterol.[47] Cholesterol biosynthesis was inhibited 84% in the preconfluent keratinocytes and 95% in the dermal fibroblasts. In another study, 25-hydroxycholesterol at 20 μg/ml concentration showed cytotoxicity to human skin fibroblasts.[48]

III. *IN VIVO* CYTOTOXICITY

In vivo studies of the effects of cholesterol oxides have been motivated primarily from interest in utilizing these compounds as hypocholesterolemic agents to lower serum cholesterol levels, because cholesterol oxides have been shown to inhibit cholesterol biosynthesis from acetate in both *in vitro* and *in vivo* studies. The recognition of the cytotoxicity of cholesterol oxides in various tissue cultures and the toxic effects in experimental animals, however, is an obstacle in their serious clinical utilization as hypocholesterolemic agents.

Nevertheless, it has been suggested that 7-ketocholesterol and 25-hydroxycholesterol be clinically evaluated in the treatment of familial hypercholesterolemia in human patients.[49,50] Chemical synthesis of the analogs of cholesterol oxides for this purpose has also been pursued.[51] Other aspects of *in vivo* studies specifically dealing with atherogenicity, mutagenicity, and carcinogenicity are discussed in the other chapters.

Early studies were conducted by Steinberg et al.,[52,53] using a group of 50 rats placed on a synthetic diet containing 1% cholest-4-en-3-one. After 8 weeks of feeding, rats became lethargic, lost appetite, and did not grow well. Twenty-two percent of rats had died and the survivors showed a weight loss of 36% as compared to controls and their serum cholesterol levels were about 40% lower than those in the controls. The adrenal hypertrophy in these animals was the most striking and reached enormous proportions, (adrenal weights 7 times those of the control group). Adrenocortical function was also sharply depressed.[54] Other studies with dogs fed 1 g of cholest-4-en-3-one for 17 d likewise revealed 30 to 45% reduction in serum cholesterol levels, and chickens fed 1% cholest-4-en-3-one showed 45 to 55% reduction in serum cholesterol levels.[55] This effect on serum cholesterol was shown to be related to the suppression of endogenous cholesterol biosynthesis from acetate. Cholest-4-en-3-one could metabolize to dihydrocholesterol (cholestanol) which was shown to be accumulated in serum, liver, and other tissues in these animals. Furthermore, cholestanol was found to be atherogenic,[56,57] and cholest-4-en-3-one was therefore determined not to be a promising hypocholesterolemic agent.

Rats fed 7-ketocholesterol at 0.5% also exhibited loss of appetite, decreased liver and body weight, and reduced cholesterol biosynthesis.[58] There were signs of hepatic toxicity and cirrhosis in rabbits, and liver necrosis in guinea pigs fed 7-ketocholesterol.[59] Electron microscopic changes in liver due to this and other cholesterol oxides were studied by Imai et al.[60] When New Zealand white rabbits were fed 2 to 5 μg/kg/day of cholesterol oxides, including 7-ketocholesterol, cholestane-3β,5α,6β-triol, cholesterol 5,6-epoxide, and 25-hydroxycholesterol for 3 weeks by gastric intubation, hepatic changes induced were diffuse vesiculation of hepatocytes and moderate reduction of hepatocellular glycogens. Cytoplasmic vacuoles were lipid droplets and autophagic vacuoles with various densities. A small number of hepatocytes had hyperplasia of the smooth endoplasmic reticulum. Occasional dead and dying hepatocytes were also seen. Injection of 7-ketocholesterol in rabbits also resulted in lowering of plasma cholesterol levels, additionally, lung granulomatous angiitis was observed.[61] In laying hens, smaller dietary levels of 0.025% 7-ketocholesterol significantly reduced hepatic HMG-CoA reductase activity and moderated the effect of cholesterol on increasing egg deposition although sterol synthesis was unchanged by 7-ketocholesterol.[62,63]

Extensive evaluation of oral administration of cholestane-3β,5α,6β-triol on acute, subacute, and chronic toxicity in mice, rats, dogs, and monkeys showed growth retardation and organ atrophy in these animals with 0.1% cholestanetriol for 2 weeks in mice, 0.1% for 5 weeks in rats, 50 to 100 mg/kg/d for 8 weeks in dogs, and 50 mg/kg/d in monkeys for 13.5 months.[64] No conspicuous toxicity in rabbits, chickens, and rats was noted, however, but their serum and liver cholesterol levels were all significantly reduced by cholestanetriol feeding.[65]

Connective tissue responses to cholesterol oxides and purified cholesterol were examined by subcutaneous injection of these sterols into the abdomens of rats for periods of time from 1 to 28 d.[11] The granulomatous inflammatory responses including lipid-filled macrophages, variable necrosis, and polymorphonuclear leukocyte infiltration were induced by these sterols in the following order of decreasing severity: cholestane-3β,5α,6β-triol, 25-hydroxycholesterol, cholesterol 5,6-epoxide, and purified cholesterol. Cholestanetriol initially produced the largest lesions, but at a later stage, the 25-hydroxycholesterol implant area became the largest.

Indirect evidence of generalized toxic responses could be demonstrated by intravenous injection of cholesterol oxidase, which catalyzes the oxidation of cholesterol to cholest-4-en-3-one, into hyperlipidemic rabbits in which aortic atheromatous lesions had been previously induced by dietary means.[21] All animals with such treatment died within hours, whereas injection of the same enzyme into normal rabbits had no apparent adverse effects. Injection of enzymatically oxidized LDL treated with cholesterol oxidase killed normal rabbits, in contrast to untreated LDL, which did not. The enzyme also proved lethal for Watanabe rabbits. Whether the toxic effect could be attributable to cholest-4-en-3-one, however, could not be certain. On the other hand, it could support the fact that endogenous peroxidation of lipids in LDL resulting in formation of oxidized LDL is not only cytotoxic, but also may play a role in the initiation of atheromatous lesions in humans.[66]

Cholesterol oxides have also been shown to be toxic to microorganisms. Cholesterol 7α-hydroperoxide, 3β-hydroxy-5α-cholest-6-ene-5-hydroperoxide, and 25-hydroxycholesterol were bacteriocidal to *Salmonella typhimurium*.[67] Cholestane-3β,5α,6β-triol and 3β-hydroxy-5α-cholestan-6-one inhibited growth of *Mycoplasma gallisepticum*.[68] Some side-chain cholesterol oxides inhibited growth of the *Saccharomyces cerevisiae* auxotroph.[69]

IV. MECHANISMS OF CYTOTOXICITY

Kandutsch et al.[70] speculated that the cytotoxicity of cholesterol oxides might be a consequence of blocked cholesterol biosynthesis. Many cholesterol oxides are potent inhibitors of cholesterol biosynthesis by suppressing HMG-CoA reductase activity. When 25-hydroxycholesterol or 7-ketocholesterol was added to serum-free medium of mouse fibroblast L-cell cultures, HMG-CoA reductase activity and the rate of sterol biosynthesis declined and were nearly completely suppressed within a period of hours. Concomitant with, or immediately following the suppression of HMG-CoA reductase, the concentration of cellular sterol and the rate of DNA synthesis also declined. At 24 to 36 h after addition of cholesterol oxides, growth of the cultures had essentially stopped, and after 48 h, the cells became rounded and detached from the surface of the flask.[71] Apparently the cholesterol oxides blocked cell division in the G_1 phase of the cell cycle.[35] Roughly coinciding with the cessatiuon of growth, the concentrations of sterol in the whole cells and in the plasma membranes declined to approximately one-half of the initial level. No significant changes in the concentration of phospholipids occurred in the blocked cells, and thus the molar ratio of sterol to phospholipids decreased from approximately 0.4 to 0.2. The decrease in the cholesterol to phospholipid ratio in the membranes could result in membrane dysfunction and eventually cell death. Reversal of the cytotoxicity and growth inhibition of cholesterol oxides by cholesterol or mevalonate strongly suggests that suppression of cholesterol biosynthesis is one of the underlying mechanisms. This mechanism is particularly evidenced with regard to growth inhibition by cholesterol oxides. Obviously cell replication requires the production of new membrane and cholesterol is required for this. It may be that the failure to produce cholesterol for membrane formation during an early stage of the cell cycle prevents further steps leading to mitosis. Some investigators, however, believe that the effect of cholesterol oxides on HMG-CoA reductase and cholesterol biosynthesis has to be dissociated from their inhibitory activity upon DNA synthesis, since the concentrations required to achieve both effects in lymphocytes are different.[72] Thus, although cholesterol biosynthesis may be prerequisite for proliferative response of lymphocytes to mitogens, it is not sufficient, and a specific cytosolic binding protein is likely to account for the effect on DNA synthesis. Moreover, other investigators insist upon the determinant role of membrane fluidity in modulating the response of lymphocytes to mitogenic stimuli and differentiation.[36]

There was also some discrepancy between the cytotoxicity and the potency of inhibition of cholesterol biosynthesis when twelve cholesterol oxides and purified cholesterol were

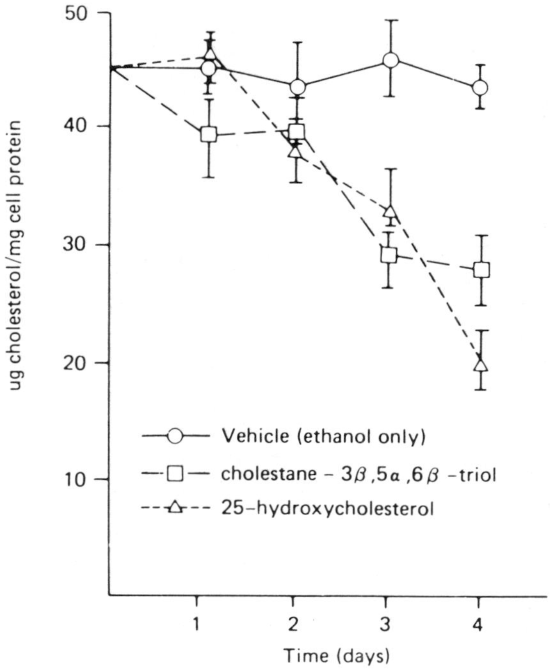

FIGURE 1. Effect of 25-hydroxycholesterol and cholestane-3β,5α,6β-triol on the cellular cholesterol content. Cells were incubated with either sterol at a final concentration of 10 μg/ml for the time periods indicated. (Reprinted from Peng, S. K., Hill, J. C., Morin, R. J., and Taylor, C. B., *Proc. Soc. Exp. Biol. Med.*, 180, 126, 1985. © by The Society for Experimental Biology and Medicine. With permission.)

tested in cultured rabbit aortic smooth muscle cells.[10] For example, 25-hydroxycholesterol was one of the most cytotoxic sterols tested as well as the most potent inhibitor of cholesterol biosynthesis. On the other hand, cholestane-3β,5α,6β-triol was also shown to have severe cytotoxic effects on the cultured aortic smooth muscle cells, but it only moderately inhibited cholesterol biosynthesis. An alternative mechanism could be a direct effect of cholesterol oxides by replacing cholesterol in membranes. It is conceivable that the stronger polar groups of cholestane-3β,5α,6β-triol on the one end of the molecule and the hydrophobic group on the other end might make it possible for this molecule to penetrate the membrane easily and thereby affect membrane fluidity, stability, and functions, resulting in toxicity. In fact, cytotoxic effects of cholestane-3β,5α,6β-triol and also cholestan-3β,5α-diol-6-one appeared much earlier than those of 25-hydroxycholesterol. The timing of the effects was considered to have been too early for a depletion of cholesterol in membrane sterol to have occurred (a cellular deficiency of cholesterol would take many hours to develop). Peng et al.[73] measured the cellular cholesterol content after cultured aortic smooth muscle cells were incubated with a 10-μg/ml concentration of cholestane-3β,5α,6β-triol or 25-hydroxycholesterol in culture medium (Figure 1). The cholesterol content steadily decreased after cells were exposed to cholestane-3β,5α,6β-triol, and membrane-associated functions were inhibited after only a 15-min period of incubation with cholestanetriol. On the other hand, cellular cholesterol content was unchanged until after 24 h of incubation with 25-hydroxycholesterol, and membrane-associated functions were not significantly altered after up to 8 h of incubation with this sterol. The cholesterol content of the cells was significantly decreased after 24 to 48 h of incubation with both sterols and the cell viability declined.

Moreover, the inhibitory effect of the cholestane-3β,5α,6β-triol could be reversed quickly after the cells were transferred to control medium (cholesterol oxide-free). Insertion of cholesterol oxides into the cell membrane, with displacement of membrane cholesterol may therefore be an additional mechanism of cytotoxicity.

V. SUMMARY

The cytotoxicity of individual cholesterol oxides varies depending upon the type of cell line used. Cholestane-3β,5α,6β-triol, 25-hydroxycholesterol, and 26-hydroxycholesterol proved to be the most toxic to virtually all the cell lines tested. These sterols inhibited cell growth, usually at 5-μg/ml concentrations, and cell viability started to decline at about 10 μg/ml after 24 h of incubation. Cholestan-3β,5α,-diol-6-one and 20α-hydroxycholesterol were the next most toxic, followed by 7-ketocholesterol, 7β-hydroxycholesterol, and 7α-hydroxycholesterol, which were toxic at higher concentrations. Cholesterol 5,6-epoxide was nontoxic to some cultured cells and mildly toxic to others. Purified cholesterol was innocuous to almost all the cell lines. Other parameters for cytotoxicity, such as a decreased [^{3}H]-thymidine incorporation into DNA, were shown to be well correlated with the degrees of cytotoxicity. Although lipid accumulation or steatosis could be another sign of cell toxicity, some of the steatotic cholesterol oxides such as the α and β isomers of cholesterol 5,6-epoxides were highly active with respect to steatosis, but were not cytotoxic. It also appeared that cholesterol oxides had more potent effects on proliferating cells in sparse cultures than on quiescent cells in confluent cultures (10 to 100 times more effective in the former than in the latter). Lymphocytes entering blastic transformation were shown to be very sensitive to 7β-hydroxycholesterol, whereas normal lymphocytes remained insensitive at 50 times higher concentrations. The composition of the culture medium may influence the cytotoxicity of cholesterol oxides. Cholesterol oxides are more toxic to cultured cells incubated in lipoprotein-depleted serum than in lipoprotein-containing fetal calf serum. In other words, lipoproteins could moderate the effects of cholesterol oxides by serving as receptors of cholesterol oxides or by competing with cholesterol oxides for membrane insertion.

The cytotoxicity of cholesterol oxides is not only concentration dependent, but also time dependent. The influence of duration of exposure to cholesterol oxides may vary depending upon the nature of the compounds. It has been repeatedly demonstrated that cholestane-3β,5α,6β-triol is the most toxic sterol in short-term incubations of one day or less. On the other hand, 25-hydroxycholesterol was the most toxic sterol after prolonged incubation for 3 d or more. These manifestations of cytotoxicity appear to have at least two separate causes. One mechanism may be incorporation of cholesterol oxides into cell membranes, directly affecting membrane fluidity, stability, and functions and thereby inducing toxicity. The long-recognized inhibition of HMG-CoA reductase activity accounts for those toxicities caused by depletion of cholesterol in dividing cells; this is usually reversible by addition of cholesterol or its precursor to the medium. Cholestane-3β,5α,6β-triol and cholestan-3β,5α-diol-6-one are very potent in those mature cells which no longer divide or do not synthesize cholesterol, such as echinocyte formation of red blood cells or cytotoxicity and cytolysis in unstimulated lymphocytes. 25-Hydroxycholesterol and 20α-hydroxycholesterol are more effective in cells undergoing proliferation, such as lymphocytes entering blastic transformation. The demonstrated cytotoxicity to Morris hepatoma HTC cells by 7β-hydroxycholesterol and the selective killing of lymphoma cells by 7β-hydroxycholesterol and of proliferating C-6 glioma cells by 25-hydroxycholesterol may have an antitumor therapeutic potential.

Another significant implication of cholesterol oxide-induced cytotoxicity is their role in the development of atherosclerotic lesions. Recent implication of oxidatively modified LDL as a primary culprit in the initiation of atherosclerosis further enhances this hypothesis. Not only are cholesterol oxides themselves cytotoxic to vascular endothelial and smooth muscle cells, but oxidatively modified LDL, either via free radical-induced lipid peroxidation or by enzymatically catalyzed oxidation with cholesterol oxidase also prove toxic to vascular cells. Furthermore, neither pure cholesterol nor untreated natural LDL shows any apparent adverse effects.

In vivo toxic responses to cholesterol oxides have been demonstrated in certain animals. These included lethargy, loss of appetite, loss of body weight, growth retardation, and organ atrophy (except adrenal glands, which showed significant hypertrophy associated with adrenocortical functional insufficiency). Their serum cholesterol levels were reduced and cholesterol biosynthesis declined. In addition, hepatic toxicity was also demonstrated in these animals. Interestingly, intravenous injection of enzymatically oxidized LDL treated with cholesterol oxidase has proven lethal to rabbits. Whether the toxic effect could be specifically attributed to cholest-4-en-3-one could not be certain, but this further supports the concept that oxidized LDL, either exogenously or endogenously produced, could be harmful to human health.

REFERENCES

1. **Peng, S. K., Taylor, C. B., Tham, P., Werthessen, N. T., and Mikkelson, B.,** Effect of autoxidation products from cholesterol on aortic smooth muscle cells, an in vitro study, *Arch. Pathol. Lab. Med.*, 102, 57, 1978.
2. **Peng, S. K., Phillips, G. A., Xia, G. Z., and Morin, R. J.,** Transport of cholesterol autoxidation products by rabbit lipoproteins, *Atherosclerosis,* 64, 1, 1987.
3. **Mitton, J. R., Scholan, N. A., and Boyd, G. S.,** The oxidation of cholesterol in rat liver subcellular particles, *Eur. J. Biochem.*, 20, 569, 1971.
4. **Sevanian, A., Stein, R. A., and Mead, J. F.,** Epoxides as products of lipid autoxidation in rat lungs, *Lipids,* 14, 634, 1979.
5. **Watabe, T. and Sawahata, T.,** Biotransformation of cholesterol to cholestane-3β,5α,6β-triol via cholesterol epoxide in bovine adrenal cortex, *J. Biol. Chem.*, 254, 3854, 1979.
6. **Hayaishi, O.,** Enzyme hydroxylation, *Annu. Rev. Biochem.*, 38, 21, 1969.
7. **Flury, F.,** über die pharmakologischen Eigenschaften einiger saurer Oxydationsprodukte des Cholesterins, *Arch. Exp. Pathol. Pharmakol.*, 66, 221, 1911.
8. **Biswas, S., MacDougall, J. D. B., and Cook, R. P.,** The effect of various steroids, mainly of the C27 series, on the growth of chick heart explants, *Br. J. Exp. Pathol.*, 45, 13, 1964.
9. **MacDougall, J. D. B., Biswas, S., and Cook, R. P.,** The effects of certain C27 steroids on organ cultures of rabbit aorta, *Br. J. Exp. Pathol.*, 46, 549, 1965.
10. **Peng, S. K., Tham, P., Taylor, C. B., and Mikkelson, B.,** Cytotoxicity of cholesterol oxidation derivatives on cultured aortic smooth muscle cells and their effect on cholesterol biosynthesis, *Am. J. Clin. Nutr.*, 32, 1033, 1979.
11. **Baranowski, A., Adams, C. W. M., Bayliss High, O. B., and Bowyer, D. B.,** Connective tissue responses to oxysterols, *Atherosclerosis,* 41, 255, 1982.
12. **Goldstein, A. L., Cox, D. C., and Comai, K.,** Effects of 25-hydroxycholesterol and cholesterol smooth muscle cell and endothelial cell growth in vitro, *Arteriosclerosis,* 5, 500a, 1985.
13. **Cox, D. C., Comai, K., and Goldstein, A. L.,** Effects of cholesterol and 25-hydroxycholesterol on smooth muscle cell and endothelial cell growth, *Lipids,* 23, 85, 1988.
14. **Naseem, S. M. and Heald, F. P.,** Cytotoxicity of cholesterol oxides and their effects on cholesterol metabolism in cultured human aortic smooth muscle cells, *Biochem. Int.*, 14, 71, 1987.
15. **Wolinsky, H., Goldfischer, S., Daly, M. M., Easak, L. E., and Coltoff-Schiller, B.,** Arterial lysosomes and connective tissue in primate atherosclerosis and hypertension, *Cir. Res.*, 36, 553, 1975.
16. **Miller, B. F. and Kiltari, H. V.,** Increased activity of lysosomal enzymes in human atherosclerotic aortas, *Exp. Mol. Pathol.*, 10, 288, 1969.
17. **Goldstein, J. L., Danna, S. E., Faust, J. R., Beavdet, A. L., and Brown, M. S.,** Role of lysosomal acid lipase in the metabolism of plasma low density lipoprotein. Observation in cultured fibroblasts from a patient with cholesteryl ester storage disease, *J. Biol. Chem.*, 250, 8487, 1975.

18. **Hu, B., Jin, D., Fan, W., Peng, S. K., and Morin, R. J.,** Effects of cholestanetriol on cytotoxicity and prostacyclin production in cultured rabbit aortic endothelial cells, *Artery,* 18, 87, 1991.
19. **Sevanian, A., Berliner, J., and Peterson, H.,** Uptake, metabolism and cytotoxicity of isomeric cholesterol-5,6,-epoxides in rabbit aortic endothelial cells, *J. Lipid Res.,* 32, 147, 1991.
20. **Sevanian, A., Peterson, H., and Coates, T.,** Cytotoxicity of cholesterol oxidation products in bovine and rabbit aortic endothelial cells. 8th Int. Symp. Atherosclerosis, Venice, Italy, Oct. 7-8, 1988, 103.
21. **Bernheimer, A. W., Robinson, W. G., Linder, R., Mullins, D., Yip, Y. K., Cooper, N. S., Seidman, I., and Uwajima, T.,** Toxicity of enzymically-oxidized low-density lipoprotein, *Biochem. Biophys. Res. Commun.,* 148, 260, 1987.
22. **Henriksen, T., Evensen, S. A., and Carlander, B.,** Injury to human endothelial cells in culture induced by low density lipoproteins, *Scand. J. Clin. Lab. Invest.,* 39, 361, 1979.
23. **Hessler, J. R., Robertson, A. L., Jr., and Chisolm, G. M.,** LDL-induced cytotoxicity and its inhibition by HDL in human vascular smooth muscle and endothelial cells in culture, *Atherosclerosis,* 32, 213, 1979.
24. **Hessler, J. R., Morel, D. W., Lewis, L. J., and Chisolm, G. M.,** Lipoprotein oxidation and lipoprotein-induced cytotoxicity, *Arteriosclerosis,* 3, 215, 1983.
25. **Morel, D. W., Hessler, J. R., and Chisolm, G. M.,** Low density lipoprotein cytotoxicity induced by free radical peroxidation of lipid, *J. Lipid. Res.,* 24, 1070, 1983.
26. **Chen, H. W., Kandutsch, A. A., and Waymouth, C.,** Inhibition of cell growth by oxygenated derivatives of cholesterol, *Nature,* 251, 419, 1974.
27. **Higley, N. A. and Taylor, S. L.,** The steatotic and cytotoxic effects of cholesterol oxides in cultured L cells, *Food Chem. Toxicol.,* 22, 983, 1984.
28. **Brown, M. S., Dana, S. E., and Goldstein, J. L.,** Cholesterol ester formation in cultured human fibroblasts, stimulation by oxygenated sterols, *J. Biol. Chem.,* 250, 4025, 1975.
29. **Sevanian, A. and Peterson, A. R.,** The cytotoxic and mutagenic properties of cholesterol oxidation products, *Food Chem. Toxicol.,* 24, 1103, 1986.
30. **Peterson, A. R., Peterson, H., Spears, C. P., Trosko, J. E., and Sevanian, A.,** Mutagenic characterization of cholesterol epoxides in Chinese hamster V79 cells, *Mutation Res.,* 203, 355, 1988.
31. **Blackburn, G. M., Rashid, A., and Thompson, M. H.,** Interactions of 5α,6α-cholesterol oxide with DNA and other nucleophiles, *J. Chem. Soc. Chem. Commun.,* 420, 1979.
32. **Brown, M. S. and Goldstein, J. L.,** Lipoprotein metabolism in the macrophage; implications for cholesterol deposition in atherosclerosis, *Annu. Rev. Biochem.,* 52, 223, 1983.
33. **Tabas, I., Weiland, D. A., and Tall, A. R.,** Unmodified low density lipoprotein causes cholesteryl ester accumulation in J744 macrophages, *Proc. Natl. Acad. Sci. U.S.A.,* 82, 416, 1985.
34. **Yachnin, S., Streuli, R. A., Gordon, L. I., and Hsu, R. C.,** Alteration of peripheral blood cell membrane function and morphology by oxygenated sterols; a membrane insertion hypothesis *Cur. Top. Hematol.,* 2, 245, 1979.
35. **Chen, H. W., Hans-Jörg, H., and Kandutsch, A. A.,** Relationship between sterol synthesis and DNA synthesis in phytohemagglutinin-stimulated mouse lymphocytes, *Proc. Nat. Acad. Sci. U.S.A.,* 72, 1950, 1975.
36. **Ip, S. H. C., Abraham, J., and Cooper, R. A.,** Enhancement of blastogenesis in cholesterol-enriched lymphocytes, *J. Immunol.,* 124, 87, 1980.
37. **Hietter, H., Bischoff, P., Beck, J. P., Ourisson, G., and Luu, B.,** Comparative effects of 7β-hydroxycholesterol towards murine lymphomas, lymphoblasts and lymphocytes: selective cytotoxicity and blastogenesis inhibition, *Cancer Biochem. Biophys.,* 9, 75, 1986.
38. **Schuh, J., Novogrodsky, A., and Haschemeyer, R. H.,** Inhibition of lymphocyte mitogenesis by autoxidized low-density lipoprotein, *Biochem. Biophys. Res. Commun.,* 84, 763, 1978.
39. **Saito, Y., Shimada, H., Imada, T., Kikuchi, T., Ikekawa, N., and Inada, Y.,** Lysis of platelets and erythrocytes by the incorporation of a unique oxygenated sterol: 22R-hydroxycholesterol, *J. Membrane Biol.,* 83, 187, 1985.
40. **Hoffman, P. C., Richman, C. M., Hsu, R. C., Chung, J., Scanu, A. M., and Yachnin, S.,** Effect of oxygenated sterol compounds on human bone marrow granulocytic progenitor cells, *Blood,* 57, 164, 1981.
41. **Yachnin, S., Chung, J., and Scanu, A. M.,** Inhibition of oxygenated sterol entry into human red cells and lymphocytes by isolated serum lipoproteins, *Biochim. Biophys. Acta,* 713, 538, 1982.
42. **Richert, L., Bergmann, C., Beck, J.-P., Rong, S., Luu, B., and Ourisson, G.,** The importance of serum lipoproteins in the cytolytic action of 7β-hydroxycholesterol on cultured hepatoma cells, *Biochem. Biophys. Res. Commun.,* 117, 851, 1983.
43. **Hietter, H., Trifilieff, E., Richert, L., Beck, J.-P., Luu, B., and Ourisson, G.,** Antagonistic action of cholesterol towards the toxicity of hydroxysterols on cultured hepatoma cells, *Biochem. Biophys. Res. Commun.,* 120, 657, 1984.
44. **Hsu, R. C., Kanofsky, J. R., and Yachnin, S.,** The formation of echinocytes by insertion of oxygenated sterol compounds in red cell membranes, *Blood,* 56, 109, 1980.

45. **Maltese, W. A., Reitz, B. A., and Volpe, J. J.,** Selective decrease of the viability and the sterol content of proliferating versus quiescent glioma cells exposed to 25-hydroxycholesterol, *Cancer Res.,* 41, 3448, 1981.
46. **Mersel, M., Hietter, H., and Luu, B.,** Differential sensitivity of heart fibroblasts and myocytes to 7β-hydroxycholesterol, in *The Heart Cell in Culture,* Vol. 3., Pinson, A., Ed., CRC Press, Boca Raton, FL, 1987, 125.
47. **Williams, M. L., Rutherford, S. L., and Feingold, K. R.,** Effects of cholesterol sulfate on lipid metabolism in cultured human keratinocytes and fibroblasts, *J. Lipid Res.,* 28, 955, 1987.
48. **Oram, J. F., Brinton, E. A., and Bierman, E. L.,** Regulation of high density lipoprotein receptor activity in cultured human skin fibroblasts and human arterial smooth muscle cells, *J. Clin. Invest.,* 72, 1611, 1981.
49. **Breslow, J. L., Lothrop, D. A., Spaulding, D. R., and Kandutsch, A. A.,** Cholesterol, 7-ketocholesterol and 25-hydroxycholesterol uptake studies and effects on 3-hydroxy-3-methylglutaryl-coenzyme A reductase activity in human fibroblasts, *Biochim. Biophys. Acta,* 398, 10, 1975.
50. **Goldstein, J. L. and Brown, M. S.,** Familial hypercholesterolemia. A genetic regulatory defect in cholesterol metabolism, *Am. J. Med.,* 58, 147, 1975.
51. **Goldstein, J. L., Faust, J. R., Dygos, J. H., Chorvat, R. J., and Brown, M. S.,** Inhibition of cholesteryl ester formation in human fibroblasts by an analogue of 7-ketocholesterol and progesterone, *Proc. Natl. Acad. Sci. U.S.A.,* 75, 1877, 1978.
52. **Steinberg, D. and Fredrickson, D. S.,** Inhibition of cholesterol biosynthesis and the problem of hypercholesterolemia, *Ann. N.Y. Acad. Sci.,* 64, 579, 1956.
53. **Steinberg, D., Fredrickson, D. S., and Avigan, J.,** Effects of Δ^4-cholestenone in animals and in man, *Proc. Soc. Exp. Biol. Med.,* 97, 784, 1958.
54. **Fredrickson, D. S., Peterson, R. E., and Steinberg, D.,** Inhibition of adrenocortical steroid secretion by Δ^4-cholestenone, *Science,* 127, 704, 1958.
55. **Tomkins, G. M., Nichols, C. W., Chapman, D. D., Hotta, S., and Chaikoff, I. L.,** Use of Δ^4-cholestenone to reduce the level of serum cholesterol in man, *Science,* 125, 936, 1957.
56. **Nichols, C. W., Lindsay, S., and Chaikoff, I. L.,** Production of arteriosclerosis in bird by the prolonged feeding of dihydrocholesterol, *Proc. Soc. Exp. Biol. Med.,* 89, 609, 1955.
57. **Nichols, C. W., Siperstein, M. D., and Chaikoff, I. L.,** Effect of dihydrocholesterol (cholestanol) administration on plasma cholesterol and atherosclerosis in the rabbits, *Proc. Soc. Exp. Biol. Med.,* 83, 756, 1953.
58. **Erickson, S. K., Cooper, A. D., Matsui, S. M., and Gould, R. G.,** 7-Ketocholesterol, its effects on hepatic cholesterogenesis and its hepatic metabolism in vivo and in vitro, *J. Biol. Chem.,* 252, 5186, 1977.
59. **Bing, R. J., Sarma, J. S. M., and Chan, S. I.,** Inhibition of cholesterol uptake by the arterial wall in the intact animals, *Artery,* 5, 14, 1979.
60. **Imai, H., Subramanyam, V., Le Quesne, P. W., and Werthessen, N. T.,** Ultrastructural changes of liver due to oxidation products of cholesterol, *Fed. Proc. Fed. Am. Soc. Exp. Biol.,* 37, 934, 1978.
61. **Santillan, G., Sarma, J. S. M., Pawlik, G., Rackl, A., Grenier, A., and Bing, R. J.,** Toxicity, pharmacokinetics, and cholesterol-inhibitory effect of 7-ketocholesterol, *Atherosclerosis,* 35, 1, 1980.
62. **Naber, E. C., Allred, J. B., Winget, C. J., and Stock, A. E.,** Effect of cholesterol oxidation products on cholesterol metabolism in the laying hen, *Poultry Sci.,* 64, 675, 1985.
63. **Vargas, R. E., Allred, J. B., Biggert, M. D., and Naber, E. G.,** Effect of dietary 7-ketocholesterol, pure, or oxidized cholesterol on hepatic 3-hydroxy-3-methylglutaryl-coenzyme A reductase activity, energy balance, egg cholesterol concentration and ^{14}C-acetate incorporation into yolk lipids of laying hens, *Poultry Sci.,* 65, 1333, 1986.
64. **Kaziwara, K., Yokotani, H., Imai, Y., Ogiwara, S., and Aramaki, Y.,** Studies of cholestane-3β,5α,6β-triol and its derivative. IV. Experimental evaluation on safety for clinical trials, acute, subacute, and chronic toxicity test using mice, rats, dogs and monkey, *Jpn. J. Pharmacol.,* 17, 340, 1967.
65. **Aramaki, Y., Kobayashi, T., Imai, Y., Kikuchi, S., Matsukawa, T., and Kanazawa, K.,** Biological studies of cholestane-3β,5α,6β-triol and its derivatives. I. Hypocholesterolimic effects in rabbits, chickens and rats on atherogenic diets, *J. Atheroscler. Res.,* 7, 653, 1967.
66. **Jurgens, G., Hoff, H. F., Chisolm, G. M., and Esterbauer, H.,** Modification of human serum low density lipoprotein by oxidation — characterization and pathophysiological implications, *Chem. Phys. Lipids,* 45, 315, 1987.
67. **Smith, L. L., Smart, V. B., and Made Gowda, N. M.,** Mutagenic sterol hydroperoxides, *Mutation Res.,* 161, 39, 1986.
68. **Herian, A. M., Kuehl, N. M., and Lee, K.,** Growth inhibition of Mycoplasma gallisepticum following membrane insertion of cholesterol autoxidation products, *J. Food Protection,* 48, 1050, 1985.
69. **Rodriguez, R. J., Arunachalam, T. A., Parks, L. W., and Caspi, E.,** Growth of a sterol auxotroph derived from Saccharomyces cerevisiae on chemically synthesized derivatives of cholesterol possessing side-chain modifications, *Lipids,* 18, 772, 1983.
70. **Kandutsch, A. A., Chen, H. W., and Heiniger, H.-J.,** Biological activity of some oxygenated sterols, *Science,* 201, 498, 1978.

71. **Kandutsch, A. A. and Chen, H. W.,** Consequences of blocked sterol synthesis in cultured cells, DNA synthesis and membrane composition, *J. Biol. Chem.*, 252, 409, 1977.
72. **Defay, R., Astruc, M. E., Rousillon, S., Descomp, B., and Crastes de Paulet, A.,** DNA synthesis and 3-hydroxy-3-methylglutaryl CoA reductase activity in PHA stimulated human lymphocytes: a comparative study of the inhibitory effects of some oxysterols with special reference to side chain hydroxylated derivatives, *Biochem. Biophys. Res. Commun.*, 106, 362, 1982.
73. **Peng, S. K., Hill, J. C., Morin, R. J., and Taylor, C. B.,** Influence of cholesterol oxidation derivatives on membrane-bound enzymes in cultured aortic smooth muscle cells, *Proc. Soc. Exp. Biol. Med.*, 180, 126, 1985.

Chapter 9

EFFECTS OF CHOLESTEROL OXIDES ON ATHEROGENESIS

Shi-Kaung Peng, Bing Hu, and Robert J. Morin

TABLE OF CONTENTS

I. INTRODUCTION

Cholesterol has been implicated in the development of atherosclerosis since 1913 when Anitschkow first induced atherosclerosis by diets containing cholesterol.[1] While there are some suggestions that hypercholesterolemia may cause arterial injury and increased endothelial turnover, the mechanism is not well characterized.[2,3] In one earlier study, hypercholesterolemia was endogenously induced in chickens on a cholesterol-free diet by the subcutaneous implantation of diethylstilbesterol.[4] These were compared with chickens with comparable levels of hypercholesterolemia (about 400 mg/dl) induced by feeding a 2% U.S.P.-grade cholesterol-containing diet. After six and one half months on these regimens, numerous large atherosclerotic lesions were observed in the thoracic aortas of the cholesterol-fed group of chickens, whereas the former group with endogenous hypercholesterolemia showed only minor degrees of atherosclerosis. Cholesterol readily oxidizes in air at room temperature.[5-8] In the preparation of cholesterol diets to induce experimental atherosclerosis, it is common practice to mix chow with a solution of cholesterol in organic solvent, then air dry it to remove the solvent, thereby exposing thin films of cholesterol to ideal conditions for autoxidation. The more extensive atherosclerosis in the cholesterol-fed chicken may well have been due to cholesterol autoxidation products in the U.S.P.-grade cholesterol. Many investigators conducting experiments on the effects of diet on the cardiovascular system have often not taken into consideration the ability of cholesterol to undergo spontaneous oxidation. In some experiments, workers have added cholesterol to the diet by heating it in air, thereby causing significant oxidation of the cholesterol. No toxicity of pure cholesterol has yet been demonstrated in either *in vitro* or *in vivo* studies. On the other hand, small amounts of cholesterol oxides have shown significant toxicity to many mammalian cell lines including vascular endothelial and smooth muscle cells.[9,12]

Atherosclerosis has been postulated to be initiated by arterial injury followed by increased vascular permeability to plasma components, including low-density lipoproteins (LDL) and chylomicron remnants, and production of chemotactic factors which enhance monocyte adhesion to endothelial cells.[3] Once penetrated inside the arterial tissue, monocytes become macrophages or scavenger cells. Monocytes have very few binding sites for native LDL, but ingest large amounts of oxidatively modified LDL via a high-affinity scavenger receptor which is not down-regulated in the presence of an excess of cholesterol.[13] Oxidatively modified LDL can be internalized by macrophages in an unregulated fashion until the cell is entirely filled with lipid to become a foam cell, or until the cell is destroyed by the accumulation.[14] Accumulation of foam cells leads to endothelial cell desquamation in vessel areas of flow turbulence. This denudation induces platelet adhesion with release of growth factor. Prostacyclin production by vascular cells may also have a significant influence on platelet adhesion. Growth factors from platelets, monocytes, or vascular cells cause intimal migration and proliferation of underlying smooth muscle cells to form fibrous plaques, the hallmark of atherosclerotic lesions. Following is a review of the effects of cholesterol oxides on each step of atherogenesis.

II. VASCULAR INJURY BY CHOLESTEROL OXIDES

In 1976, Imai et al.[15] reported that purification of a stock of 5-year-old U.S.P. cholesterol largely prevented its atherogenicity when fed to rabbits, whereas administration of the concentrated impurities present in U.S.P. cholesterol produced severe arterial damage, with smooth muscle cell death and focal intimal edema. In short-term experiments, the rabbits were given by gastric gavage either 1 g/kg of newly purchased or 5-year-old U.S.P. cholesterol or 250 mg/kg of the concentrated impurities of cholesterol. One group of controls received vehicle only and another group received 250 mg/kg of purified cholesterol. All

TABLE 1
Average Frequency per 100 Nucleated Cells of Cellular Debris and Degenerated Aortic Cells in Control and Experimental Rabbits (means ± SEM)

Treatment	No. of aminals	Percentage of aggregate debris	Percentage of degenerated cells	Average cell count per animal
Controls	15	0.03 ± 0.01	0.61 ± 0.14	498 ± 28
Old USP-grade cholesterol, 1 g/kg	5	0.70 ± 0.57[a]	7.60 ± 0.24[b]	654 ± 103
New USP-grade cholesterol, 1 g/kg	5	0.22 ± 0.08[b]	4.38 ± 0.54[b]	616 ± 22
New USP-grade cholesterol, 250 mg/kg	3	0.07 ± 0.06	1.63 ± 0.75[a]	439 ± 19
Cholesterol purified via dibromination, 250 mg/kg	4	1.20 ± 0.95[a]	0.35 ± 0.13	542 ± 95
Concentrate, 250 mg/kg	7	5.59 ± 1.99[b]	7.28 ± 2.86[b]	451 ± 47
Concentrate, 10 mg/kg	4	1.05 ± 0.38[b]	1.70 ± 0.67[b]	459 ± 68

[a] $0.05 > p > 0.01$
[b] $p > 0.01$

Reprinted from Taylor, C. B., Peng, S. K., Werthessen, N. T., Tham, P., and Lee, K. T., *Am. J. Clin. Nutr.*, 32, 40, 1979. With permission.

rabbits were killed and aortas were studied by light and electron microscopy 24 h after their single gastric gavage feeding. Significant increases of degenerative and necrotic arterial smooth muscle cells were observed in aortas of the group fed 1 g/kg of newly purchased, 5-year-old U.S.P.-grade cholesterol, and 250 mg/kg of the concentrated impurities of cholesterol when compared with the control group (Table 1). Aortas of the group fed purified cholesterol showed no significant differences from the control group. These impurities of cholesterol have been shown to be largely composed of cholesterol oxides.[16] The major components were identified to be 25-hydroxycholesterol, cholestane-3β,5α,6β-triol, cholesterol 5,6-epoxide, 7-ketocholesterol, and 7α- and 7β-hydroxycholesterol.

To determine the *in vivo* effects of these compounds, the concentrated impurities of cholesterol were separated by thin-layer chromatography into three bands. Since oral administration may lead to uncertainties associated with biotransformation, retention, excretion, and absorption, the intravenous route was used.[17] The compounds in each band were eluted with ethanol, emulsified in physiological saline, and then injected into New Zealand white rabbits. Aorta were examined 24 h after the injections. Two of the bands showed significant angiotoxicity; only the components of the middle band, identified as sterol hydroperoxides, showed no angiotoxicity. Synthetic cholesterol oxides known to be present in the concentrate of cholesterol impurities were also tested. The test substances were injected intravenously into rabbits once daily for 3 days through an ear vein. At 24 hours, 2 weeks, or 10 weeks after the last injection, the rabbits were sacrificed. At 24 hours after the last injection with the concentrate of cholesterol oxides, their aortas showed increased smooth muscle cell death. At 2 weeks, their aortas contained areas of hypocellularity and cellular debris. Repair, in the form of partial restoration of the media, and intimal fibromuscular thickening, was demonstrable at 10 weeks. An average of three mound-like fibromuscular plaques were distributed at regular intervals along the entire descending thoracic aorta. Unlike the usual cholesterol-induced lesion in the rabbit, these plaques had no foam cells or stainable lipids. Major and minor branches of pulmonary arteries also responded to intravenously injected cholesterol oxides by undergoing grossly visible segmental thickening at 24 hours after the last injection. The thickening was due to a series of changes ranging from the initial inflammation to subsequent repair by fibromuscular thickening. These pulmonary arterial responses provided a rapid bioassay procedure for arterial wall injury and repair. Neither purified cholesterol nor the ethanol vehicle induced changes in the major or minor branches of the pulmonary arteries. Cholestane-3β,5α,6β-triol and 25-hydroxycholesterol were equipotent, with visible responses in the major branches at doses of 5 mg/kg. Cholesterol 5,6-epoxide and 7-ketocholesterol did not affect the major branches at doses of 10 mg/kg although at 5 mg/kg, the minor branches reacted.

The morphological changes on the endothelial surface of the aorta could be better appreciated by scanning electron microscopy (SEM). Peng et al.[18] studied three groups of New Zealand White rabbits given intravenously either 25 mg/kg of 25-hydroxycholesterol, cholestane-3β,5α,6β-triol or vehicle only. The sterols were dissolved in 0.3 ml ethanol, then added to 3 ml of the rabbits' own serum. Twenty-four hours after injection, rabbits were sacrificed. The aortic endothelial surfaces from rabbits receiving 25-hydroxycholesterol showed numerous balloon-like protrusions and crater-like defects, as well as adherence of circulating formed elements. In contrast, aortas of the control group showed very few lesions (Figure 1). Aortas from rabbits given cholestane-3β,5α,6β-triol had similar but more frequent lesions (Figure 2) when compared with those of the 25-hydroxycholesterol group and control group (Table 2). Microthrombi were occasionally found (Figure 3). Transmission electron microscopic studies also showed intracytoplasmic vacuoles and diffuse subendothelial edema in the aortas of the two groups receiving the cholesterol oxides (Figure 4). These balloon-like protrusions and crater-like defects observed by SEM appeared to represent initial endothelial injury induced by cholesterol oxides.

Matthias et al.[19] administered cholestane-3β,5α,6β-triol orally to rats in different doses (100 to 200 mg/d) and for varying length of times (5 to 119 d) and demonstrated an angiotoxic effect of cholestanetriol. The rat aorta is normally resistant to atherosclerosis. Cholestanetriol caused alterations in the media of the aorta, representing a toxic cell damage. These were discernible as wrinkling of the nuclear membranes, and cell alterations such as nuclear pyknosis and vacuolar degenerations of the cytoplasm, followed by smooth muscle cell death and necrosis, sometimes of groups of cells in the media. Similar effects on the endothelium were observed ultrastructurally with cells having degenerated over large areas and desquamated areas partly coated with fibrin and platelets. The medial defects were largely repaired by fibrotic materials 4 months later. Cholesterol, when administered in the same doses as cholestanetriol, did not cause similar endothelial or medial toxic alterations of aorta, but caused a numeric atrophy of the smooth muscle cells of the media with replacement by collagen fibers after 4 months.

Beside cholestane-3β,5α,6β-triol and 25-hydroxycholesterol, which have been repeatedly identified as the most toxic as well as possibly the most atherogenic cholesterol oxides, 7-ketocholesterol has also been demonstrated to be angiotoxic when fed to chicks at 2% of the diet.[20] Although dietary 7-ketocholesterol treatment had little effect on plasma and liver lipid metabolism, aortas from chicks treated with 7-ketocholesterol contained numerous dead and dying smooth muscle cells. Chicks receiving dietary cholesterol in amounts equal to or greater than the amount of 7-ketocholesterol did not demonstrate significant smooth muscle degeneration. After eight weeks, a combined diet of 1% cholesterol plus 10 to 20 mg/d of 7-ketocholesterol produced no greater amounts of aortic lipid deposition, or smooth muscle cell degeneration than in a comparable group supplemented with 7-ketocholesterol alone. The abdominal aorta was always more severely damaged than the ascending arch or thoracic aorta in chicks fed 7-ketocholesterol.

III. EFFECTS OF CHOLESTEROL OXIDES ON VASCULAR PERMEABILITY

The vascular endothelium acts as a selectively permeable barrier to plasma components. Since cholesterol oxides have been shown to be injurious to the endothelium both *in vitro*

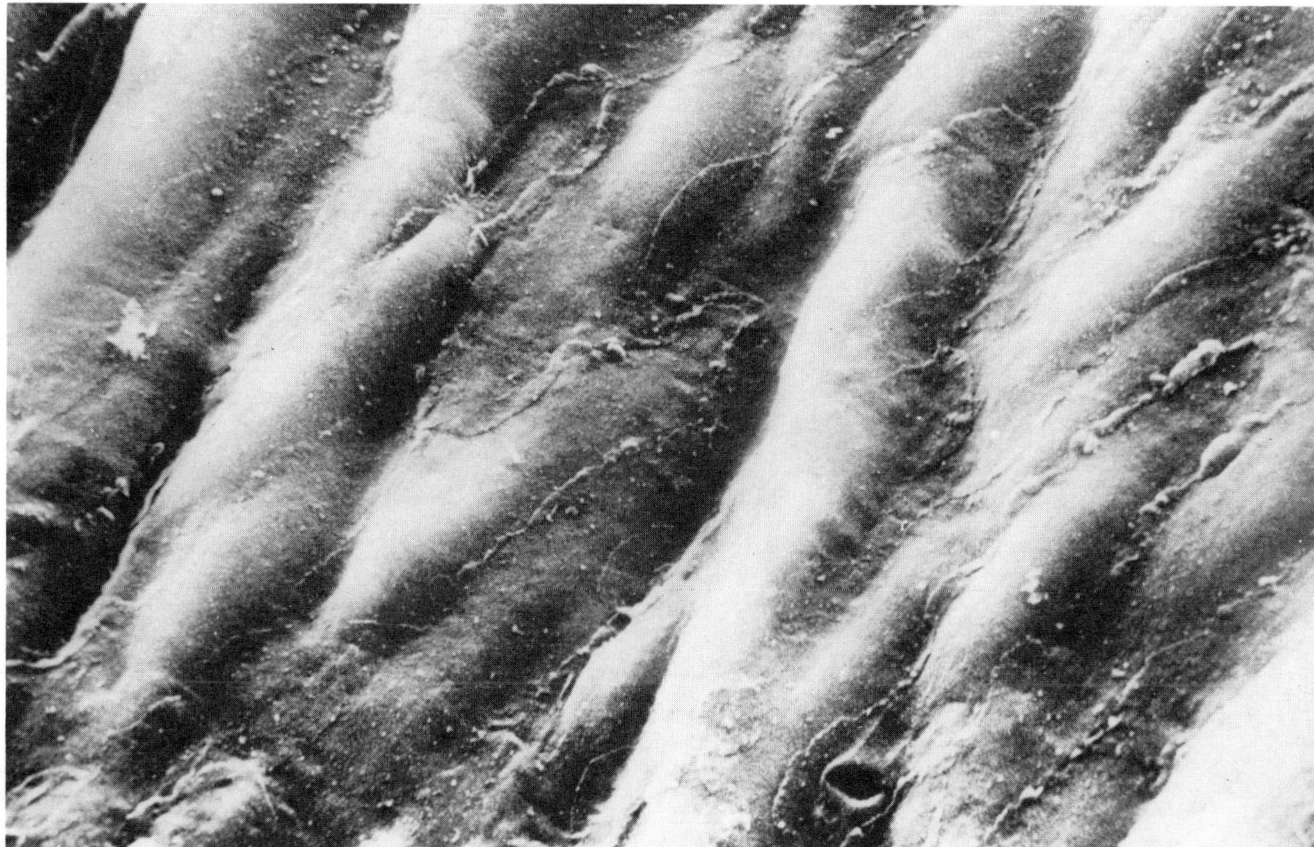

FIGURE 1. Aorta of control rabbit showing undulating liminal surface, discernible intercellular borders and ovoid protrusions of nuclei. A small crater is seen at the right lower corner. Magnification × 2000. (Reprinted from Peng, S. K., Taylor, C. B., Hill, J. C., and Morin, R. J., *Atherosclerosis,* 54, 121, 1985. With permission.)

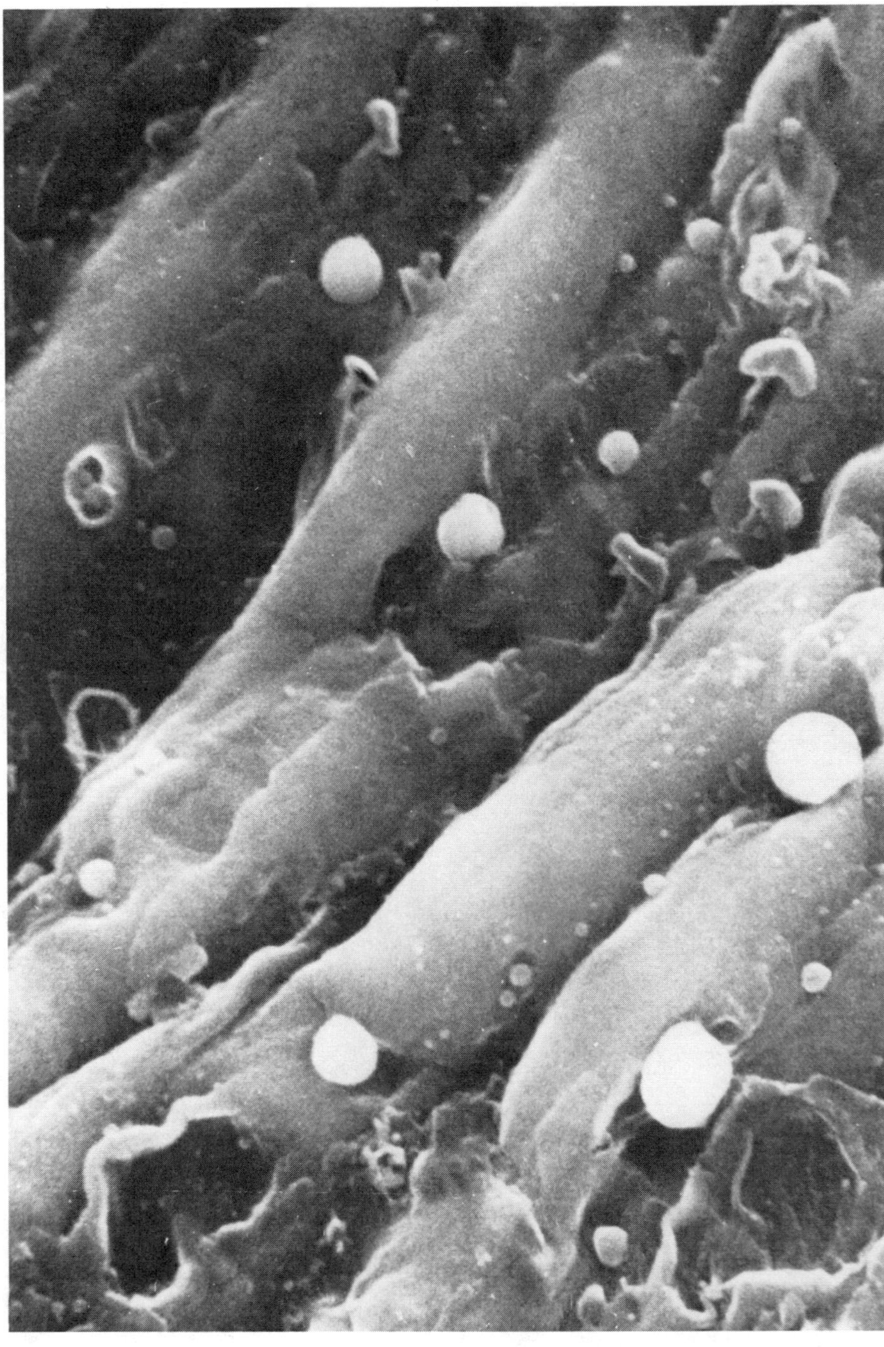

FIGURE 2. Aorta of rabbit receiving cholestane-3β,5α,6β-triol showing the presence of numerous blebs and craters on the luminal surface. Many circulating formed elements are seen adhering on the surface. Magnification × 4700.

and *in vivo,* this may disturb endothelial integrity resulting in alteration of the barrier function of the vascular endothelium and allowing increased penetration of plasma components, such as LDL or chylomicron remnants. Henning and Boissonneault[21] had studied the effect of cholesterol and cholesterol oxides on albumin transfer across vascular endothelium mono-

TABLE 2
Comparison of the Number of Craters and Balloons in the Aorta of 3 Groups

Group	Crater	Balloon
Control	2.0 ± 0.4[a]	0.8 ± 0.3
25-OH cholesterol	6.6 ± 0.6[b]	2.2 ± 0.7[c]
Cholestane-triol	13.6 ± 1.5[b]	4.5 ± 1.3[b]

[a] Each number represents the average number ± SEM of craters or balloons larger than 5 μg in diameter for the 6 animals in 10 random fields of 40,000 μm^2 (at × 500 magnification).
[b] Significant difference ($p < 0.01$) to control.
[c] Borderline significant difference ($p < 0.05$) to control.

Reprinted from Peng, S. K., Taylor, C. B., Hill, J. C., and Morin, R. J., *Atherosclerosis,* 54, 121, 1985. With permission.

layers cultured on micropore filters. Exposure to cholestane-3β,5α,6β-triol significantly increased albumin transfer across endothelial monolayers. The effect of cholestanetriol on endothelial cell barrier function was time and concentration dependent, with maximum albumin transfer being reached at 20 μ*M* and after a 24-h exposure. Pure cholesterol, on the other hand, did not affect albumin transfer at concentrations as high as 130 μ*M*.

Permeability to ions has also been observed to be increased by cholesterol oxides. Increased permeability to calcium and rubidium results from exposure of P815 cells or L-cells to 25-hydroxycholesterol.[22,23] Liposomes containing 25-hydroxycholesterol also show increased permeability to calcium.[24] The 7-oxygenated derivatives, 7-ketocholesterol and 7α- and 7β-hydroxycholesterol, however, reduce liposomal permeability to glucose.[25] The formation of echinocytes upon exposure of erythrocytes to cholesterol oxides[26] may also be a result of the induced changes in ionic permeability.

IV. ALTERATION OF LDL RECEPTOR FUNCTION BY CHOLESTEROL OXIDES

LDL is metabolized in many cells throughout the body, including fibroblasts, smooth muscle cells, lymphoid cells, and endothelial cells, by the LDL receptor pathway.[27] It has been demonstrated that absent or defective LDL receptors will result in hypercholesterolemia, increased accumulation of cholesterol in macrophages, and a predisposition to early atherosclerosis.[28] Normally cells can regulate the number of cell-surface LDL receptors, thus controlling the rate of entry of cholesterol needed for membrane synthesis. Since cholesterol oxides have been shown to alter many membrane-related functions, and LDL receptors are located in the membrane, it is conceivable that cholesterol oxides may also affect LDL receptor function, leading to derangement in the metabolism of LDL. 25-Hydroxycholesterol was reported by Brown and Goldstein[29] to decrease receptor-mediated LDL uptake in human fibroblasts. Lorenzo et al.[30] also demonstrated that 26-hydroxycholesterol, which is an intermediate in the synthesis of bile acids and is present in the serum of normal adults,[31] inhibited the specific binding, uptake, and degradation of ^{125}I-LDL in human fibroblasts. The effect was dose dependent and significant suppression occurred at a concentration of 1 μ*M*, not greatly different from concentrations observed in human serum (0.23 to 0.64 μ*M*).[31] This effect involved a reduction of LDL-binding sites with no change in their affinity for the ligand. Freshly crystalized cholesterol was less efficient in reducing cellular degradation

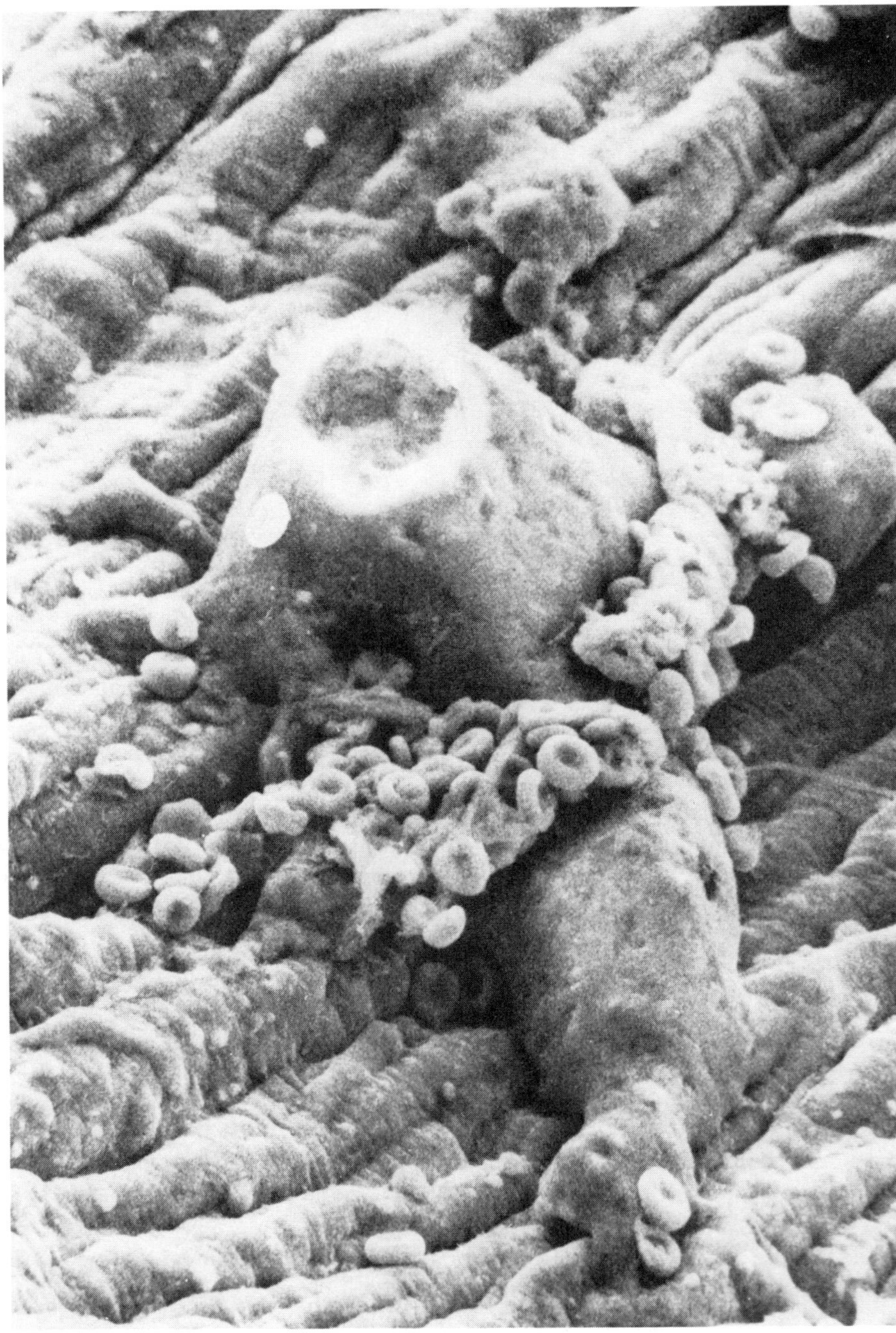

FIGURE 3. Aorta of rabbit receiving cholestane-3β,5α,6β-triol showing giant balloons with conical protrusions surrounded by microthrombi and a giant crater formed over the swollen endothelium. Magnification × 1200. (Reprinted from Peng, S. K., Taylor, C. B., Hill, J. C., and Morin, R. J., *Atherosclerosis,* 54, 121, 1985. With permission.)

of ^{125}I-LDL. The vascular smooth muscle cells which are involved in atherogenesis also showed similar responses. Rabbit aortic smooth muscle cells were preincubated in 5% lipoprotein-deficient medium for 24 h and then incubated with 1 to 5 μg/ml of either pure cholesterol, 25-hydroxycholesterol, 7-ketocholesterol, cholestane-3β,5α,6β-triol, or

FIGURE 4. Transmission electron microscopy of aorta of rabbit receiving 25-hydroxycholesterol showing subendothelial edema and intracytoplasmic vacuoles in endothelium. Magnification × 7100.

cholesterol 5α,6α-epoxide for 12 to 24 h. The uptake of ^{125}I-LDL after 25-hydroxycholesterol or cholesterol 5,6-epoxide at 1 μg/ml for 12 h of incubation was suppressed to 142 ± 65 cells bound ^{125}I-LDL μg/mg protein and 169 ± 70, respectively, and after cholestane-3β,5α,6β-triol or 7-ketocholesterol at 5 μg/ml for 12 h was 158 ± 40 and 157 ± 29, respectively when compared with controls at 219 ± 89. Similar suppression of ^{125}I-LDL degradation by cholesterol oxides was also observed.

On the other hand, exposure of bovine endothelial cell cultures to 25-hydroxycholesterol (50 to 100 μg/ml) resulted in a 5- to 10-fold increase in cell-surface binding sites for HDL.[32] This increase in HDL-binding sites was dependent on time and temperature. After a 48-h exposure to the 25-hydroxycholesterol, a maximal increase in HDL binding could be observed, and newly induced binding sites disappeared rapidly once 25-hydroxycholesterol was removed from the medium. HDL-binding sites were capable of binding LDL. When LDL bound to 25-hydroxycholesterol-induced newly synthesized HDL-binding sites, the LDL was no longer internalized by its own receptors, resulting in a 90% reduction of LDL degradation. It is possible, therefore, that cholesterol oxides not only reduce the number of cell-surface LDL receptors but also stimulate synthesis of cell-surface binding sites for HDL, resulting in reduction of LDL uptake and degradation.

It appears that cholesterol oxides coordinately regulate cholesterol metabolism in cells, since LDL receptor function and cholesterol biosynthesis are similarly inhibited by cholesterol oxides. 25-hydroxycholesterol is the most potent inhibitor. 7-Ketocholesterol, cholestane-3β,5α,6β-triol, and cholesterol 5,6-epoxide all show comparable degrees of inhibition of cholesterol biosynthesis and LDL receptor function. It has been postulated that cholesterol oxides bind to specific cytosolic receptor proteins which may participate in the feedback repression of enzymes of cholesterol biosynthesis and the LDL receptor.[33] Although exact mechanisms of how the oxysterol-binding protein influences the genetic regulation of enzymes in the cholesterol biosynthetic pathway and synthesis of cell surface receptor for LDL have not yet been elucidated, the promoters for the three sterol-repressed genes, the LDL receptor,[34] hydroxymethylglutaryl coenzyme A (HMG-CoA) reductase,[35] and HMG-CoA synthase[36] all contain a related octanucleotide designated as the sterol regulatory element. Cholesterol oxides may simultaneously suppress the gene promoter activities and thereby reduce the transcription of mRNA for enzymes in the cholesterol biosynthesis pathway and the LDL receptor.

V. CHOLESTERYL ESTER ACCUMULATION AND FOAM CELL FORMATION

Another possible mode of action of cholesterol oxides in atherogenesis might be at the stage of cholesteryl ester accumulation and foam cell formation. The LDL receptor activity is so closely regulated that even prolonged incubation of human monocyte/macrophages *in vitro* with high concentrations of native LDL does not produce deposition of cholesteryl ester in these cells.[37] Oxidatively modified LDL, however, can be ingested by monocytes in large quantities via a high-affinity scavenger cell receptor, which in contrast to the LDL receptor is not subject to down-regulation by the intracellular content of cholesterol,[15] and this may in part account for the accumulation of cholesteryl ester in macrophage-derived foam cells. The sources of the oxidatively modified LDL could be either exogenous from LDL-containing cholesterol oxides or endogenously produced by free radical-induced oxidation of LDL in the vessel wall.[38]

In atherosclerotic lesions, in addition to monocyte/macrophages, arterial smooth muscle cells also show cytoplasmic cholesteryl ester accumulation. Lipoproteins from plasma or other sources enter cell lysosomes where cholesteryl esters are hydrolyzed and the free cholesterol subsequently resynthesized into lipid droplets of cholesteryl esters by the enzyme acyl CoA cholesterol acyltransferase (ACAT). Cholesteryl esters in lipid droplets are continuously being acted on by hydrolytic enzymes. Free cholesterol derived from this hydrolysis has the ability to migrate through the cell membrane and is available for transport out of the cell by HDL. Incubation of monolayers of cultured human fibroblasts with cholesterol oxides including 25-hydroxycholesterol, 7-ketocholesterol, or 6-ketocholesterol markedly enhanced the rate at which the cells esterified their endogenous cholesterol and produced an increase in the cellular content of cholesteryl esters.[39] The enhanced esterification capacity was associated with an increase in the activity of ACAT. Incubation of cells for 5 h with 5 μg/ml of 25-hydroxycholesterol produced an 8-fold increase in the specific activity of this enzyme. Since the cholesterol oxides also suppressed the activity of HMG-CoA reductase, this suggests that the processes of cholesteryl ester formation and cholesterol synthesis in human fibroblasts are regulated in a reciprocal manner by coordinate changes in the activities of these two membrane-bound enzymes. Drevon and Evensen[40] had also shown a 6- to 46-fold increase in intracellular cholesterol esterification in response to 25-hydroxycholesterol and LDL in normal and in lecithin:cholesterol acyltransferase-deficient fibroblasts. In cultured mouse L fibroblast cells, incubation with 25-hydroxycholesterol or cholestane-3β,5α,6β-triol resulted in intracellular lipid accumulation.[41] The cholesteryl ester content of cultured hepatocytes from rabbits fed oxidized cholesterol derivatives was greatly increased as compared to hepatocytes from control or pure cholesterol-fed rabbits.[42] Morin and Peng[43] examined the effects of cholesterol oxides on enzymes that participate in accumulation and removal of cellular cholesteryl esters. Rabbit aortic smooth muscle cells in culture were preincubated for 24 h in lipoprotein-deficient media and then incubated for 15 min, 1 or 5 h with 5-μg/ml media of either 25-hydroxycholesterol, cholestane-3β,5α,6β-triol, cholesterol, or ethanol control. After one hour of incubation, ACAT activity showed a fourfold increase in the 25-hydroxycholesterol group (7.63 ± 2.24% esterified per mg protein per hour as compared to the control 1.96 ± 1.27%). Cholestane-3β,5α,6β-triol did not affect ACAT activity after any of the incubation time intervals. The activities of acid cholesteryl ester hydrolase and neutral cholesteryl ester hydrolase were not significantly affected by either 25-hydroxycholesterol or cholestane-3β,5α,6β-triol at any time interval. Enhancement of ACAT by 25-hydroxycholesterol without change in cholesteryl ester hydrolase may predispose the cholesteryl ester accumulation by arterial smooth muscle cells.

VI. EFFECT OF CHOLESTEROL OXIDES ON PROSTACYCLIN PRODUCTION AND PLATELET ADHESION

Prostacyclin (PGI_2) is synthesized primarily by endothelial cells and is essential for maintenance of vascular integrity,[44] and therefore, may play a role in the development of atherosclerosis.[45] Since cholesterol oxides have been shown to alter endothelial cell functions, they may also alter PGI_2 production. Human umbilical vein endothelial cells in culture were incubated with either pure cholesterol, 25-hydroxycholesterol, or cholestane-3β,5α,6β-triol at 10 μg/ml of culture media for 12 h. PGI_2 production measured by radioimmunoassay of 6-keto $PGF_{1\alpha}$, the stable metabolite of PGI_2, was inhibited 46% by cholestanetriol and 60% by 25-hydroxycholesterol.[46] Pure cholesterol had no inhibitory effect. The antioxidant, α-tocopherol (vitamin E) appeared to prevent the inhibitory effects of cholesterol oxides on PGI_2 production. When α-tocopherol, 50 μg/ml, was added to the culture media, cholestanetriol showed no inhibitory effects and 25-hydroxycholesterol inhibited PGI_2 production by only 21%. Decreased PGI_2 production by endothelial cells was also shown to enhance platelet adhesion to endothelial cells.[47] A platelet adhesion assay was performed on cultured human umbilical vein endothelial cells preincubated with 50 μg/ml of either cholestane-3β,5α,6β-triol, 25-hydroxycholesterol, or 7-ketocholesterol for 12 h. The original media covering endothelial monolayers were then replaced by media containing ^{111}In-labeled platelets.

Exposure to cholestane-3β,5α,6β-triol increased platelet adhesion by 104% when compared with controls, 25-hydroxycholesterol by 54%, and 7-ketocholesterol by 37%. Pure cholesterol at the same concentration had no effect on platelet adhesion. Cholesterol oxides, therefore, appeared not only to be able to suppress PGI_2 production but also to enhance platelet adhesion to endothelium with subsequent release of growth factors which may lead to subintimal migration and proliferation of smooth muscle cells.[3,48]

Effects of cholesterol oxides on prostaglandin metabolism using bovine aortic smooth muscle cells in culture were also studied.[49] PGI_2 synthesis was measured by two different methods, i.e., radioactive labeling of cells with ^{14}C-arachidonic acid and radioimmunoassay of 6-keto $PGF_{1\alpha}$. Toxicity was assessed by lactic dehydrogenase (LDH) activity measurement in the incubation medium, and 20 μg/ml was found to be a toxic dose for all the cholesterol oxides tested, regardless of their morphological changes in the cultures. Effects of cholesterol oxides on PGI_2 synthesis varied depending on the structure of cholesterol oxides. Cholesterol oxidized on the rings including cholestane-3β,5α,6β-triol and cholesterol 5α,6α-epoxide elicited a marked and significant reduction of PGI_2 synthesis only at toxic dose of 20 μg/ml and no effect at lower doses. Side-chain oxidation products such as 25-hydroxycholesterol and 22R-hydroxycholesterol both induced a biphasic effect. They inhibited PGI_2 synthesis and conversion of arachidonate into oxygenated derivatives in a dose-dependent manner at low doses. When present at 20 μg/ml in the incubation medium, they enhanced PGI_2 release. 20α-Hydroxycholesterol and 22S-hydroxycholesterol also inhibited PGI_2 production and arachidonate metabolism at low doses, in a dose-dependent manner, but at 20 μg/ml they did not elicit any enhancement of PGI_2 which remained reduced at this dose. Hypercholesterolemia has often been shown to be associated with an alteration in arachidonate metabolism and particularly a reduction of PGI_2 synthesis.[50] It has also been shown that cholesterol from LDL, once internalized into the cells, can be oxidized by the Cytochrome p-450-mediated mechanism, therefore, the presence of cholesterol oxides inside the cells is well recognized,[51] which may in part be responsible for the reduction of PGI_2 synthesis. On the other hand, Martinez-Sales et al.[52] demonstrated that the aortas of rats fed a diet enriched with 2% cholesterol autoxidation products for 24 h prior to sacrifice showed a slight increase in both lipid peroxidase and prostacyclin production. They believed the results suggested that the increase in the prostacyclin production might be a protective compensatory mechanism in response to vessel wall injury.

VII. EXPERIMENTAL ATHEROSCLEROSIS INDUCED BY CHOLESTEROL OXIDES

Cook et al. were the first to use cholesterol oxides to induce experimental atherosclerosis in 1968.[53] They tested 103 cholesterol derivatives on organ cultures of chick hearts[54] and subsequently on rabbit aortas.[55] Cholestane-3β,5α,6β-triol was found to be the most toxic. This compound, therefore, was chosen to feed to New Zealand white male rabbits, in doses of 0.1% of a standard commercial diet, corresponding to about 30 mg/kg body weight per day. The duration of feeding ranged from 27 to 350 d. Lesions were observed in the aorta, starting with deposition of sudanophilic material, progressing through various pathological changes in the tunica intima and tunica media, and terminating with the aorta showing considerable fibrosis and calcification. They pointed out that lesions induced in the rabbit aorta by feeding a high cholesterol diet bear only a partial resemblance to human atherosclerosis. In the human type, not only are lipid deposits present, but there are other changes such as disruption of elastic tissue, alteration in ground substance, tissue necrosis, and calcification. They claimed that the changes induced by prolonged feeding of relatively small amounts of cholestane-3β,5α,6β-triol produced lesions in rabbits with a closer resemblance to those of human atherosclerosis. More recently, Taylor et al.[56] studied the effect of

TABLE 3
Summary of Atherosclerotic Scores[b] in Each Group of Monkeys

Group	Control	Cholesterol	Cholestane-3β,5α,6β-triol
Aortic arch	0.6 ± 0.4	1.0 ± 0.3	2.2 ± 0.2[a]
Thoracic aorta	0.4 ± 0.2	0.4 ± 0.2	2.1 ± 0.2[a]
Abdominal aorta	0.5 ± 0.2	0.6 ± 0.2	2.6 ± 0.3[a]
Grand total	1.5 ± 0.8	2.0 ± 0.7	6.9 ± 0.7[a]

Note: All data expressed as mean ± S.E.M.

[a] Significant difference when compared with control and cholesterol groups ($p < 0.01$).
[b] Atherosclerotic scores graded on 0 to 4 scale as follows:

Grade 0	No intimal thickening
Grade 1	Lesion showing 1 to 3 cell layers
Grade 2	Lesion showing 4 cell layers to 50% of aortic wall thickness
Grade 3	Lesion exceeding 50% of aortic wall thickness
Grade 4	Lesion exceeding the full thickness of original aortic wall

cholestane-3β,5α,6β-triol on atherogenesis using 15 squirrel monkeys fed one of the following diets for 10 months: (1) control cholesterol-free diet; (2) 0.5% cholesterol diet; and (3) control diet plus a daily dose of cholestane-3β,5α,6β-triol 5 mg/kg body weight. Serum cholesterol was elevated only in the cholesterol-fed group (279 ± 14 mg % vs. 178 ± 11 mg % control). Light microscopic study revealed multifocal and marked intimal thickening in cholestanetriol-fed monkeys and much less in control and cholesterol-fed monkeys (Table 3). Ultrastructurally, the lesions showed proliferation of smooth muscle cells, macrophages, occasional foam cells, numerous cell debris, and stroma with accumulation of collagen fibers, elastica, and calcium cyrstals. Crater-like defects and areas of desquamation with occasional microthrombus formation were seen on SEM. Calcium and phosphorus were markedly increased by X-ray microanalysis. Control and cholesterol-fed monkeys showed no significant arterial damage or calcium deposition except some foam cells seen in the latter group. It is concluded that cholestane-3β,5α,6β-triol is more atherogenic than pure cholesterol. Jacobson et al.[57] used white Carneau pigeons gavage-fed either 0.05% pure cholesterol or 0.05% pure cholesterol with trace levels of cholestane-3β,5α,6β-triol for three months. These are amounts similar to the estimated U.S. dietary intake levels. Aortic lipids, aortic calcium, and coronary artery histopathology were assessed. Aortic lipids showed no significant differences between the two groups, but aortic calcium accumulation in the cholesterol with cholestanetriol group was 42% greater than in the cholesterol-fed group ($P < 0.02$). Coronary artery atherosclerosis, as measured by percent mean luminal stenosis, was 5.23 ± 5.4% in the cholesterol with triol group as compared to 2.80 ± 1.4% in the cholesterol-fed group, a difference of 87% ($P < 0.01$). These results showed that dietary exposure to low levels of cholestane-3β,5α,6β-triol is atherogenic to a greater degree than exposure to pure cholesterol alone. The data suggest that cholestanetriol is an atherogenic risk factor in doses approximating U.S. dietary exposure to cholesterol and its auto-oxides. Furthermore, the results call for reinterpretation of data from studies using animal models fed U.S.P. cholesterol and of human epidemiologic data not accounting for cholesterol oxide intake. The atherogenicity of 25-hydroxycholesterol was also studied using squirrel monkeys which were fed 5 mg/kg/d of 25-hydroxycholesterol for 2 to 8 months.[58] Arteriosclerosis induced by 25-hydroxycholesterol was visible by light microscopy as multifocal intimal thickening of the aortas. Ultrastructurally, there was evidence of arterial injury manifested

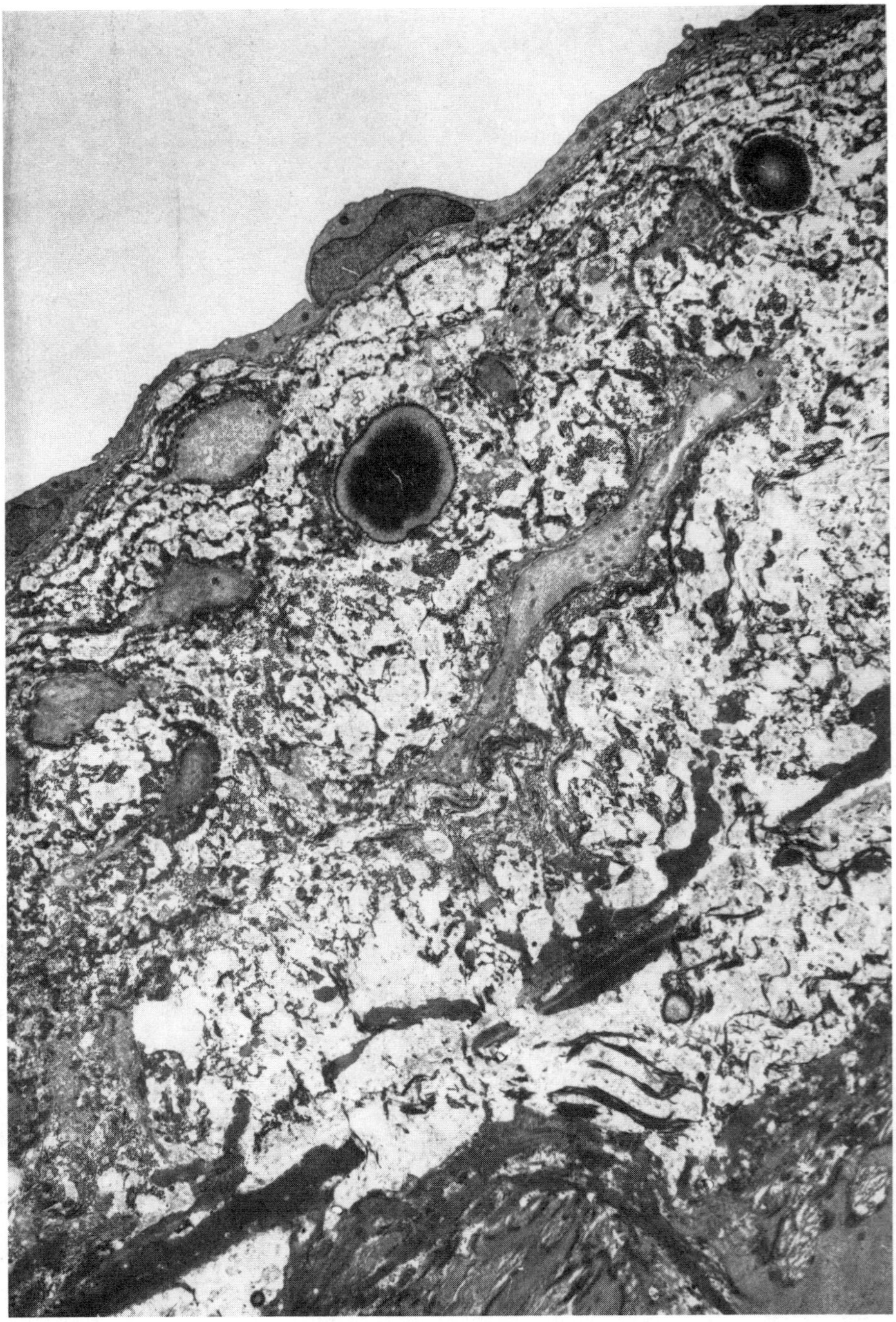

FIGURE 5. Aorta of monkey fed 25-hydroxycholesterol for 4 months showing intimal thickening composed of many degenerative myointimal cells, abundant collagen fibers and scattered calcium crystals. Magnification × 5900.

as edema, degenerating endothelial and smooth muscle cells, frequent deposits of cellular debritus, and calcium crystals with proliferative smooth muscle cells and macrophages in the subendothelial space, similar to those seen in human arteriosclerotic lesion (Figures 5 and 6).

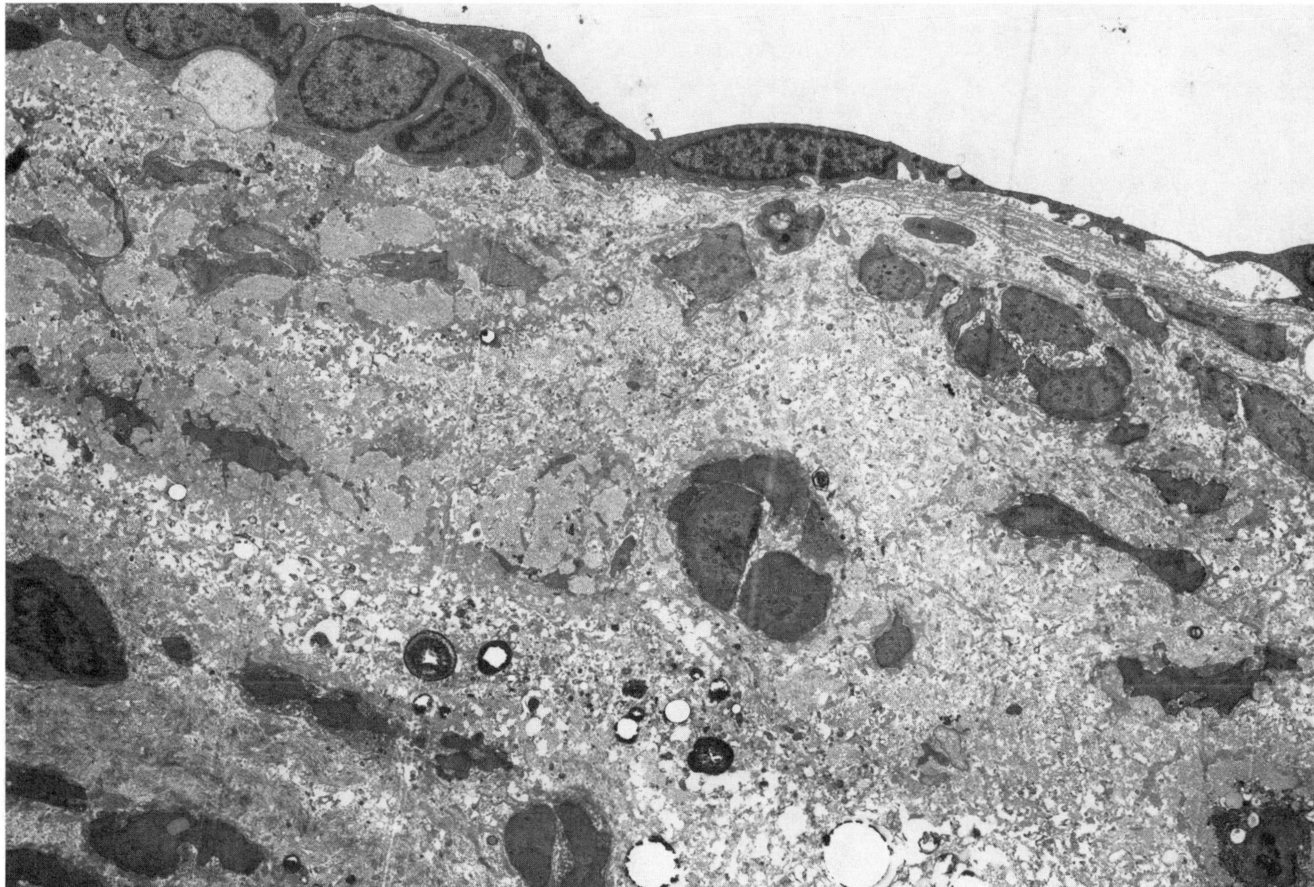

FIGURE 6. Aorta of monkey fed 25-hydroxycholesterol for 8 months showing a large amount of cell debris, calcium crystals, disrupted membranes, and amorphous granular materials. There are some macrophages and numerous myointimal cells. Magnification × 12,000.

In addition to the injury and repair hypothesis of atherogenesis, Benditt has proposed that atherosclerotic lesions are monoclonal, arising from single-stem cells.[59] Certain cholesterol oxides are more mutagenic than cytotoxic. Interestingly, serial studies by Lee et al.[60-62] demonstrated the ability of cholesterol oxide to produce monotypic lesions of atherosclerosis. Hybrid hares (*Lepus europaeus* males × *Lepus timidus* females) that display glucose-6-phosphate dehydrogenase (G-6-PD) mosaicism in the normal state, when fed a high-cholesterol diet developed ditypic atherosclerotic lesions. No monotypic samples were found. In contrast, hybrid hares fed 25-hydroxycholesterol or cholestane-3β,5α,6β-triol in addition to a high-cholesterol diet developed monotypic foci, all of the *timidus* type, in atherosclerotic lesions and normal aortic tissue. These observations suggest that the change from ditypism to monotypism induced by the cholesterol oxides was the result of progressive elimination of the *europaeus* type to leave only the *timidus* G-6-PD, which is consistent with the selective advantage of one phenotype over the other. On the other hand, the ratio of *europaeus* phenotype to *timidus* phenotype in cultured hare skin fibroblasts appeared to be shifted by exogenous 25-hydroxycholesterol or ultraviolet light irradiation in favor of the *europaeus* phenotype, perhaps because the *timidus* phenotype is more sensitive to toxic effect.[63-65] These results tend to support the monoclonal hypothesis of atherogenesis.

Despite extensive evidence that cholesterol oxides are associated with atherosclerosis, some experiments suggest otherwise. A comparative study of the atherogenic effects of cholesterol and cholesterol oxides was performed on New Zealand white rabbits by Higley et al.[66] Carefully purified cholesterol was fed to rabbits as a supplement for 11 weeks at 166/mg/kg/d and the resulting aortic lesions were compared with those from animals supplemented with oxidized cholesterols only or with a mixture of cholesterol and oxidized cholesterols. Microscopically, there were no statistically significant differences between any of the groups in the numbers of histochemically defined lesion. Macroscopically, however, the cholesterol-fed animals exhibited more severe arterial lesions than the animals fed cholesterol-free oxidized cholesterols or a mixture of cholesterol and oxidized cholesterols. The divergent results of this study may in part be explainable by the use of refluxing in toluene to oxidize the cholesterol. This method resulted in the absence of the highly toxic cholestane-3β,5α,6β-triol and relatively low levels of another toxic sterol, 25-hydroxycholesterol. It has been shown that in rabbits fed a diet containing 1% pure cholesterol for 2 months, that the plasma 5,6-oxygenated cholesterols increase remarkably, i.e., the concentration of 5α,6α-epoxide, 5β,6β-epoxide, and cholestane-3β,5α,6β-triol in the plasma were elevated to 2.6, 2.3, and 4.6 times and those in the liver to 7.5, 6.0, and 5.0 times the normal levels, respectively.[67] Unfortunately, the comparative study by Higley et al. did not measure the levels of cholesterol oxides in the plasma or tissue, and it is conceivable that the plasma levels of cholestane-3β,5α,6β-triol in rabbits fed pure cholesterol may be higher than in those fed oxidized cholesterols containing no cholestanetriol. Further investigations to correlate plasma levels of cholesterol oxides with severity of atherosclerosis or cardiovascular disease may be helpful to reconcile these results. Yet another viewpoint, proposed by Krut[68,69] is that solubilizing effects of cholesterol oxides on cholesterol if applicable *in vivo* might actually solubilize cholesterol from tissue deposits in atherosclerotic lesions. The effect of these cholesterol oxides on solubilization and clearance of cholesterol *in vivo* was examined in rats by subcutaneously implanting tablets made of cholesterol, cholesterol plus oxidation products, and both with phosphatidylcholine. Tablets containing cholesterol oxides were cleared rapidly and tablets without cholesterol oxides showed no clearance at all. Baranowski et al.,[70] however, demonstrated that cholesterol oxides caused more tissue necrosis with inflammatory responses and granuloma formation than pure cholesterol when injected subcutaneously in rats. The severity of responses were in the following descending order: cholestane-3β,5α,6β-triol, 25-hydroxycholesterol, cholesterol 5,6-epoxide, oxidized cholesterols = cholesterol. The discrepancy of the results of these two studies, could again be explained in part by the absence of the highly toxic cholestane-3β,5α,6β-triol and 25-hydroxycholesterol in Krut's experiment in which oxidized cholesterols contained only 7-ketocholesterol, and 7α- and 7β-hydroxycholesterol. Furthermore, atherosclerosis is more closely related to inflammatory processes than physical processes.

Epidemiologic evidence implicating cholesterol oxides as atherogenic agents comes from two reports of Indian immigrants to London and the West Indies who are afflicted with a higher-than-expected mortality from atherosclerosis but who are not characterized by the usual high risk factors associated with such disease frequency[71,72] These people continue to consume large amounts of ghee, the clarified butter used extensively for Indian cooking and which contains up to 12.3% of its sterols as cholesterol oxides, including 7α- and 7β-hydroxycholesterol, 25-hydroxycholesterol, cholesterol 5,6-epoxide, and 20α-hydroxycholesterol. The high level of cholesterol oxides in ghee was strongly suggested to be the cause of the high incidence of this human atherosclerosis.

VIII. EFFECTS OF ANTIOXIDANTS ON ATHEROSCLEROSIS

In recent years it has become apparent that oxidatively modified LDL plays an important role in the initiation and progression of atherosclerosis.[73] The oxidatively modified form, but not native LDL, is recognized by the scavenger receptor and could, therefore, be taken up rapidly by tissue macrophages to generate the fatty-streak lesion of atherosclerosis. LDL degradation in fatty-streak lesions reflects primarily degradation by macrophages, whereas LDL degradation in nonlesioned areas represents primarily degradation by endothelial cells and smooth muscle cells. It has been reported that treatment of Watanabe heritable hyperlipidemic (WHHL) rabbits with probucol, which is an antioxidant, can prevent progression of atherosclerosis.[74] A complication in the evaluation of the mechanism behind the antiatherogenic effect of probucol *in vivo,* however, is that this drug has an intrinsic hypolipidemic effect, which in itself could be antiatherogenic. This issue was addressed by Carew et al.[75] who demonstrated that a considerable part of the antiatherogenic effect of probucol obtained by treatment of WHHL rabbits was unrelated to the hypolipidemic properties of the drug. They have shown that the rate of degradation of LDL in the macrophage-rich fatty-streak lesions of the LDL receptor-deficient (WHHL) rabbit treated with probucol (1% by weight in the diet) was reduced to about one half of that in the lesions of receptor-deficient rabbits not given probucol (control group matched for plasma cholesterol levels by lovastatin treatment). In contrast, the rates of degradation in the nonlesioned areas of aorta were no different in probucol-treated and control animals. After about 33 weeks of treatment, the severity of aortic atherosclerosis in the probucol-treated rabbits was much less than in untreated rabbits and most importantly, was significantly less than that in the lovastatin-treated (control) group in which plasma cholesterol levels were the same as (or lower than) in the probucol-treated rabbits. Björkhem et al.[76] used an antioxidant butylated hydroxytoluene (BHT) that has no hypolipidemic effect to investigate whether or not antioxidants can protect against atherosclerosis. Rabbits fed a 1% cholesterol diet with or without BHT developed typical atherosclerotic lesions. The addition of BHT gave higher levels of total cholesterol (+ 40%), triglyceride (+ 250%), LDL and VLDL (very-low-density lipoproteins) in plasma. Despite the lower plasma lipid levels, the degree of atherosclerosis of the aortic surface was considerably higher in rabbits fed cholesterol than in the group treated with cholesterol and BHT. The mean atherosclerotic involvement was 18.6 ± 4.4% in the former group and 5.9 ± 1.7% in the latter group ($p = 0.02$). In all animals, there was a high correlation between the area of the arterial lesion and cholesterol content ($r = 0.96$). Serum levels of cholesterol autoxidation products (7-ketocholesterol and cholesterol 5α,6α-epoxide) were lower in the group of rabbits treated with BHT ($p < 0.005$). Serum levels of vitamin E were slightly higher in the BHT group. Furthermore, Chait et al.[77] have demonstrated that a free

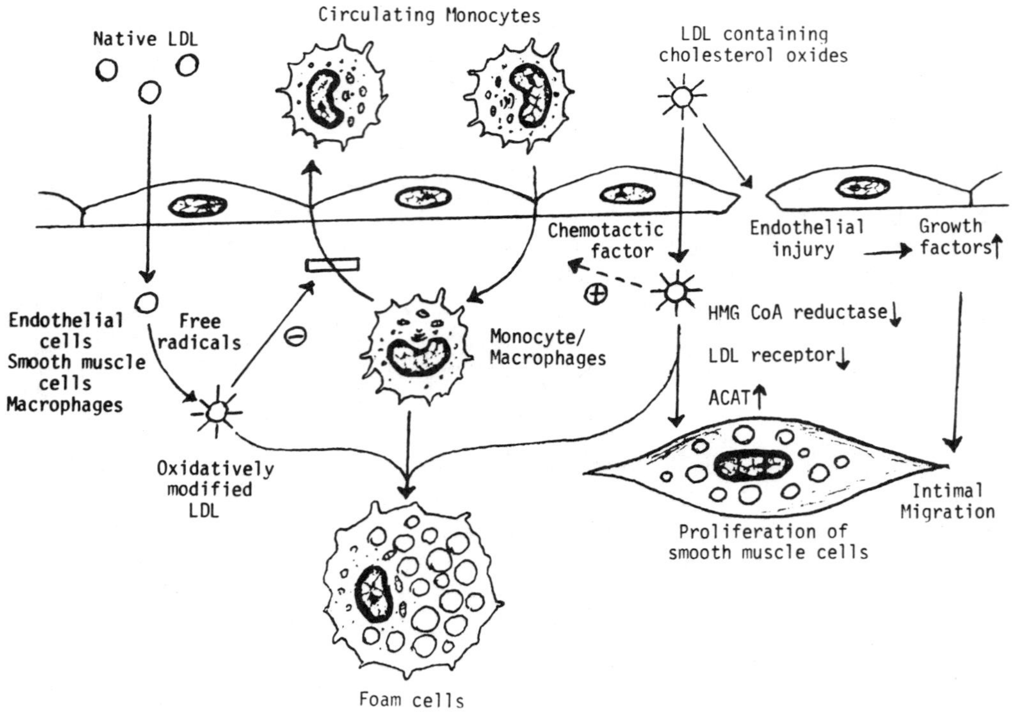

FIGURE 7. Summary of how cholesterol oxides could contribute to atherogenesis.

radical mechanism involving superoxide is involved in the oxidative modification of LDL and *in vitro*-produced oxidatively modified LDL has shown to be quite high in cholesterol oxidation products.[78] If *in vivo*-produced oxidatively modified LDL contains significant levels of cholesterol oxides, the cholesterol oxides present could explain the fact that oxidatively modified LDL is far more cytotoxic than native LDL. In addition, oxidatively modified LDL, but not native LDL, is also a chemoattractant with the ability to recruit circulating monocytes as well as a potent inhibitor of the motility of the resident macrophage.[79] These properties may provide the basis for possible mechanisms whereby oxidatively modified LDL could be intimately involved in both arterial injury and atherosclerotic plaque formation.[73]

IX. CONCLUSIONS

Studies implicating cholesterol oxides in the initiation and progression of atherosclerosis have been conducted for more than two decades but the matter still remains unresolved. The possible role of cholesterol oxides in atherogenesis can be summarized in Figure 7. Crucial to the involvement of cholesterol oxides with atherosclerosis is the presence of cholesterol oxides in human blood, aortic tissue, and atherosclerotic plaques. In this regard, the evidence is ample. Gray and Lawrie[80] have suggested that the level of cholesterol 5α,6α-epoxide in human serum may be related to the severity of atherosclerosis, based on the detection of very high concentrations (250 to 3250 μg/dl) of cholesterol 5α,6α-epoxide in type II hypercholesterolemic patients, as compared to controls with levels of less than 5 μg/dl serum. Brooks et al.[81] have analyzed human plasma or serum and identified the presence of cholesterol oxides, including cholesterol 5α,6α-epoxide, 7-ketocholesterol, and 26-hydroxycholesterol. These cholesterol oxides were predominantly located in VLDL and LDL. Exogenous 7α-hydroxycholesterol, 25-hydroxycholesterol, and 20α-hydroxycholesterol appeared to preferentially distribute into LDL and HDL of human serum.[82] In squirrel monkeys, it appeared that exogenous 25-hydroxycholesterol was selectively present in VLDL and LDL and in relatively low levels in HDL.[83] In rabbits fed cholesterol oxidation products, cholestane-3β,5α,6β-triol, 7α- and 7β-hydroxycholesterol, and 7-ketocholesterol were selectively transported in VLDL, whereas most of the 25-hydroxycholesterol was present in LDL; HDL contained only minute amounts of cholesterol oxides.[84] Analysis of tissue from atherosclerotic human aortas have demonstrated the presence of 24-, 25- and 26-hydroxycholesterol, 7α- and 7β-hydroxycholesterol, 7-ketocholesterol, and a range of other minor sterol constituents.[85,86] Furthermore, the increased accumulation of 26-hydroxycholesterol and its fatty acyl esters in atherosclerotic human aorta was shown to be proportional to advancing age and severity of atherosclerosis,[87,88] and therefore, may play a role in the progression of atherosclerotic plaques. Cholestane-3β,5α,6β-triol has also been found in atheromatous human arteries.[87] More recently, cholesterol 5,6-epoxides and 25-hydroxycholesterol were again demonstrated in human aortic tissue and plaques.[89] Most recently, Sevanian et al.[90] analyzed the cholesterol and cholesterol oxides contents in serum and aortas obtained from four experimental groups of New Zealand white rabbits: (1) normotensive/normolipemic, (2) hypertensive/normolipemic, (3) normotensive/hyperlipemic, (4) hypertensive/hyperlipemic, and found that the levels of cholesterol oxides, particularly cholesterol epoxides and cholestanetriol, were markedly increased in hypertensive/hyperlipemic rabbits. In fact, the severity of hypertension conbined with hyperlipemia was much better correlated with the proportion of cholesterol oxides in serum lipids than with the level of cholesterol. The severity of disease in the experimental rabbits correlated with markedly higher amounts of cholesterol epoxides (expressed as a percent of cholesterol) when compared to normal animals. Interestingly, cholesterol oxides were substantially elevated in the hypertensive/normolipemic rabbits, despite the fact that cholesterol levels were within normal limits. The levels of cholesterol and cholesterol oxides in the aortic arch essentially reflected those measured in serum. It should be noted that cholesterol 5β,6β-epoxide was the major cholesterol oxide identified in nearly all samples, suggesting that a substantial portion of the cholesterol oxides arose via free radical chain propagation reactions. The sources of elevated cholesterol oxides in blood and tissue are not exactly known since these compounds are present in foods and can thus be ingested and distributed to various tissues. They may also be generated endogenously via free radical and enzymatic reactions, however, the relative contribution from these two recognized sources remains uncertain. Many arguments that link cholesterol to human atherosclerosis may be equally valid for cholesterol oxides present in foods, blood, and aortic tissue and atherosclerotic plaques. Perhaps a more balanced judgment could be achieved if there were epidemiological data examining relationships between atherosclerosis or cardiovascular disease with plasma levels of cholesterol oxides as independent risk factors.

REFERENCES

1. **Anitschkow, N.,** Uber die veranderungen der kaninchenaorta by experimenteller cholesterinsteatose, *Beitr. Pathol. Anat. Allg. Pathol.,* 56, 379, 1913.
2. **Florentin, R. A., Nam, S. C., Lee, K. T., and Thomas, W. A.,** Increased ^{3}H-thymidine incorporation into endothelial cells of swine fed cholesterol for 3 days, *Exp. Mol. Pathol.,* 10, 250, 1969.
3. **Ross, R.,** The pathogenesis of atherosclerosis — an update, *N. Engl. J. Med.,* 314, 488, 1986.
4. **Chaikoff, I. L., Lindsay, S., Lorenz, F. W., and Entenman, C.,** Production of atherosclerosis in the aorta of the bird by administration of diethylstilbesterol, *J. Esp. Med.,* 88, 373, 1948.
5. **Beckwith, A. L. J.,** The oxidation of crystalline cholesterol, *Proc. Chem. Soc.,* 194, 1958.
6. **Fiority, J. A. and Sims, R. J.,** Autoxidation products of cholesterol, *J. Am. Oil Chem. Soc.,* 44, 221, 1967.

7. **Horvath, C.,** Quantitative determination of cholesterol in autoxidation mixtures by thin layer chromatography, *J. Chromatogr.*, 22, 52, 1966.
8. **Smith, L. L., Matthews, W. S., Price, J. C., Bachman, R. C., and Reynolds, B.,** Thin layer chromatographic examination of cholesterol autoxidation, *J. Chromatogr.*, 27, 187, 1976.
9. **Peng, S. K., Taylor, C. B., Tham, P., Werthessen, N. T., and Mikkelson, B.,** Effect of auto-oxidation products from cholesterol on aortic smooth muscle cells — an in vitro study, *Arch. Pathol. Lab. Med.*, 102, 57, 1978.
10. **Peng, S. K., Tham, P., Taylor, C. B., and Mikkelson, B.,** Cytotoxicity of cholesterol oxidation derivatives on cultured aortic smooth muscle cells and their effect on cholesterol biosynthesis, *Am. J. Clin. Nutr.*, 32, 1033, 1979.
11. **Naseem, S. M. and Heald, F. P.,** Cytotoxicity of cholesterol oxides and their effects on cholesterol metabolism in cultured human aortic smooth muscle cells, *Biochem. Int.*, 14, 71, 1987.
12. **Cox, D. C., Comai, K., and Goldstein, A. L.,** Effects of cholesterol and 25-hydroxycholesterol on smooth muscle cell and endothelial cell growth, *Lipids,* 23, 85, 1988.
13. **Goldstein, J. L., Ho, Y. K., Basu, S. K., and Brown, M. S.,** Binding site on macrophages that mediated uptake and degradation of acetylated low density lipoprotein, producing massive cholesterol deposition, *Proc. Natl. Acad. Sci. U.S.A.*, 76, 333, 1979.
14. **Brown, M. L. and Goldstein, J. L.,** Lipoprotein metabolism in the macrophage; implications for cholesterol deposition in atherosclerosis, *Annu. Rev. Biochem.*, 52, 223, 1983.
15. **Imai, H., Werthessen, N. T., Taylor, C. B., and Lee, K. T.,** Angiotoxicity and atherosclerosis due to contaminants of U.S.P. grade cholesterol, *Arch. Pathol. Lab. Med.*, 100, 565, 1976.
16. **Taylor, C. B., Peng, S. K., Werthessen, N. T., Tham, P., and Lee, K. T.,** Spontaneously occurring angiotoxic derivatives of cholesterol, *Am. J. Clin. Nutr.*, 32, 40, 1979.
17. **Imai, H., Werthessen, N. T., Subramanyam, V., Lequosne, P. W., Soloway, A. H., and Kanisawa, M.,** Angiotoxicity of oxygenated sterols and possible precursors, *Science,* 207, 651, 1980.
18. **Peng, S. K., Taylor, C. B., Hill, J. C., and Morin, R. J.,** Cholesterol oxidation derivatives and arterial endothelial damage, *Atherosclerosis,* 54, 121, 1985.
19. **Matthias, D., Becker, C. H., Godicke, W., Schmidt, R., and Ponsold, K.,** Action of cholestane-3,5,6-triol on rats with particular reference to the aorta, *Atherosclerosis,* 63, 115, 1987.
20. **Toda, T., Leszczynski, D., and Kummerow, F.,** Angiotoxic effects of dietary 7-ketocholesterol in chick aorta, *Paroi Arterielle,* 7, 167, 1981.
21. **Henning, B. and Boissonneault, G. A.,** Cholestane-3β,5α,6β-triol decreases barrier function of cultured endothelial cell monolayers, *Atherosclerosis,* 68, 255, 1987.
22. **Biossonneault, G. A. and Heiniger, H.-J.,** 25-Hydroxycholesterol-induced elevations in ^{45}Ca uptake: permeability changes in P815 cells, *J. Cell Physiol.*, 125, 4715, 1985.
23. **Chen, H. W., Heiniger H.-J., and Kandutsch A. A.,** Alteration of ^{86}RB^{+} influx and efflux following depletion of membrane sterol in L-cells, *J. Biol. Chem.*, 253, 3180, 1978.
24. **Holmes, R. P. and Yoss, N. L.,** 25-Hydroxysterols increase the permeability of liposomes to Ca^{2+} and other actions, *Biochim. Biophys. Acta.*, 770, 15, 1984.
25. **Theunissen, J. J. H., Jackson, R. L., Kempen, H. J. M., and Demel, R. A.,** Membrane properties of oxysterols, interfacial orientation, influence on membrane permeability and redistribution between membranes, *Biochim. Biophys. Acta,* 860, 66, 1986.
26. **Hsu, R. C., Kanofsky, J. R., and Yachnin, S.,** The formation of echinocytes by the insertion of oxygenated sterol compounds into red cell membranes, *Blood,* 56, 109, 1980.
27. **Goldstein, J. L.. and Brown, M. S.,** The low density lipoprotein pathway and its relation to atherosclerosis, *Annu. Rev. Biochem.*, 46, 897, 1977.
28. **Goldstein, J. L., Kita, T., and Brown, M. S.,** Defective lipoprotein receptor and atherosclerosis, *N. Engl. J. Med.*, 309, 288, 1983.
29. **Brown, M. S. and Goldstein, J. L.,** Regulation of the activity of the low density lipoprotein receptor in human fibroblasts, *Cell,* 6, 307, 1975.
30. **Lorenzo, J. L., Allorio, M., Bernini, F., Corsini, A., and Fumagalli, R.,** Regulation of low density lipoprotein metabolism by 25-hydroxycholesterol in human fibroblasts, *Fed. Eur. Biochem. Soc.*, 218, 77, 1987.
31. **Javitt, N. B., Kok, E., Burstein, S., Cohen, B., and Kutscher, J.,** 26-hydroxycholesterol. Identification and quantification in human serum, *J. Biol. Chem.*, 256, 12644, 1981.
32. **Tauber, J. P., Goldminz, D., and Gospodarowicz, D.,** Up-regulation in vascular endothelial cells of binding sites of high density lipoprotein induced by 25-hydroxycholesterol, *Eur. J. Biochem.*, 119, 327, 1981.
33. **Taylor, F. R., Saucier, S. E., Shown, E. P., Parish, E. J., and Kandutsch, A. A.,** Correlation between oxysterol binding to a cytosolic binding protein and potency in the repression of hydroxymethylglutaryl coenzyme A reductase, *J. Biol. Chem.*, 259, 12382, 1984.

34. **Dawson, P. A., Hofmann, S. L., Van der Westhuyzen, D. R., Südhof, T. C., Brown, M. S., and Goldstein, J. L.,** Sterol-dependent repression of low density lipoprotein receptor promoter mediated by 16-bse pair sequence adjacent to binding site for transcription factor spl, *J. Biol. Chem.*, 263, 3372, 1988.
35. **Osborne, T. F., Gil, G., Goldstein, J. L., and Brown, M. S.,** Operator constitutive mutation of 3-hydroxy-3-methyglutaryl coenzyme A reductase promoter abolishes protein binding to sterol regulatory element, *J. Biol. Chem.*, 263, 3380, 1988.
36. **Smith, J. R., Osborne, T. F., Brown, M. S., Goldstein, J. L., and Gil, G.,** Multiple sterol regulatory elements in promoter for hamster 3-hydroxy-3-methyglutaryl coenzymes A synthase, *J. Biol. Chem.*, 263, 18480, 1988.
37. **Shechter, I., Fogelman, A. M., Haberland, M. E., Seager, J. Hokom, M., and Edwards, P. A.,** The metabolism of native and malondialdehyde-altered low density lipoproteins by human monocyte-macrophages, *J. Lipid Res.*, 22, 63, 1981.
38. **Morel, D. W., DiCorleto, P. E., and Chisolm, G. M.,** Endothelial and smooth muscle cells alter low density lipoprotein in vitro by free radical oxidation, *Arteriosclerosis*, 4, 357, 1984.
39. **Brown, M. S., Dana, S. E., and Goldstein, J. C.,** Cholesterol ester formation in cultured human fibroblasts, stimulation by oxygenated sterols, *J. Biol. Chem.*, 250, 4025, 1975.
40. **Drevon, C. A. and Evensen, S. A.,** Demonstration of intact intracellular cholesterol esterification and normal low-density lipoprotein pathway in fibroblasts from a patient with lecithin:cholesterol acyltransferase deficiency, *Biochim. Biophys. Acta*, 794, 479,, 1984.
41. **Highley, N. A. and Taylor, S. L.,** The steatotic and cytotoxic effects of cholesterol oxides in cultured L cells, *Food Chem. Toxicol.*, 22, 983, 1984.
42. **Kosykh, V. A., Lankin, V., Podrez, E. A., Dmitriy, K. N., Volgushev, S. A., Victorov, A. V., Repin, V. S., and Smirnov, V. N.,** Very low density lipopritein secretion by cultured hepatocytes of rabbits receiving old commercial and purified cholesterol, *Lipids*, 24, 109, 1989.
43. **Morin, R. J. and Peng, S. K.,** Effects of cholesterol oxidation derivatives on cholesterol esterifying and cholesteryl ester hydrolytic enzyme activity of cultured rabbit aortic smooth muscle cells, *Lipids*, 24, 217, 1989.
44. **Weksler, B. B., Marcus, A. J., and Jaffe, E. A.,** Synthesis of prostaglandin I_2 (prostacyclin) by cultured human and bovine endothelial cells, *Proc. Natl. Acad. Sci. U.S.A.*, 74, 3922, 1977.
45. **Moncada, S.,** Prostacyclin and arterial wall biology, *Arteriosclerosis*, 2, 193, 1982.
46. **Peng, S. K., Hu, B., Brubaker, D., Marcus, C., and Morin, R. J.,** Cholesterol oxides enhance platelet adhesion to endothelial cells in vitro, *FASEB J.*, 4, 3733, 1990.
47. **Hu, B., Peng, S. K., Brubaker, D., and Morin R. J.,** Influence of cholesterol oxides and antioxidants on prostacyclin production by cultured endothelial cells, *FASEB J.*, 4, 3734, 1990.
48. **Ross, R.,** Platelet-derived growth factor, *Annu. Rev. Med.*, 38, 71, 1987.
49. **Seilan, C. and Dubuquoy, C.,** Effects of oxysterols on arachidonic acid metabolism and prostacyclin synthesis in bovine smooth muscle cells in culture, *Prostaglandins, Leukotrienes and Essential Fatty Acids*, 39, 11, 1990.
50. **Gupta, A., Sexton, R. G., and Rudney, H.,** Modulation of regulatory oxysterol formation and low density lipoprotein suppression of 3-hydroxy-3-methylgutaryl coenzyme A (HMG-CoA) reductase activity by ketoconazole, *J. Biol. Chem.*, 261, 8348, 1986.
51. **Pomerantz, K. B. and Hajjar, D. P.,** Cholesteryl ester accumulation reduces prostacyclin and PGE_2 synthesis by aortic smooth muscle cells, *Fed. Proc. Fed. Am. Soc. Exp. Biol.*, 46, 972, 1987.
52. **Martinez-Sales, V., Fornas, E., and Camanas, A.,** Prostacyclin production and lipid peroxidation in the aorta of rats fed with cholesterol oxidation products, *Artery*, 12, 213, 1983.
53. **Cook, R. P., MacDougal, J. D. B., and Biswas S.,** Experimental atherosclerosis in rabbits after feeding cholestane-triol, *Br. J. Exp. Pathol.*, 49, 265, 1968.
54. **Biswas, S., MacDougal, J. D. B., and Cook, R. P.,** The effect of various steroids, mainly of the C27 Series, on the growth of chick heart explants, *Br. J. Exp. Pathol.*, 45, 13, 1964.
55. **MacDougal, J. D. B., Biswas, S., and Cook, R. P.,** The effects of certain C27 steroids on organ cultures of rabbit aortas, *Br. J. Exp. Pathol.*, 46, 549, 1965.
56. **Taylor, C. B., Peng, S. K., Safarik, J., Kapuscinska, B., Bochenek, W., and Hill, J.,** Experimental atherosclerosis in squirrel monkey by feeding cholestane-3β-5α,6β-triol, *Fed. Proc. Fed. Am. Soc. Exp. Biol.*, 42, 789, 1983.
57. **Jacobson, M. S., Price, M. G., Shamoo, A. E., and Heald, F. P.,** Atherogenesis in White Carneau pigeons, Effects of low-level cholestane-triol feeding, *Atherosclerosis*, 57, 209, 1985.
58. **Peng, S. K., Taylor, C. B., Safarik, J., Hill, J., and Mikkelson, B.,** Arteriosclerosis induced by 25-hydroxycholesterol in squirrel monkeys, *Fed. Proc. Fed. Am. Soc. Exp. Biol.*, 41, 452, 1982.
59. **Benditt, E. P.,** The monoclonal theory of atherogenesis, *Atherosclerosis Rev.*, 3, 77, 1978.
60. **Lee, K. T., Thomas, W. A., Janakidevi, K., Kroms, M., Reiner, J. M., and Borg, K. Y.,** Mosaicism in female hybrid hares heterozygous for glucose-6-phosphate dehydrogenase (G-6-PD). I. General properties of a hybrid hare model with special reference to atherogenesis, *Exp. Mol. Pathol.*, 34, 191, 1981.

61. **Janakidevi, K., Lee, K. T., Krom, M., Imai, H., and Thomas, W. A.,** Mosaicism in female hybrid hares heterozygous for glucose-6-phosphate dehydrogenase. VI. Production of monotypicism on the aortas of 4 of 10 mosaic hares fed cholesterol oxidation products, *Exp. Mol. Pathol.,* 41, 354, 1984.
62. **Lee, K. T., Janakidevi, K., Kroms, M., Schmee, J., and Thomas, W. A.,** Mosaicism in female hybrid hares heterozygous for glucose-6-phosphate dehydrogenase. VII. Evidence for selective advantage of one phenotype over the other in ditypic samples from aortas of hares fed cholesterol oxidation products, *Exp. Mol. Pathol.,* 42, 71, 1985.
63. **Janakidevi, K., Lee, K. T., Thomas, W. A., Reiner, J. M., and Murray, C. D.,** Mosaicism in female hybrid hares heterozygous for glucose-6-phosphate dehydrogenase (G-6-PD). II. Changes in the ratios of G-6-PD types in skin fibroblast cultures carried through multiple passages, *Exp. Mol. Pathol.,* 34, 202, 1981.
64. **Murray, C. D., Lee, K. T., Thomas, W. A., Reiner, J. M., and Janakidevi, K.,** Mosaicism in female hybrid hares heterozygous for glucose-6-phosphate dehydrogenase (G-6-PD). III. Changes in the ratios of G-6-PD types in skin fibroblast cultures exposed to 25-hydroxycholesterol, *Exp. Mol. Pathol.,* 34, 209, 1981.
65. **Janakidevi, K., Lee, K. T., Thomas, W. A., Kroms, M., and Murray, C. D.,** Mosaicism in female hybrid hares heterozygous for glucose-6-phosphate dehydrogenase. V. The recovery of DNA synthesis of hare fibroblasts after ultraviolet irradiation, *Exp. Mol. Pathol.,* 41, 344, 1984.
66. **Higley, N. A., Berry, J. T., Taylor, S. L., Porter, J. W., Dziuba, J. A., and Lalich, J. J.,** Comparative atherogenic effects of cholesterol and cholesterol oxides, *Atherosclerosis,* 62, 91, 1986.
67. **Watabe, T., Isobe, M., and Danai, M.,** Cholesterol diet increases plasma and liver concentrations of cholesterol epoxides and cholestane-triol, *J. Pharm. Dyn.,* 3, 553, 1980.
68. **Krut, L. H.,** Solubility of cholesterol in vitro promoted by oxidation products of cholesterol, *Atherosclerosis,* 43, 95, 1982.
69. **Krut, L. H.,** Clearance of subcutaneous implants of cholesterol in the rat promoted by oxidation products of cholesterol, *Atherosclerosis,* 43, 105, 1982.
70. **Baranowski, A., Adams, C. W. M., Bayliss-High, O. B., and Bowyer, D. B.,** Connective tissue responses to oxysterols, *Atherosclerosis,* 41, 255, 1982.
71. **Jacobson, M. S.,** Cholesterol oxides in Indian ghee: possible cause of unexplained high risk of atherosclerosis in Indian immigrant populations, *Lancet,* ii, 656, 1987.
72. **Beckles, G. L. A., Kirkwood, B. R., and Carson, D. C.,** High total and cardiovascular disease mortality in adults of Indian descent in Trinidad, unexplained by major risk factors, *Lancet,* i, 1298, 1986.
73. **Steinberg, D., Parthasarathy, S., Carew, T. E., Khoo, J. C., and Witztum, J. L.,** Beyond cholesterol: modification of low density lipoprotein that increases its atherogenicity, *N. Engl. J. Med.,* 320, 915, 1989.
74. **Kita, T., Nagano, Y., Yokode, M., Ishii, K., Kume, N., Ooshima, A., Yoshida, H., and Kawai, C.,** Probucol prevents the progression of atherosclerosis in Watanabe heritable hyperlipidemic rabbit, an animal model for familial hypercholesterolemia, *Proc. Natl. Acad. Sci. U.S.A.,* 84, 5928, 1987.
75. **Carew, T. E., Schwenke, D. C., and Steinberg, D. A.** Antiatherogenic effect of probucol unrelated to its hypochelesterolemic effect: evidence that antioxidants in vivo inhibit low density lipoprotein degradation in macrophage-rich fatty streaks and slow the progression of atherosclerosis in the Watanabe heritable hyperlipidemic rabbit, *Proc. Natl. Acad. Sci. U.S.A.,* 84, 7725, 1987.
76. **Björkhem, I., Henriksson-Freyschuss, A., Breuer, O., Diczfalusyz, U., Berglund, L., and Henriksson, P.,** The antioxidant butylated hydroxytoluene protects against atherosclerosis, *Arteriosclerosis and Thrombosis,* 11, 15, 1991.
77. **Chait, A., Heinecke, J. W., Hiramatsu, K., Baker, L., and Rosen, H.,** Superoxide-mediated modification of LDL by cells of the arterial wall, *Atherosclerosis,* 7, 491, 1986.
78. **Emanuel, H. A.,** Cholesterol oxidation products in lipoproteins, M.S. thesis, University of Minnesota, 1989.
79. **Quinn, M. T., Pasthasarathy, S., Fong, L. G., and Steinberg, D.,** Oxidatively modified low density lipoprotein: a potential role in recruitment and retention of monocyte/macrophages during atherogenesis, *Proc. Natl. Acad. Sci. U.S.A.,* 84, 2995, 1987.
80. **Gray, M. F. and Lawrie, T. D. V.,** Isolation and identification of cholesterol α-oxide and other minor sterols in human serum, *Lipids,* 6, 836, 1971.
81. **Brooks, C. J. W., McKenna, R. M., Cole, W. J., MacLachlan, J., and Lawrie, T. D. V.,** "Profile" analysis of oxygenated sterols in plasma and serum, *Biochem. Soc. Trans.,* 11, 700, 1983.
82. **Streuli, R. A., Chung, J., Scanu, A. M., and Yachnin, S.,** Serum lipoproteins modulate oxygenated sterol insertion into human red cell membranes, *Science,* 212, 1294, 1981.
83. **Peng, S. K., Taylor, C. B., Haung, W. Y., Hill, J. C., and Mikkelson, B.,** Distribution of 25-hydroxycholesterol in the lipoprotein and its role in atherogenesis. A study in squirrel monkeys, *Atherosclerosis,* 41, 395, 1982.
84. **Peng, S. K., Phillips, G. A., Xia, G. Z., and Morin, R. J.,** Transport of cholesterol autoxidation products by rabbit lipoproteins, *Atherosclerosis,* 64, 1, 1987.

85. **Brooks, C. J. W., Steel, G., Gilbert, J. D., and Harland, W. A.,** Characterization of a new group of polar sterol esters from human atherosclerotic plaques, *Atherosclerosis,* 13, 223, 1971.
86. **Harland, W. A., Gilbert, J. D., Steel, G., and Brooks, C. J. W.,** The occurrence of a new group of polar sterol esters in various stages of human atherosclerosis, *Atherosclerosis,* 13, 239, 1971.
87. **Smith, L. L. and van Lier, J. E.,** Sterol metabolism. IX. 26-Hydroxycholesterol levels in the human aorta, *Atherosclerosis,* 12, 1, 1970.
88. **Fumagalli, R., Galli, G., and Urna, G.,** Cholesterol and 26-hydroxycholesterol in normal and atherosclerotic human aorta, *Life Sci.,* 10, 25, 1971.
89. **Jacobson, M. S., Nair, P. P., Naseem, S. M., Smith, T. D., and Heald, F. P.,** Oxygenated cholesterols: new risk factors for atherosclerosis?, *Pediatr. Res.,* 15, 535, 1981.
90. **Sevanian, A., Peterson, H., and Coates, T.,** Cytotoxicity of cholesterol oxidation products in bovine and rabbit aortic endothelial cells, 8th Int. Symp. Atherosclerosis, Venice, Italy, Oct. 7-8, 1988, 103.

Chapter 10

CHOLESTEROL OXIDATION AND CANCER

Robert J. Morin, Bing Hu, Shi-Kaung Peng, and Alex Sevanian

TABLE OF CONTENTS

I. HISTORICAL — EXPERIMENTAL CARCINOGENESIS

The earliest experimental evidence for the possible carcinogenicity of cholesterol oxidation products originated from studies by Bischoff and Rupp[1] in which ovariectomized mice were injected subcutaneously with either crude or purified preparations of 10 mg/mouse of progesterone in sesame oil. The incidence of cancer was found to be 32% in mice receiving the crude progesterone vs. 0% for those receiving purified progesterone. Analysis of the crude progesterone revealed that each mouse had received 20 mg of cholesterol oxidation products. These studies led to a series of further experiments in which individual cholesterol oxidation products were administered to mice and rats to determine their possible contributory effects to carcinogenicity.

When 4-cholestene-3,6-dione was injected subcutaneously in sesame oil to male Marsh mice at 20 mg/mouse, this produced a 34% incidence of local sarcomas.[2] In castrated female mice, however, the same treatment resulted in only a 3% incidence of local sarcomas.[2] Other studies by Hieger[3] reported no induction of sarcomas in mice using this compound. These studies were all complicated by use of a natural oil as a vehicle and uncertain purity of the cholesterol oxide used.

A 30% incidence of tumors in female mice was reported by Koch et al.[4] after injection of 5α-hydroperoxy-6-cholestene-3β-ol in olive oil, significantly higher than with vehicle alone. Male mice appeared to be resistant to this compound. The authors suggested that this oxide may have acted as a co-carcinogen in the female mice, augmenting the effects of endogenous carcinogens.

Using 6β-hydroxy-4-cholestene-3-one injected subcutaneously in sesame oil into male and castrated female Marsh mice, Bischoff reported a 15 to 19% incidence of local sarcomas.[2] Other investigators reported an incidence of 5 to 20% local sarcomas in three other strains of mice using similar experimental procedures with an oily vehicle.[5] When Evans rats were injected subcutaneously with an aqueous suspension of 6β-hydroxy-4-cholestene-3-one, no local cancers were produced.[5]

The oxide 6β-hydroperoxy-4-cholestene-3-one has been more extensively studied than the above compounds. Castrated male Marsh mice showed a 9 to 59% incidence of local sarcomas when the oxide was injected in either sesame oil or triolein.[2,6] Some of these experiments were complicated by a high mouse early mortality rate. This oxide was also carcinogenic in C57 and BALB mice[6,7] and in Evans female rats[8] when injected subcutaneously in oils containing triolein. When injected as an aqueous emulsion in Evans female rats there was no significant increase in local sarcomas as compared to controls.[5]

Cholesterol 5α,6α-epoxide (cholesterol α-oxide) is the only cholesterol oxidation product that has induced significant carcinogenesis without an oily vehicle. When aqueous suspensions of this oxide in saline, albumin, or serum were injected into female Marsh mice, there were 18 to 25% incidences of local sarcomas vs. 0% in controls.[5] Intratesticular injections also produced local sarcomas in Marsh mice but not in Evans rats.[8]

Other cholesterol oxides tested were 5-cholestene-3-one and 3,5-cholestadiene, neither of which were proven to have carcinogenic effects.[5]

II. ROLE OF CHOLESTEROL OXIDATION PRODUCTS IN SITE-SPECIFIC CARCINOGENESIS

A. ULTRAVIOLET RADIATION-INDUCED SKIN CARCINOGENESIS

There is abundant evidence that UV exposure is an etiologic factor in human skin cancers.[9] In 1933 Roffo suggested that production of cholesterol oxidation products by UV exposure of the skin might be the carcinogenic mechanism.[10] When skin from Albino hairless mice was incubated with ^{14}C-cholesterol and then exposed to UV radiation from a Mercury arc lamp for 15, 30, or 60 min, production of cholesterol α-epoxide could be demonstrated by gas chromatographic analysis.[11] The amounts of this compound produced were proportional to the UV exposure time. Cholesterol α-epoxide had also previously been demonstrated in human skin after UV radiation.[12,13] In a further study, human skin biopsy specimens were labeled *in vitro* with ^{14}C-cholesterol and then irradiated under a Mercury arc lamp.[14] Lipids were extracted and separated by silicic acid thin-layer and column chromatography, and identified by gas chromatography, revealing the production of ^{14}C-labeled cholestane-3β,5α,6β-triol, 7α- and 7β-hydroxycholesterol, 7-ketocholesterol, and cholesterol 5α,6α-epoxide. The rate of formation of these compounds was proportional to the time of UV exposure. In further experiments to determine the significance of these effects and their possible prevention, the diets of female hairless mice were supplemented with the antioxidant nutrients ascorbate, DLα-tocopherol, and glutathione, plus the synthetic antioxidant butylated hydroxytoluene (BHT).[15] After two weeks, the total water-soluble antioxidant content of the skin had increased by 54%, at which time the *in vitro* skin production of UV-induced cholesterol α-oxide from ^{14}C-labeled cholesterol was inhibited by more than 90%. At 4 to 12 weeks, skin antioxidant content remained fairly constant at 18% above control levels, and cholesterol α-epoxide production remained approximately 50% inhibited. To determine the association of this inhibitory effect to skin cancer, in another series of experiments female Albino hairless mice were irradiated with UV light.[16] The diet of one group was supplemented with the antioxidants ascorbate, DLα-tocopherol, glutathione, and BHT. At the end of 22 weeks, 30% of mice on the control diet had developed squamous cell carcinomas of the skin, whereas only 7% of mice on the antioxidant-supplemented diet had developed these skin cancers. To further explore the relationship between cholesterol α-oxide and skin cancer, the time course of biochemical and carcinogenic changes were studied.[17] Skin cholesterol α-epoxide levels began to rise 4 weeks after UV radiation of hairless mice, and reached peak levels at 10 weeks. Skin tumors began to appear at 15 weeks, and were present in greater than 90% of animals at 24 weeks (Figure 1). The enzyme cholesterol α-epoxide hydrase was identified in the skin, and was demonstrated to convert the epoxides to less reactive dihydrodiols. The activity of cholesterol α-epoxide hydrase was noted to gradually increase 8 weeks after UV radiation, and reached maximum levels at 15 weeks, after which it remained higher than controls through 20 weeks. The close association between the peak levels of cholesterol α-epoxide levels and the increased hydrase activity immediately following this suggested that this enzyme was substrate inducible. The increased hydrase activity explains why the epoxide levels decreased after 15 weeks. Apparently tumor induction had already occurred at this time, and detoxification of the carcinogen did not decrease tumor incidence.

Mechanisms for the possible carcinogenic effects of cholesterol α-epoxide in human skin fibroblasts in culture were explored in a later study by Parsons and Goss,[18] in which nuclear changes induced by UV radiation alone were compared to those induced by cholesterol α-epoxide alone. Cholesterol α-epoxide was found to induce similar degrees of chromosome damage and similar stimulation of DNA repair synthesis as did low doses of UV radiation. Simultaneous treatment with both UV and cholesterol α-epoxide produced additive chromosome-damaging effects but no additive effects on DNA repair synthesis. Pure cholesterol had no effect on these processes. UV exposure was found to produce single-strand DNA breaks, whereas cholesterol α-epoxide did not. DNA damage, as detected by inhibition of thymidine incorporation, also resulted from UV exposure only (in other tissues the epoxide does decrease thymidine incorporation). Although these experiments demonstrated the above carcinogen-like effects, the fibroblasts were not transformed to neoplastic cells even with combined UV plus cholesterol α-epoxide treatment. This is not unexpected, as there is no demonstrated correlation between repair, cytotoxicity, and transformation.

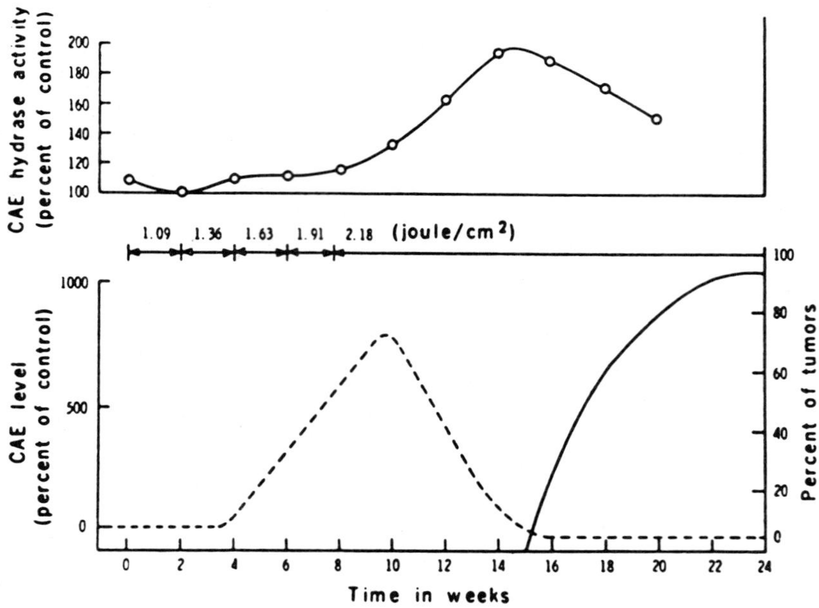

FIGURE 1. Relation between cholesterol α-epoxide (CAE) hydrase activity and CAE levels to onset of tumors induced by UV light. The upper curve illustrates CAE hydrase activity in skin of mice irradiated with UV light as a percentage of nonirradiated controls. Enzyme activity was calculated from the conversion rate of CAE to triol per milligram of protein. Each point on the curve represents the mean value from triplicate incubations. The standard deviation of the percentage of conversion within each triplicate set of incubations did not exceed ± 0.63. Average control activity was 0.16 nmol of CAE per mg. of protein per 2 h. In the lower curves, the dashed line represents levels of CAE in the skin of irradiated mice as a percentage of controls; the solid line and right ordinate represent the percentage of animals bearing tumors. The middle abscissa shows the regimen of UV radiation. Exposure levels were increased at 2-week intervals to compensate for epidermal thickening. (Reprinted from Chan, J. T. and Black, H. S., *Science,* 186, 1216, 1974. Copyright, American Association for the Advancement of Science, 1974. With permission.)

B. CHOLESTEROL OXIDES AND COLON CANCER

Epidemiologic data indicate that those populations ingesting high fat-containing diets, such as the U.S. and Sweden, have a higher incidence of colon cancer than areas such as Japan and Puerto Rico, where diets are lower in fat.[19,20] In a large Swedish population, higher levels of serum cholesterol and β lipoproteins were associated with higher rates of colon cancer.[21] High-fat diets have been demonstrated to lower intestinal redox potential and to increase fecal excretion of cholesterol and bile acid metabolites.[22] Patients with adenomatous polyps of the colon (a precancerous lesion) or those with colon cancer were found to excrete higher levels of cholesterol metabolites and bile acids than normal American or Japanese controls or patients with other types of digestive disorders (Table 1).[23] Among the fecal neutral sterols that were increased were cholestane-3β,5α,6β-triol (cholestanetriol) and coprostanone (which is produced by bacterial oxdiation of the 3β-hydroxyl group of cholesterol to a ketone). Patients with chronic ulcerative colitis, which is also associated with an increased risk of colon cancer, were found to have increased fecal excretion of cholestanetriol, cholesterol, coprostanol, and coprostanone[24] (Table 1). In the three primary groups at high risk for colon cancer, therefore, those with adenomatous polyps showed both increased fecal cholesterol metabolites and bile acid metabolites; those with familial polyposis showed increased cholesterol excretion only, but no increase in bile acid metabolites; and those with ulcerative colitis showed increases in all fecal neutral sterols, but not bile acids.

TABLE 1
Fecal Neutral Sterols in Patients With Colon Cancer, Adenomatous Polyps, Ulcerative Colitis, Other Digestive Diseases, and Healthy Controls

	Colon cancer	Adenomatous polyps	Ulcerative colitis	Other digestive diseases	Controls
Cholesterol	12.6 ± 1.5[a]	6.4 ± 0.8[b]	12.5 ± 2.9[a]	3.3 ± 0.5[c]	3.2 ± 0.4[c]
Coprostanol	18.7 ± 1.8[a]	19.6 ± 3.2[a]	26.6 ± 7.0[a]	11.9 ± 1.5[b]	12.9 ± 0.8[b]
Coprostanone	3.9 ± 0.4[a]	4.0 ± 1.0[a]	2.0 ± 0.5[b]	1.8 ± 0.2[b]	1.9 ± 0.2[b]
Cholestane-3β,5α,6β-triol	0.2 ± 0.02[a]	0.2 ± 0.02[a]	0.3 ± 0.02[b]	0.05 ± 0.01[c]	0.06 ± 0.01[c]
Total neutral sterols	35.4 ± 1.7[a]	30.2 ± 3.1[a]	41.4 ± 7.0[a]	17.1 ± 1.5[b]	18.1 ± 0.9[b]
β-Sitosterol	2.5 ± 0.4[a]	2.3 ± 0.3[a]	2.3 ± 0.5[a]	2.5 ± 0.5[a]	2.5 ± 0.3[a]
Campesterol	4.3 ± 0.4[a]	5.0 ± 1.5[a]	5.7 ± 0.8[a]	3.9 ± 1.4[a]	4.6 ± 0.9[a]

Note: Mean mg/g dry feces ± standard errors of means. Those values for each compound not sharing a common superscript letter are significantly different ($p < 0.05$).

Reprinted from Reddy, S., Martin, C. W., and Wynder, E. L., *Cancer Res.*, 37, 1697, 1977; Sloga, T. J., Sinak, A., and Boatwell, R. K., Eds., *Carcinogenesis*, Vol. 2, *Mechanisms of Tumor Promotion and Carcinogenesis,* Raven Press, New York, 1978. With permission.

The common factor, therefore, appears to be cholesterol metabolites. It is possible that the increased cholesterol that accompanies increased dietary fat is converted to cholesterol epoxide or further carcinogenic metabolites by intestinal colonic enzymes or by intestinal bacteria. To test this hypothesis, Reddy and Watanabe[25] investigated the possible promoting effects of cholesterol epoxide and cholestanetriol as well as cholesterol and lithocholic acid on *N*-methyl-*N*′-nitro-*N*-nitrosoguanidine (MNNG)-induced colon cancer in standard and germ-free female rats. No colonic tumors were induced by any of the sterols or by lithocholic acid in the absence of MNNG. The incidence of colon adenocarcinomas was no greater in rats receiving both MNNG and cholesterol epoxide, cholestanetriol, or cholesterol as compared to MNNG alone in either the standard or germ-free rats. Lithocholic acid was the only agent that increased the incidence of adenocarcinomas when administered with MNNG in both models of rats.

The role of cholesterol oxidation products in colon cancer appears to be as yet unresolved, and requires further studies with other experimental models.

C. CHOLESTEROL OXIDES AND BREAST CANCER

The incidence of breast cancer is also higher in those populations ingesting high-fat diets, although there is no apparent correlation between serum cholesterol levels and breast cancer.[26] Analysis of breast fluid aspirates by Petrakis et al.[28] revealed that 45% of women had detectable amounts of cholesterol α- and β-epoxides. The ratio of α- and β-epoxide isomers was later determined to be 1:3. Total cholesterol epoxide levels ranged from 0 to 31.5 mg/dl breast fluid. High levels of cholesterol epoxides were correlated with high levels of both free and total cholesterol. In a subsequent study by Gruenke et al.[27] the levels of cholesterol β-epoxide were found to be higher in nulliparous women than in those who were lactating, parous, or had previously breast fed. The lower levels in parous women remained low for 2 or more years following parturition or lactation. The authors postulated that the ductal epithelium of these women may have had a lesser cumulative exposure to the possible carcinogenic effects of cholesterol epoxide, and that this may explain the previously observed lower risk of breast cancer in women who have had a full-term pregnancy. No correlation was noted between either breast fluid cholesterol or cholesterol β-epoxide levels and serum cholesterol, which may explain the lack of correlation between breast cancer incidence and serum cholesterol level. In a more recent study, Wrensch et al.[29] found that cholesterol and

TABLE 2
Breast Fluid Cholesterol β-Epoxide Levels According to Breast Disease Status

	N	Median μg/dl	% Of women with undetectable levels
Controls	125	447	28
Nonproliferative	14	483	21
Proliferative without atypia	24	780	8
Proliferative with atypia	15	506	7
Fibroadenoma	8	654	13

Reprinted from Wrench, M. R. et al., *Cancer Res.*, 49, 2168, 1989. With permission.

cholesterol β-epoxide levels increased with age, were higher in white than in non-white women, and were higher in smokers. Women with benign breast disease, whether simple hyperplasia or associated with atypia, had higher breast fluid concentrations of both cholesterol and cholesterol β-epoxide. Epithelial hyperplasia and hyperplasia with atypia have been associated with 1.5 to 7× increased risk of breast cancer. Progressively increasing concentrations of cholesterol and cholesterol β-epoxides were observed in women with normal epithelium, hyperplasia, and hyperplasia with atypia, respectively. The incidence of all types of benign breast cancer was 8.5 times more common in women who had detectable levels of cholesterol β-epoxide in their breast fluid as compared to those with no detectable amounts (Table 2). The authors considered it unlikely that the increased breast fluid cholesterol and epoxides were the result rather than the cause of the proliferative disease, because a small proportion of women with normal epithelium had concentrations of these compounds in the same range as those with proliferative disease. The main difference between normals and those with benign breast disease was that detectable levels of cholesterol epoxides were present in a much higher proportion of the latter group. As well as cholesterol β-epoxide, the authors indicated that similar associations as described above were found for breast fluid cholesterol α-epoxide and cholestanetriol. The ratio of β/α epoxide concentrations of 3:1 suggests the possibility that these epoxides may be an indicator of general lipid peroxidative processes occurring in the breast, which result in the injurious and possibly carcinogenic effects on the ductal epithelium.

D. PROSTATIC HYPERTROPHY AND CANCER

Analysis of prostatic secretions from 25 normal subjects by Sporer et al.[30] revealed a substantial cholesterol content of 893 ± 133 μg/ml and the presence of cholesterol epoxides in 1/3 of all prostatic secretions. Mean cholesterol epoxide concentration was 22.6 ± 3.3 μg/ml. In 29 surgical patients requiring primary prostactectomy for benign prostatic hypertrophy, total cholesterol epoxide concentration was significantly higher at 33.7 ± 7.9 μg/ml. Analysis of extracts of prostate lobes removed at surgery revealed a cholesterol content double that of normal prostate tissue. The concentration of cholesterol α-epoxide was 120 ± 23.4 μg/g dry tissue, and for β-epoxide was 9.9 ± 5. No differences in epoxide concentrations were noted between benign prostatic tissue lobes and lobes containing prostate cancer. It had previously been found that prostate adenomas and benign prostatic hypertrophy tissue contain more cholesterol in the glandular epithelial acinus than does normal prostatic tissue,[31] and it is possible that peroxidation of cholesterol in this area may be associated with subsequent development of prostatic carcinoma.

III. POSSIBLE MECHANISMS OF CARCINOGENESIS BY CHOLESTEROL OXIDATION PRODUCTS

A. MUTAGENIC EFFECTS

Smith et al.[32] found that air-aged samples of reagent-grade cholesterol had mutagenic effects when tested with 5 strains of *Salmonella typhimurium* using the Ames method. Although pure cholesterol was nonmutagenic, exposure of pure cholesterol to heating in air at 70°C or to ^{60}Co γ-radiation resulted in production of mutagenic compounds. Testing of cholesterol 5α,6α-epoxide and cholesterol 5β,6β-epoxide failed to produce mutagenic effects. A previous study had also demonstrated negative results with the α-epoxide.[33] A number of other individual cholesterol oxidation products were also found to be nonmutagenic to *S. typhimurium*.[31,32]

Using a different assay system involving induction of 8-azaguanine-resistant mutants in V79 Chinese lung fibroblasts in culture, Sevanian and Peterson found that cholesterol α-epoxide had significant, although relatively weak mutagenic effects.[34] The mutation rate with the α-epoxide was 4.6 to 11.8 × higher than the spontaneous mutation rate. Cholesterol β-epoxide was reported to be more mutagenic than the α isomer.[35] Cholestanetriol was not mutagenic except at very high concentrations, although this compound was more toxic and suppressed DNA synthesis more than the epoxide. Since the isomeric cholesterol epoxides are converted to cholestanetriol by epoxide hydrolase, it was postulated that a low level of activity of this enzyme might predispose to mutagenicity.

The mutagenic effects of cholesterol α-epoxide and β-epoxide were linearly related to dosage in the range of 0 to 12.4 μ*M*, above which a plateau was observed. More recently the mutagenic properties of cholesterol β-epoxide were examined in V79 cells.[36] Significant dose-dependent increases in mutant frequencies were found over the same dose ranges as the α-epoxide, although the β-epoxide was approximately three times more potent than the α-epoxide. It is possible that at the higher treatment concentrations (>12.4 μ*M*) the epoxide hydrolase activity generated toxic levels of cholestanetriol from both epoxide isomers, thus limiting the capacity of the cells to mutate. This is further supported by studies which showed that the metabolism of the isomeric cholesterol epoxides by epoxide hydrolase converted approximately 5 to 15% of the epoxides, administered at their optimal mutagenic concentrations, to cholestane triol over a 24-hour interval. The limited mutagenic dose ranges for the epoxides, as well as differential toxicities, corresponded to their relative rates of hydration, where the β-epoxide was metabolized approximately twice as rapidly as the α-epoxide to cholestane triol.[36] Another cholesterol oxide, 7-ketocholesterol, was found not to be significantly mutagenic to the V79 cells.[35]

B. TRANSFORMATION AND DNA EFFECTS

An *in vitro* carcinogenesis assay system based upon morphological transformation rates of Syrian Golden hamster embryo cells was employed by Kelsey and Pienta[37,38] to study possible transforming effects of sterols. In this system, cholesterol α-epoxide consistently resulted in transformations when added to the cultures over the concentration range of 54 to 2,000 nmol. Cholesterol was inactive as a transforming agent. In a subsequent study, cholestanetriol was also found to be mutagenic in this system, at only a slightly lower efficiency than cholesterol α-epoxide. Using mouse C3H 10T1/2 cells, Raaphorst et al.[39] found that both cholesterol α- and β-epoxides induced significant rates of morphological transformation. Transformation was greater with the β-epoxide than with the α isomer. The magnitudes of transformation were directly proportional to increasing dosage of the oxides and increasing exposure time. At the highest concentration and longest exposure time to the α- and β-epoxides, the level of transformation was equivalent to that produced by 3.5 Gy of X-radiation. Since the β isomer is the predominant form of epoxide produced *in vivo* by nonenzymatic processes such as autoxidation and hydroperoxidation of cholesterol, and the β isomer was more potent in this system, the authors postulated that this could be associated with significant biological damage.

Cholesterol epoxides have been found to have a high affinity for DNA, and can form strong complexes *in vitro* with calf thymus DNA.[40] The binding of α- and β-epoxides to either calf thymus DNA, or DNA isolated from V79 cells following treatments in culture, was highly resistant to organic solvent extraction. However, the epoxides appeared to be either noncovalently bound or formed unstable adducts with a high affinity for DNA.[36] The difference in mutagenic potency for the isomeric epoxides contrasted with the lack of difference in DNA binding or the stability of the adducts, suggesting that the basis for mutagenesis may not simply be due to covalent binding. This may in part account for their significant mutagenic and transforming effects. In lymphocyte cultures, cholesterol β-epoxide has been observed to induce sister chromatid exchange,[41] which is another indication of possible DNA alteration induced by these oxides. The evidence for direct DNA effects must, nevertheless, be taken with some caution. The endpoints of unscheduled DNA synthesis, chromosomal abberations, expression of 6-thioguanine-resistant phenotype, and induction of morphological transformation, are indirect measures of potential DNA damage. These measurements may be misinterpreted in the light of membrane-damaging effects that many of the cholesterol oxidation products elicit. Therefore, (1) alterations in uptake and incorporation of precursors in the measurement of unscheduled DNA synthesis; (2) release in dying cells of nonspecific nucleases which break chromosomes at cytotoxic concentrations; and (3) suppression of the HGPRT gene, thus conferring the 6-thioguanine phenotype, are all sources of misinterpretation for direct DNA effects which can arise from "epigenetic phenomena". The induction of *in vitro* transformation at frequencies much higher than the mutation frequency in the same cells suggests that an epigenetic mechanism could be involved in the transformation phenomenon.[42] Inhibition of metabolic cooperation by cholesterol epoxides has been observed at doses considerably lower than the mutagenic dose,[43] which suggests a membrane-associated mechanism for *in vitro* transformation. Since inhibition of intercellular cooperation is a property exhibited by many tumor promoters,[44] the transforming capability of cholesterol epoxides may reside in the capacity to clonally amplify initiated cells.

C. IMMUNOSUPPRESSIVE EFFECTS

Almost total inhibition of plaque-forming cell (PFC) response was induced by 5×10^{-7} *M* 25-hydroxycholesterol added in phospholipid liposomes during the first 24 h of mouse spleen cell cultures.[45] DNA synthesis, as measured by ^{3}H-thymidine uptake after mitogen stimulation of human peripheral blood lymphocytes *in vitro,* was inhibited by 2 μg/ml 25-hydroxycholesterol[46] and in another study by 3.9×10^{-6} *M* 25-hydroxycholesterol and by 7.4×10^{-6} *M* of 20α-hydroxycholesterol.[47] In the latter study, the effect was apparent only after 40 h in culture and could be prevented by removal of the cholesterol oxides from the media within 18 h. The generation of tumor-specific cytotoxic T-lymphocytes (CTL) has been found by Spangrunde et al. to be inhibited by some cholesterol oxides.[48] Addition of 7α- and 7β-hydroxycholesterol to cultures of HC57BL16 spleen cells resulted in a moderate, dosage-dependent inhibition of CTL generation. The most potent sterols were 20α-hydroxycholesterol and 25-hydroxycholesterol, which totally inhibited CTL generation at concentrations below 2.5 μ*M*. Cholesterol α-epoxide did not show significant inhibitory effects. The inhibitory effects appeared to occur during the differentiation processes leading to CTL generation, and not directly upon the function of the CTLs.

Using mixed lymphocyte cultures of B6 and C3H spleen cells, it was determined that 20α-hydroxycholesterol and 25-hydroxycholesterol also had the highest inhibitory effects on cell proliferation. 7α- and 7β-hydroxycholesterol were moderately inhibitory, whereas cholesterol α-epoxide had no effect.

In another model system, the effects of the cholesterol oxides on CTL generation in draining popliteal lymph nodes (DLN) was studied after immunization of mice with syngenic UV-induced 58a tumor cells. Subsequent CTL activity assay revealed inhibitory effects of 7α- and 7β-hydroxycholesterol at 2.5-μ*M* concentrations, but at 0.5-μ*M* concentrations

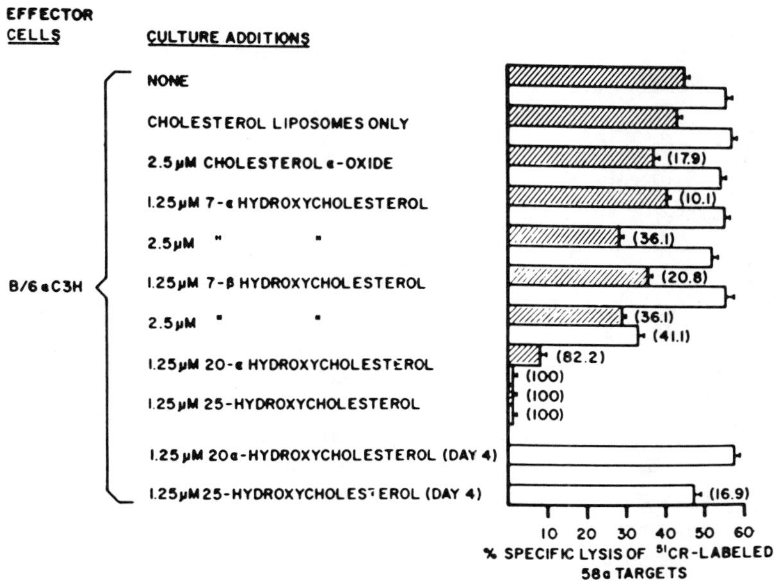

FIGURE 2. Inhibitory effects of liposomes containing various oxygenated sterols on the *in vitro* generation of allogeneic cytotoxic lymphocytes. Liposomes were added to MLC cultures at the onset of the 5-d incubation period to the indicated final concentrations. Cultures were harvested and assayed for CTL activity in a 6-h ^{51}Cr release assay using 58a, a UV-induced C3H tumor, as the target. Effector to target (E:T) ratio was 25:1 ▨, experiment 1: □, experiment 2. 25-Hydroxycholesterol (1.25 μ*M*)-containing liposomes were also added to some of the cultures in experiment 2 on day 4. This was done to establish whether oxysterols at concentrations known to inhibit CTL differentiation were capable of exerting a direct effect on CTL function (percentage of inhibition). (Reprinted from Spangrude, G. J., Sherris, D., and Daynes, R. A., *Transplantation,* 33, 482, 1982. © by Williams & Wilkins, 1982. With permission.)

only 20α-hydroxycholesterol and 25-hydroxycholesterol were inhibitory (Figure 2). Cholesterol α-epoxide again showed no inhibitory effects.

When 25-hydroxycholesterol was added to the syngenic tumor-stimulated lymphocyte cultures on day 0 and the CTL activity assayed on day 4, a marked inhibition of activity was found. Addition of the sterol on day 2 had no effect on activity assayed on day 4. Addition of the 25-hydroxycholesterol in the form of lecithin-cholesterol liposomes produced 10 times greater inhibition than addition of the sterol in its soluble form.

In further experiments, lymphocyte proliferation and antitumor CTL activity were simultaneously measured in the antigen-sensitized DNL cells. 25-hydroxycholesterol at 2.5 μ*M* completely inhibited both processes, whereas 7α- and 7β-hydroxycholesterol at that concentration inhibited CTL generation without influencing cell proliferation.

The process of transformation of DLN cells to functional CTL cells after antigenic stimulation takes 4 to 5 days. Addition of 7α- or 7β-hydroxycholesterol on day 4 had little effect on CTL activity assayed on day 5. A similar sequence for 25-hydroxycholesterol produced only a small increase in inhibition on day 5. These results suggested that the differentiation period is when the lymphocytes are most sensitive to oxysterol-mediated inhibition of CTL activity.

The most potent inhibitors of both CTL activity and lymphocyte proliferation, 25-hydroxycholesterol and 20α-hydroxycholesterol, are also the strongest inhibitors of endogenous cholesterol biosynthesis.[48] It has also been demonstrated that normal activity of hydroxymethyl glutaryl coenzyme A (HMG-CoA) reductase, the rate-limiting enzyme in cholesterol biosynthesis, is required for normal cell proliferation.[49] Although inhibition of

PTC response in cultured mouse spleen cells could not be reversed by addition of mevalonate,[45] inhibition of DNA synthesis in mitogen-stimulated lymphocytes by 25-hydroxycholesterol could be reversed by addition of cholesterol or mevalonate,[46] and inhibition by either 25-hydroxycholesterol or 20α-hydroxycholesterol could also be reversed by mevalonate.[47] The growth inhibition of 3T6 and SV-3T3 tumor derivatives of fibroblasts in culture by 25-hydroxycholesterol was attributed to inhibition of HMG-CoA reductase activity.[48] The mechanism of the latter effect was elucidated by Boissonneault and Heiniger who observed a 10-fold increase in calcium uptake by P815 mastocytoma cells after incubation with 25-hydroxycholesterol.[50] This resulted in a decrease in DNA synthesis to a level at which DNA replication was not possible, with subsequent cell death. It has previously been shown that very specific cell calcium shifts are required for normal progression of the cell cycle,[51-54] and that large sustained surges in calcium concentration result in inhibition of DNA synthesis.[55]

Plasma membrane cholesterol reduction in the lymphocyte cultures, therefore, is probably responsible for the reduced cell proliferation and inhibition of lymphocyte function. Induction of CTLs specific for tumor antigens is an important mechanism in the rejection of tumors, and this could be impaired *in vivo* by cholesterol oxides. Exposure of animals to whole body UV radiation for 6 days resulted in a systemic defect in splenic antigen-presenting cell functions.[47] This is probably related to the epidermal inflammatory response, with oxidized cholesterol products binding to circulating lymphocytes, which then migrate to other lymphoid organs.

IV. SUMMARY

Earlier studies demonstrated that a number of cholesterol oxidation products induced sarcomas and other tumors when injected subcutaneously to animals in an oil vehicle, but only cholesterol α-epoxide was carcinogenic in aqueous suspension. Subsequently, experimental evidence was obtained for a role of cholesterol α-epoxide in skin cancer. Exposure of human and mouse skin fibroblasts to UV radiation induced the formation of significant levels of cholesterol epoxide. These levels could be reduced, and squamous cell carcinomas in mice prevented, by supplementation of the diets with autoxidant nutrients prior to UV exposure. Colon cancer is also etiologically linked to cholesterol oxidation products. Higher than normal levels of cholestanetriol have been found in patients with colon cancer and also in those with precancerous disorders such as adenomatous polyps and ulcerative colitis. Higher than normal levels of cholesterol α-epoxide have been found in breast fluid aspirates of women with benign breast disease, with or without atypical hyperplasia of the epithelium, and this may be a factor in the increased incidence of breast cancer associated with hyperplasia. Similarly, the observed increased levels of cholesterol α- and β-epoxides in prostatic fluid of men with benign prostatic hypertrophy may be associated with subsequent development of prostate cancer.

Cholesterol α-epoxide has been found to be mutagenic to fibroblasts in culture and to induce morphological transformation in hamster embryo cells and in mouse C3H cells. Cholesterol β-epoxide seems to possess greater *in vitro* mutagenicity and transforming potency than the isomer. Other cholesterol oxides, such as 25-hydroxycholesterol and 20α-hydroxycholesterol are potent suppressors of generation and proliferation of tumor-specific cytotoxic T-lymphocytes.

Although investigations into the role of cholesterol oxidation products in cancer are still in the early stages, evidence to date indicates a potentially significant role in the induction of some types of cancer.

REFERENCES

1. **Bischoff, F. and Rupp, J. J.,** The production of a carcinogenic agent in the degradation of cholesterol to progesterone, *Cancer Res.,* 6, 403, 1946.
2. **Bischoff, F.,** Cocarcinogenic activity of cholesterol oxidation products and sesame oil, *J. Natl. Cancer Inst.,* 19, 977, 1957.
3. **Hieger, I.,** Carcinogenesis by cholesterol, *Br. J. Cancer,* 13, 439, 1959.
4. **Koch, R., Schenck, G. O., and Neumuller, O.-A.,** Moglichkeiten des Eingreifens sensibilisierender und desensibilisierender zusate in der strahlenchemie und-biologie, *Strahlentherapie,* 124, 626, 1964.
5. **Bischoff, F.,** Carcinogenic effects of steroids, *Adv. Lipid Res.,* 7, 165, 1969.
6. **Bryson, G. and Bischoff, F.,** Triolein as a cocarcinogen, *Fed. Proc. Fed. Am. Soc. Exp. Biol.,* 23, 106, 1964.
7. **Hieger, I.,** Cholesterol as carcinogen, *Br. J. Cancer,* 16, 716, 1962.
8. **Bischoff, F.,** Carcinogenesis through cholesterol and derivatives, *Progr. Exp. Tumor Res.,* 3, 412, 1963.
9. **Urbach, F.,** Ultraviolet radiation: interaction with biological molecules, in *Cancer, a Comprehensive Treatise,* Becker, F. F., Ed., Plenum Press, New York, 1975, 441.
10. **Roffo, A. H.,** Heliotropism of cholesterol in relation to skin cancer, *Am. J. Cancer,* 17, 42, 1933.
11. **Black, H. S. and Douglas, D. R.,** A model system for the evaluation of the role of cholesterol α-oxide in ultraviolet carcinogenesis, *Cancer Res.,* 32, 2630, 1972.
12. **Black, H. S. and Lo, W.-B.,** Formation of a carcinogen in human skin irradiated with ultraviolet light, *Nature,* 234, 306, 1971.
13. **Black, H. S. and Douglas, D. R.,** Formation of a carcinogen of natural origin in the etiology of ultraviolet light-induced carcinogenesis, *Cancer Res.,* 33, 2094, 1973.
14. **Lo, W. B. and Black, H. S.,** Formation of cholesterol-derived photoproducts in human skin, *J. Invest. Dermatol.,* 58, 278, 1972.
15. **Lo, W. B. and Black, S. B.,** Inhibition of carcinogen formation in skin irradiated with ultraviolet light, *Nature,* 246, 489, 1973.
16. **Black, H. S. and Chan, J. T.,** suppression of ultraviolet light-induced tumor formation by dietary antioxidants, *J. Invest. Dermatol.,* 65, 412, 1975.
17. **Chan, J. T. and Black, H. S.,** Skin carcinogenesis: cholesterol-5α,6α-epoxide hydrase activity in mouse skin irradiated with ultraviolet light, *Science,* 186, 1216, 1974.
18. **Parsons, P. G. and Goss, P.,** Chromosome damage and DNA repair induced in human fibroblasts by UV and cholesterol oxide, *Aust. J. Exp. Biol. Med. Sci.,* 56, 287, 1978.
19. **Wynder, E. L.,** The epidemiology of large bowel cancer, *Cancer Res.,* 35, 3388, 1975.
20. **Wynder, E. L., Kajitani, T., Ishikawa, S., Dodo, H., and Takano, A.,** Environmental factors of cancer of the colon and rectum, *Cancer,* 23, 1210, 1969.
21. **Tornberg, S. A., Holm, L.-E., Carstensen, J. M., and Eklund, G. A.,** Risks of cancer of the colon and rectum in relation to serum cholesterol and beta-lipoprotein, *N. Engl. J. Med.,* 315, 1629, 1986.
22. **Hill, M. J.,** Colon cancer: a disease of fibre depletion or of dietary excess?, *Digestion,* 11, 289, 1974.
23. **Reddy, B. S. and Wynder, E. L.,** Metabolic epidemiology of colon cancer, *Cancer,* 39, 2533, 1977.
24. **Reddy, S., Martin, C. W., and Wynder, E. L.,** Fecal bile acids and cholesterol metabolites of patients with ulcerative colitis, a high-risk group for development of colon cancer, *Cancer Res.,* 37, 1697, 1977.
25. **Reddy, S. and Watanabe, K.,** Effect of cholesterol metabolites and promoting effect of lithocholic acid in colon carcinogenesis in germ-free and conventional F344 Rats, *Cancer Res.,* 39, 1521, 1979.
26. **Hiatt, R. A., Friedman, G. D., Bawol, R. D., and Ury, H. K.,** Breast cancer and serum cholesterol, *J. Natl. Cancer Inst.,* 68, 885, 1982.
27. **Gruenke, L. D., Wrensch, M. R., Petrakis, N. L., Miike, R., Ernster, V. L., and Craig, J. C.,** Breast fluid cholesterol and cholesterol epoxides: relationship to breast cancer risk factors and other characteristics, *Cancer Res.,* 47, 5483, 1987.
28. **Petrakis, N., Gruenke, L. D., and Craig, J. C.,** Cholesterol and cholesterol epoxides in nipple aspirates of human breast fluid, *Cancer Res.,* 41, 2563, 1981.
29. **Wrensch, M. R., Petrakis, N. L., Gruenke, L. D., Miikke, R., Ernster, V. L., King, E. B., Hauck, W. W., Craig, J. C., and Goodson, W. H.,** III, Breast fluid cholesterol and cholesterol β-epoxide concentrations in women with benign breast disease, *Cancer Res.,* 49, 2168, 1989.
30. **Sporer, A., Brill, D. R., and Schaffner, C. P.,** Epoxycholesterols in secretions and tissues of normal benign, and cancerous human prostate glands, *Urology,* 20, 3, 1982.
31. **Swyer, G. I. M.,** The cholesterol content of normal and enlarged prostate, *Cancer Res.,* 2, 372, 1942.
32. **Smith, L. L., Smart, V. B., and Ansari, G. A. S.,** Mutagenic cholesterol preparations, *Mutation Res.,* 68, 23, 1979.
33. **Kadis, B.,** Steroid epoxides in biological systems: a review, *J. Steroid Biochem.,* 9, 75, 1978.
34. **Sevanian, A. and Peterson, A. R.,** Cholesterol epoxide is a direct-acting mutagen, *Proc. Natl. Acad. Sci. U.S.A.,* 81, 4198, 1984.

35. **Sevanian, A. and Peterson, A. R.,** The cytotoxic and mutagenic properties of cholesterol oxidation products, *Food Chem. Toxicol.,* 24, 1103, 1986.
36. **Peterson, A. R., Peterson, H., Spears, C. P., Trosko, J. E., and Sevanian, A.,** Mutagenic characterization of cholesterol epoxides in Chinese hamster V79 cells, *Mutat. Res.,* 203, 355, 1988.
37. **Kelsey, M. L. and Pienta, R. J.,** Transformation of hamster embryo cells by cholesterol-α-epoxide and lithocholic acid, *Cancer Lett.,* 6, 143, 1979.
38. **Kelsey, M. L. and Pienta, R. J.,** Transformation of hamster embryo cells by neutral sterols and bile acids, *Toxicol. Lett.,* 9, 177, 1981.
39. **Raaphorst, G. P., Azzam, E. I., Langlois, R., and Van Lier, J. E.,** Effect of cholesterol and epoxides on cell killing and transformation, *Biochem. Pharmacol.,* 36, 14, 2369, 1987.
40. **Blackburn, G. M., Rashid, A., and Thompson, M. H.,** Interaction of 5,6α-cholesterol oxide with DNA and other nucleophiles, *J. Chem. Soc. Chem. Commun.,* 420, 1979.
41. **Petrakis, N. L., Dupuy, M. E., Lee, R. E., Lyon, M., Mack, C. A., Gruenke, L. D., and Craig, J. C.,** Mutagens in nipple aspirates of breast fluid, *Banbury Rep.,* 17, 67, 1982.
42. **Tso, P. O. P., Barrett, J. C., and Moyzis, R.,** The relationship between neoplastic transformation and the cellular genetic appartus, in *The Molecular Biology of the Mammalian Genetic Apparatus,* Tso, P. O. P., Ed., Elsevier/North Holland, Amsterdam, 1977, 241.
43. **Chang, C. C., Jone, C., Trosko, J. E., Peterson, A. R., and Sevanian, A.,** Effects of cholesterol epoxides on the inhibition of intercellular communication and on mutation induction in Chinese hamster V79 cells, *Mutat. Res.,* 206, 471, 1988.
44. **Jone, C., Trosko, J. E., Aylsworth, C. F., Parker, L., and Chang, C. C.,** Further characterization of the in vitro assay for inhibitors of metabolic cooperation in the Chinese hamster V79 cell line, *Carcinogenesis,* 6, 361, 1985.
45. **Humphries, G. M. K. and McConnell, H. M.,** Potent immunosuppression by oxidized cholesterol, *J. Immunol.,* 122, 121, 1979.
46. **Curtiss, L. K. and Edgington, T. S.,** Differences in the characteristics of inhibition of lymphocyte stimulation by 25-hydroxycholesterol and by the immunoregulatory serum lipoprotein LDL-In, *J. Immunol.,* 125, 1470, 1980.
47. **Yachnin, S. and Hsu, R.,** Inhibition of human lymphocyte transformation by oxygenated sterol compounds, *Cell. Immunol.,* 51, 42, 1980.
48. **Spangrude, G. J., Sherris, D., and Daynes, R. A.,** Inhibitory effects of various oxygenated sterols on the differentiation and function of tumor-specific cytotoxic T lymphocytes, *Transplantation,* 33, 482, 1982.
49. **Larsson, O. and Zetterberg, A.,** Kinetics of G_1 progression in 3T6 and SV-3T3 cells following treatment by 25-hydroxcholesterol, *Cancer Res.,* 46, 1233, 1986.
50. **Boissonneault, G. A. and Heiniger, H.-J.,** 25-hydroxycholesterol-induced elevation in ^{45}Ca uptake: correlation with depressed DNA synthesis, *J. Cell. Physiol.,* 120, 151, 1984.
51. **Freedman, M. H., Raff, M. C., and Gomperts, B.,** Induction of increased calcium uptake in mouse T lymphocytes by concanavalin A and its modulation by cyclic nucleotides, *Nature,* 255, 378, 1975.
52. **Hesketh, T. R., Smith, G. A., Moore, J. P., Taylor, M. V., and Metcalfe, J. C.,** Free cytoplasmic calcium concentration and the mitogenic stimulation of lymphocytes, *J. Biol. Chem.,* 253, 4876, 1983.
53. **Whitfield, J. F., MacManus, J. P., Rixon, R. H., Boynton, A. L., Youdale, T., and Swierenga, S.,** The positive control of cell proliferation by the interplay of calcium ions and cyclic nucleotides: a review, *In Vitro,* 12, 1, 1976.
54. **Maino, V. C., Green, N. M., and Crumpton, M. J.,** The role of calcium ions in initiating transformation of lymphocytes, *Nature,* 251, 324, 1974.
55. **Letvin, N. L., Greene, M. I. J., Benacerraf, B., and Germain, R. N.,** Immunologic effects of whole body ultraviolet (UV) irradiation. Selective defect in vitro splenic adherent cell function, *Proc. Natl. Acad. Sci. U.S.A.,* 77, 1881, 1980.

INDEX

A

B

C

D

E

F

G

I

K

L

Q

R

S

T

U

V

W

X